Richard Howard

About the Author

JOHN SEDGWICK is the author of the novels *The Dark House* and *The Education of Mrs. Bemis,* and has written extensively for *The Atlantic Monthly, GQ, Newsweek,* and many other magazines. He lives in Cambridge, Massachusetts.

IN MY BLOOD

IN MY BLOOD

Six Generations
of Madness and Desire
in an American Family

John Sedgwick

HARPER PERENNIAL

NEW YORK • LONDON • TORONTO • SYDNEY

HARPER ● PERENNIAL

Portraits, silhouette, and other images of Mumbet and the Sedgwick family are reproduced courtesy of the Sedgwick Family Society & Trust.

Photograph of Mumbet's gravestone reproduced courtesy of Zachary Kahn Photography.
All photographs digitized by Zachary Kahn Photography.

Photograph of Edie Sedgwick and Andy Warhol is reproduced courtesy of Burt Glinn/
Magnum Photos.

Sedgwick family tree was rendered by Susan W. Gilday and reproduced by permission.

Grateful acknowledgment is hereby made to the Massachusetts Historical Society for permission to quote from letters and other material in its Sedgwick Family Papers 1717–1946,
and to the Stockbridge Library for permission to quote from materials in its historical
room.

Portions of chapters 1 ("My Fall') and 43 ("Our Interior Weather") originally appeared in
somewhat different form in *GQ*.

A hardcover edition of this book was published in 2007 by HarperCollins Publishers.

FIRST HARPER PERENNIAL EDITION PUBLISHED 2008.

Designed by Joseph Rutt

The Library of Congress has catalogued the hardcover edition as follows:

Sedgwick, John.
    In my blood : six generations of madness and desire in an American family /
John Sedgwick.—1st ed.
        xi, 414 p. : ill. ; 24 cm.
    Includes bibliographical references (p. [387]-399) and index.
    ISBN: 978-0-06-052159-2
    ISBN-10: 0-06-052159-7
    1. Sedgwick, John, 1954– —Mental health. 2. Depressed persons—United
States—Biography. 3. Sedgwick family. 4. Family—Mental health—United
States. 5. Depressed persons—Family relationships—United States. I. Title.
RC537.S43 A3 2007
616.85'270092 22   2006049636

ISBN: 978-0-06-052167-7 (pbk.)

08 09 10 11 12 DIX/RRD 10 9 8 7 6 5 4 3 2 1

To the memory of my parents,

Robert Minturn Sedgwick (1899–1976)
and
Emily Lincoln Sedgwick (1913–2003)

And what thoughts or memories would you guess were passing through my mind on this extraordinary occasion? ...I shall be frank and tell you what it was, though the confession is a shameful one. I was thinking, "So, I'm Emperor, am I? What nonsense! But at least I'll be able to make people read my books now." ...I was thinking too, what opportunities I should have, as Emperor, for consulting the secret archives and finding out just what happened on this occasion or on that. How many twisted stories still remained to be straightened out. What a miraculous fate for a historian.

—Robert Graves, *I, Claudius*

CONTENTS

PART THREE

THE LEGACY DEFINED

PART FOUR

THE PRICE OF LEGACY

PART FIVE

WHAT REMAINS

The Genealogy of the Sedgwick Family

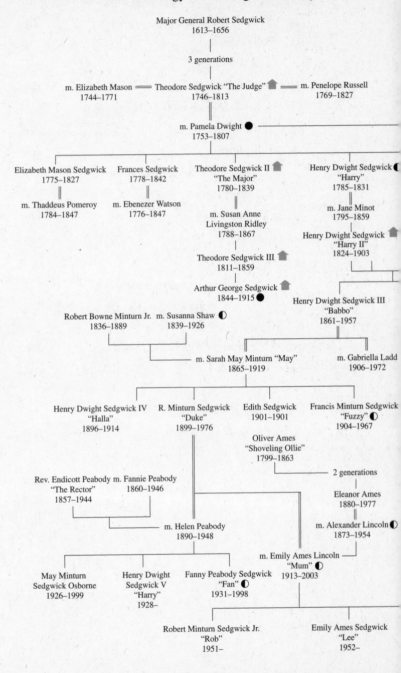

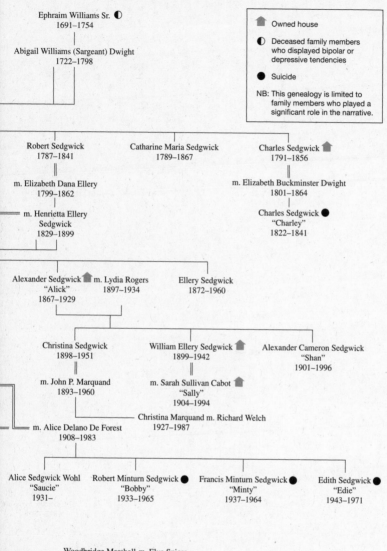
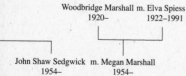

PART ONE

PRELIMINARIES

MY FALL

In the fall of the millennial year of 2000, *my* fall, I was up on the third floor of my house, and I was pacing like a wild man, each step a drumbeat that pounded inside my skull. "*I* can't do this, I *can't* do this, I can't *do* this, I can't do *this*," I chanted over and over. Each time I'd stress a different word, as if these were lines from some demonic Dr. Seuss poem, but the meaning was the same: *I can't go on like this. Not the way I'm feeling.* I was pouring sweat; my pulse thudded in my ears. My eyes jumped from the pine floor to the white wall to the open door to the window. Seeing, but not taking in. The room, the world, was senseless to me; it had no form, no order, certainly no purpose. It seemed alien, frightening, just as I did. I was a stranger to myself, a crazed weirdo who'd leapt into my clothes, taken over my body, seized my brain.

At that point, I'd gone three weeks without a solid night's sleep, but I was more wired than exhausted. I might have been a jungle warrior, ready to jump at the sound of a twig snapping. I'd stopped eating, pretty much, since I'd decided I wasn't worth food. In the mirror I could almost see my eye sockets hollowing, as if, any minute, my bones might burst through the skin. Thoughts hurtled through my head like meteors, burning out before I could quite track them.

"*I* can't do this. I *can't* . . ."

I'd been toying with death for a while by then, almost daring myself to take a suicidal plunge. To feel nothing—feel nothing forever. I craved that. In my scarce moments of calm contemplation, I pondered various ways of bringing about my own demise. It was a comfort, like the prospect of a cool drink on a broiling hot day. Hanging myself, blowing my brains out—such acts seemed not at all ghoulish.

Most of all I wanted to take a long fall from a high place. I'd always had a fear of heights, but I started to think that was actually an attraction. A few days before, I'd stood by the bannister on the second floor, lifted a foot onto the railing, and hopped up a little, to see what it might be like to hurtle downward to the first floor like Primo Levi. It wasn't much of a drop from there, barely a dozen feet, and I'd probably have crashed down onto the front hall table without much harm. But now, on the third floor, as I paced about the room, I kept returning to the window. From there, it was a long way down, a good forty feet to a concrete walkway. Such a plunge seemed so right. I was falling, so I should fall.

I reached for the window, flipped the latch.

THE PROXIMATE CAUSE, as the lawyers say, was the two Ambien sleeping pills I'd taken the night before. I was desperate for sleep, but the bed was hell for me. As I lay there, I felt a prickling heat all over me, as if my body were being licked all over by infernal flames. Breathe deep, just breathe deep, my wife, Megan, sleepily counseled, having conquered insomnia this way during her two pregnancies. But I spent most nights twisting about in agony, trying to find a spot of coolness on the rumpled, sweat-soaked sheets on my side of the bed. I got good at judging the time by the shade of gray on the ceiling, the rate of the cars passing by the street out front.

My brother, Rob, no stranger to sleep troubles as a harried New York lawyer, recommended the Ambien to me as if it were a hot stock. "No side effects," he assured me. "Every lawyer I know is on it."

"Including you?" I asked.

"Of course!" He gave a throaty chuckle.

He's my older brother. Tall and energetic, he's almost invariably cheerful, and he made the pills seem cheerful, too.

I scored an Ambien prescription through a doctor friend. In retrospect, she should probably have asked me a few more questions, but at the time I was really glad she didn't, since I didn't have many good answers. I hurried off to the pharmacy like a junkie, sure that happy, sleep-filled nights were soon to be mine.

That night, I moved upstairs to the guest bedroom on the third floor, since I didn't want to disturb Megan any more with my writhing.

I took the pill, then lay back on the bed, eager for the letting-go. But the pill didn't give me the milky calm I'd expected; if anything it made me feel alert, as if I should be doing quadratic equations, composing Elizabethan sonnets, inventorying my sins. So I took another, which set my thoughts racing even faster; I felt my heart rate rise. I didn't take another. Sleep, even the notion of it, fled. I didn't close my eyes the whole night, just lay there staring in terror at the ceiling until morning. Then I got up and went nuts.

As I say, the Ambien was the proximate cause. But there were others. I'd recently placed my mother in a locked ward at McLean Hospital for her fourth hospitalization for major depression, a disease that she'd been fighting since college. Always a tender person, she'd become increasingly frail with age, both emotionally and physically. After my father's death in 1976, she'd had trouble adjusting to the solitude, the exposure, that had come once her big bear of a husband was no longer around to protect her.

It was during hospitalization number three that I'd had the bright idea of writing a novel about her. Not her exactly, but someone like her, an elderly Bostonian patient, proud but broken, at an old-line mental hospital that, like McLean, had seen better days. The thought came to me in a rush as I sat with her in a dingy office while a psychiatric nurse did a brisk intake exam, asking about her mental history. As my mother haltingly answered the questions, her eyes going everywhere except to the woman's face, my mind flitted off like a little bird, perching somewhere above me to take in this miserable scene from a safe remove, where it all seemed rather interesting, and not at all as dismal as it actually felt.

In a fit of magical thinking, I imagined my fiction might cure her. Once I plunged into the novel, though, I found it difficult to cure even the woman in the book. The fictional psychiatric resident to whom I'd entrusted her care was exhausting herself in her patient's service—just as I was in my mother's—and, again like me, she was becoming overly identified. Progress on the book faltered as I began to feel overwhelmed by what I had undertaken both in literature and in life. Then, in the pivotal summer of 2000, my mother collapsed again, breaking up into the shards of worry that I myself was coming to know only too well. And she returned to the place where my story had started, to McLean.

As it happened, the previous novel, my first, had come out just three days before. A psychological thriller, it was itself a crazy sort of book—I can see that now—about a man who likes to follow people in his car. It was supposed to take me to a literary height that I imagined was my destiny, my proper place. I come from a line of writers, after all. My grandfather Henry Dwight Sedgwick, a dapper man who wore knickers until his death at ninety-five, wrote a shelfful of learned biographies of historical figures like the Stoic philosopher Marcus Aurelius, the Spanish conqueror Hernán Cortés, and George Sand's lover, the poet Alfred de Musset. Henry's more robust brother Ellery Sedgwick had owned and edited the *Atlantic Monthly* for three decades. And there were many others, going back generations to the country's first prominent female novelist, the sensitive and autocratic Catharine Maria Sedgwick, four of whose books are still in print. Every Sedgwick, it appeared, had his novel, or hers. Sometimes intimidated, sometimes bored, I'd never been able to get through a single one of them, but, enshrined in bookcases at Sedgwick houses throughout the country—whether it be in my half brother Harry Sedgwick's brownstone on the Upper East Side, my late avuncular cousin Cabot Sedgwick's sprawling ranch in Arizona, my cousin Tod Sedgwick's manse in Georgetown, or the Pomona, California, ranch house of my distant cousin Dennis Sedgwick, the faithful caretaker of a Web site devoted to family lore and genealogy—they seemed to set a certain imposing standard.

For generations, Sedgwicks had commanded the nation's attention. Our family's founder, the imperious Theodore Sedgwick, had served in the House of Representatives in Washington's presidency, and rose to Speaker in John Adams's; his grandson befriended Tocqueville, argued the Amistad case. More recently, my tragic cousin Edie was the sixties It girl before she crashed, and my niece Kyra is a gifted actor married to Kevin Bacon, a fellow movie star.

All in all, it gave me a need to measure up. Up, that was the whole idea. Height was what mattered. Elevation.

The truth was, I both knew these ancestors, and didn't. For the most part, the dead ones were merely "back there." Occasionally an exotic figure from the past, like my jovial cousin Shan, a former *New York Times* war correspondent, would flit into my presence and, with a lot of wild gestures, tell stories that lit up the image of these Sedgwicks of yore. But mostly they seemed to exist in some dim mausoleum of family memory, a place lit by flickering votary candles, where only the truly devout would ever visit.

But in that millennial year, I'd turned forty-six, tipping toward fifty, itself a kind of continental divide between youth, and—God, what? And I'd started to brood a bit about my losses. I'd married early, at twenty-six, had two delightful daughters, a thriving writing career, a big house in the leafy Boston suburb of Newton. I was a squash fanatic, full of health, had plenty of friends. There was nothing seriously wrong with this picture.

Yet I had reached the age my mother was when I first began to know her. She was forty when I was born, the last of her three children. My father had been fifty-five, and died when I was a senior in college. And in the months before my fall, I found myself glancing backward more, wondering about life from their perspective. Ruminating. Little things would catch me up. I found my father's old double-edged razor on a remote shelf in the family summer house in New Hampshire. When I held it in my hand, I was surprised by its heft, its age. I thought of his scratchy cheek when he kissed me—on the mouth, well into my teens—me on tiptoe, craning my neck upward as he leaned down. And his whole elaborate morning ritual, standing before the mirror, slathering hair tonic onto twin hair brushes, then sweeping his hair back on either side, with two hands. Prowling in a closet, I found the double-breasted Groton School blazer he had always worn so proudly at Sunday lunch, like the emblazoned armor of a knight. It hung limp off a wire hanger now, powdered with mildew, nibbled by moths. It all seemed so long ago.

OTHER CALAMITIES PILED up that fall. I'd overinvested an inheritance I'd received from my father in the Nasdaq stocks that were all the rage—frenzies that he, as an investment adviser, would have scrupulously avoided. It felt as if large chunks of my body were being cut out of me as they plummeted. Our older daughter, Sara, a high school soccer star, was coming off a year lost to knee surgery; when she went down hard again during one of her first games back, I watched, stricken. The injury proved to be only a nasty ankle sprain, but I took it like a spear in my chest. In a rare dream from this period, I was standing in a swamp when, below me in the murk, an alligator sliced my legs off just below the knee.

Pinpricks started at the back of my neck, spread across my shoulders, and then slithered down my arms. Mail piled up. Phone calls went unreturned. Bills went unpaid. I lost eight, ten, twelve pounds. My pants started hanging off me; there was no hole on my belt tight enough. I couldn't concentrate on anything—couldn't read a book, barely glanced at the newspaper. Exercise became a strain, conversation a chore, sex out of the question. And as the insomnia deepened, I developed nervous tics—a regular shudder under my left eye, a quivering on the left side of my neck. I thought increasingly, obsessively, about heights, about falling, being brought low. I could almost hear the wind whistling in my ears as I plunged.

It was in this state that I took the Ambien. And that's what brought me to the window. I unlocked it, drew my hand back, and stared down at the ground below: the grass that was starting to fade as winter came on, the rock garden where a few hardy chrysanthemums still bloomed. In his memoir of depression, William Styron tells of a moment when hearing a bit of Brahms's Alto Rhapsody, with its many childhood associations, saved him from doing himself in. I didn't hear any music; the room was silent except for my breathing. Still, I did not raise the sash, draw up the screen, hurl myself out. A failure of will, I guess. Or was it a triumph?

Instead, I called the psychologist I had been consulting intermittently, and he referred me to a psychopharmacologist, whom I saw that afternoon. She made the diagnosis: I was suffering from a depression that, while "major," was not "severe." I think the difference was that I was able to haul myself to her office. And she offered me a cure, putting me on Prozac. For now, I will skip over the details of my months as yet another citizen of Prozac nation, except to say that the cheery little pills worked possibly too well.

COMMON AS DEPRESSION is, I felt mine was custom-made—it seemed to exaggerate the me-ness of me. But at moments it hinted at something beyond me, too, something vast and timeless. I first perceived it as a kind of augury late in the pivotal summer of 2000, after I put my mother in McLean but still some weeks before I started to fall apart. My wife, two daughters, and I were at the family farmhouse in New Hampshire. When the first cool gusts of autumn pushed through the trees, I found myself unusually sensitive to the swirls of chilly air as they tossed the reddening maple leaves and spread cattails across the cooling water of the pond. As if by primitive glands I had never been aware of possessing before, I sensed active menace in those icy gusts of wind. Sure enough, as the fall deepened, the days shortened, the light dimmed, the leaves flared and fell, and the warm summer gave way to what seemed like endless cold, gray drizzle, my inner weather turned bleak, too.

That was the start of it, and, oddly, that is the part of it that stays with me most sharply, the dividing line between before and after, then and now. It was a wind that had been pushing through my world for a long time, but I had never noticed.

It took me nearly a year before I felt like myself again after the depression, and then, when the new novel I'd struggled with so mightily came out, I felt a kind of echo of it once more. Since then, with the aid of a Jungian therapist, I've spent a good deal of time thinking about my fall, its causes and meaning. No doubt it expressed a kind of mourning, just as Freud would suppose, in my case mourning for the death not of a corporeal person, but of the person I'd hoped to become, the stunning success, the one who could stand confidently with the ancestors whose exploits, while mysterious to me, still hung over my head, the one who was worthy of my father's pride.

But I kept returning to those swirling gusts of chilly air.

A year after my fall, my half brother Harry Sedgwick—now in his seventies, a full generation older than me, but still remarkably youthful—called to say that he was coming down with what I'd had. I could hear it in his voice, which had sunk into a deep bass, and the words came very slowly. He, too, had suffered a blow to his self-esteem. A "private venture capitalist," as he called himself, he'd been searching for a big hit since he'd backed an invention called Trig-a-tape for marking the price on grocery items in the 1960s, and now, with a recession deepening, the economic times seemed against him. "Well," he said sadly, his voice gravelly. "I think we've got the family disease."

He'd been speaking to a psychiatrist who'd been astounded by the high incidence of mood disorders in his pedigree. At that point, I'd thought of any genetic defect in a far more limited way, as something that had come down through my mother, whose father before her had been almost morbidly depressed.

"From the Sedgwick side?" I asked.

"Oh, yes. God!" And he started to tick off the names. "Think of Edie." She'd died of a heroin overdose. "And Bobby, Minty." Her brothers, both suicides. "And Fuzzy." My father's brother Francis, he meant, their demonic father. Even Dad had succumbed to a heavy period of gloom toward the end of his life. "And Babbo—didn't I tell you?"

That was my grandfather, the aforementioned Henry Dwight Sedgwick. "What?" I asked. "He nearly killed himself over a love affair."

"Really?"

And so it went, he said, back and back.

It made me wonder—what *did* lie back there in the shadows? Where had I come from? Who were my people? What legacy had they left me?

As I rolled the term "family disease" around in my head, I thought Harry had it wrong: It wasn't the family *disease,* but the *family* disease, an excessive absorption with kinsmen—their standards, attitudes, beliefs—which in turn intensified any underlying psychiatric predisposition to overdo. Certainly, there were few families that I knew who were so intensely loyal; to my father, an elderly third cousin twice removed was a brother, if not a twin. The family feeling is like patriotism or fandom, bubbling up from somewhere deep inside, enveloping us all in a tight web of common expectations, histories, self-conceptions, delusions, moods, talents, networks, loves, frames of reference, objects of veneration, landscapes, and desires that, like a cocoon, gives us our shape as it gives us our shelter.

Or can these things be traced back? Is there a source that lies behind us, an unseen force moving, as it were, through the trees, that touches us all? Sedgwick traits did not start with us, the living Sedgwicks, but are of long heritage. For us, it seems, the future *is* the past. We become what we were, our character forged by the generations that have gone before.

And if I were to try to find the source of who we are, where would I look? Where does the wind come from? Did I need to search the Yorkshire moors, from where the very first Sedgwicks sprang into the historical records in the fifteenth century? Or the places the Sedgwicks *made*? The first Sedgwick came to Charlestown, Massachusetts, in 1635. His sons cleared the New England forest, and sent their grandsons into Washington's army to create a new nation, and into the first government to establish its politics. And ever since, Sedgwick descendants have fought in the country's wars, served in its administrations, and more. Despite fits of derangement, and worse, they have helped create national fashions, set off national tempests, well through the rebellious sixties and into the present. It was here that I needed to look, if only because it was here where I could see. For the Sedgwick story is the story of America.

OUR GRAVEYARD

A memory.

STIFF, TALL, ANCIENT, proper, my father—at sixty-one, old enough to
be my grandfather—lowers the back of the station wagon with a screech
of rusty hinges and stuffs in the suitcases, the coats, the athletic gear (the
football pint-sized, to fit our small hands), and, finally, me. It's October
1960, the fall of the Kennedy-Nixon race, and some of the last days of my
being the Lone Ranger, with a six-gun routinely lashed to my leg. I'm six.
The youngest and by far the smallest of the three kids, I climb into the
back with theatrical resentment.

Through the suitcases, I can see my father climb in beside my brother
Rob in the backseat. Unable to turn his head because of an old football
injury, Dad leaves the driving to Mum while he plays the federal marshal
beside his prisoner, Rob, who is about half an inch from flying off the
handle. My mother stamps out her last Kent on the driveway, then picks
up the butt and drops it in the car ashtray as she slips in behind the wheel
beside my sister Lee, who is at eight excessively well behaved, a little princess. Then Mum fires up the Ford, puts it in first with a rumble of the
gears, and with a lurch, we're off for the weekend in Stockbridge.

To the Sedgwick house. My father's house, or so I think of it. And, as

far as I know, his father's house, and his father's, and back and back, until
the very beginning of the Sedgwicks. Even at six, I know this is an infinitely bigger deal than going to Barnstable, our shingled summer house
on the Cape, with its breezeway and view of the ocean.

The new Massachusetts Turnpike makes it a straight shot from Boston
to the Berkshires, and a lot shorter drive than it used to be, but it still
seems interminable to me.

The engine starts to strain as the road rises little by little, and then, out
the window, past my shoulder as I lie there, a glinting lake appears, and
then hills against the darkening sky, hills that rise and spread out under
the evening clouds. One last jerk of the wheel and a splashing sound from
under the car as the tires kick up pebbles from the driveway, and we're
there.

The cold grabs my nose and cheeks and settles on the backs of my
hands. Around me, the hemlocks rise up black against the evening sky.
But I feel lifted up, as if I'm older and more important just by being here.
I turn to the house. Massive and broad, with stout columns holding up the
front portico, a pair of thick chimneys spewing smoke into the night air,
and a heavy front door, it inspires grown-up manners. Rob turns polite,
and Lee talkative.

For a second, the house stands in quiet like a church. But then the door
opens with a thunk of its brass knocker, and the storm door bangs, and a
flood of pint-sized relations spill down the steps toward us. Sally, Christopher, Sam...the children of my half sister May, so I'm their uncle, which
is the funniest thing, since they're all about my age.

"They're here! They're here!" they shout, racing toward us, these
protocousins, like puppies in their eagerness, grabbing bags, surrounding us, tugging us all—even my father, who is not usually tugged—back
in toward the house, where their parents, already deep into cocktails,
await.

Inside, the ceilings are high, the hallways long. Photographs and oil
paintings of dead ancestors stare down at us, the parquet floors groan, the
evening light flickers with the spirits of the dead. The grown-ups call it
the Old House, but familiarly, as they might say "old Tom," meaning the
town's aging mailman, Tom Carey. And it's true, even the air smells old—
a mingling of wood smoke and rotting books and the penetrating damp

that drifts down from the trees or up from the Housatonic River, which wanders through the meadow below the house.

Tall people stoop to crush me in their arms. Big, grinning Harry, my half brother, tenor-voiced and full of zest, with his brisk but tiny new wife, Patsy. My half sister Fan, or, sometimes, Fan-tan, the youngest of my three half siblings; she's permanently single, mannish in her energy, and loud, and there's so much of her; she surrounds everyone with talk. And my half sister May, her skin so white it's almost blue, her manner stiffly proper (a demon for correct English), regal, despite being an up-country mother of all those children; and her gigantic husband, Erik, his head so high up, every doorway is a mortal hazard. Finally, there's Gabriella, my dear step-grandmother, loved by all the young. She'd married my grandfather Babbo when she was forty-four, and he ninety-one. Now fifty-one, she's been a widow for three years, but a merry one, a gazelle, with her high-jumper's legs and a motor scooter to zip her around town. The winter before, she took me skiing and coaxed me onto a single-seat J-bar to ride up the hill. She went first, but when I followed on the next one, the bar vaulted me into the air and landed me upside down in a snowbank. Her smile faded only for a second before she snowplowed down to me, scooped me up, and remounted another J-bar to guide me back up the hill between her knees. That was a wonderful place to be, my skis inside hers.

Cocktails over, we sit down for dinner around the big table in the dining room, the family silver glinting in the candlelight, wine bottles everywhere, and much shrieking laughter and lots of stories that are not meant for the young. As the grown-ups take their coffee, Gabriella leads us kids into the book-lined study where Babbo, and so many other Sedgwicks, wrote their books. She spreads us all out on the bearskin rug in front of the sputtering fire, a woodsy scent all through the room, and as the room's silver wallpaper—originally the lining of tea boxes—gleams in the firelight, she reads to us from *Winnie the Pooh,* allowing one or another of us to turn the pages. As she reads, she turns all of Stockbridge, if not the entire world, into the Thirteen-Acre Wood of Pooh Corner.

In the morning, May shocks us by sleeping till noon with Erik in the huge bed upstairs, while the rest of us pair off—me with Christopher— for baths in the big claw-footed tubs on either side of the upstairs hall. Afterward, I play on the antique rocking horse in the old laundry room; it seems like a real pony to me. Then a picnic in a meadow in nearby Tyringham, where Harry thrills everyone but my father by careening about our picnic sites in his Volkswagen Bug, his head and much of his body thrust up through the open sunroof, whooping wildly as he drives the car with his fingertips.

Later, Gabriella leads a treasure hunt in the apple orchard out front of the house, seeking walnuts she's tucked in the crooks of the towering spruces that shade the driveway. Then, as the sun drops behind the hills and a chill wind comes up, she leads us back into the house and upstairs, out an attic window, and onto the rooftop, where we send off wooden helicopters into the breeze.

SUNDAY MORNING, KNOWING we're not much for church, Gabriella takes us down Main Street to the town cemetery to play hide-and-seek among the gravestones. The leaves skitter about; the air crackles. We are all having a wonderful time, leaping and shouting and dashing about, until a groundskeeper emerges from his little guardhouse, furious. "What do you think you're doing there?" he growls.

"We're playing hide-and-seek," Gabriella explains, as if nothing could be more natural.

"Not here, you don't. This is a cemetery, you know, not a playground."

"All right, then," Gabriella tells him. "We'll go play in *our* graveyard."

Our graveyard? This is news. Gabriella marches us all down the cemetery's gravel drive, a goose with her goslings. To the right from there is a line of high Norwegian spruces; beyond it, slashes of light angle down. She slips between the trees, and we all follow across a mat of pine needles that silences our steps. It's like a moment from a fairy tale when the children enter some mysterious kingdom. All about us are graves—graying, moss-covered, cloaked in time. They seem, in fact, less like gravestones than like actual beings, green-bearded druids who've been turned to stone. It's so quiet here.

Something else is strange. The stones are not in straight rows, but in circles. Concentric circles, in fact, all of them ringed about a pair of granite monuments that rise up in the middle—one a soaring obelisk, the other a more demure urn. My eye is drawn to the sculpture of a dog in that first circle, and to a cast-iron structure housing full-sized statues of two boys who seem to be about my age. What are they doing there? We are full of questions. Gabriella points to the two central monuments, and says they mark the graves of the two founders of our line of Sedgwicks. "That's Theodore Sedgwick," she says, pointing, as if the stone itself were him. "We call him the Judge. He started the family, don't you know? He built our house." And that one? "That's Pamela, the mother of all his children." She points to the other monuments circled around them. "These are the children, don't you see?" She stops. "All except this one." She pauses before a rectangular slab. "This is a black woman, Mumbet, a former slave. The Judge freed her, and she came to work for the family." Gabriella turns once more, with a wave of the hand. "And those grave-stones there, they are their children. And the ones behind—those are theirs." And on she goes, out five circles altogether. Babbo is there by the graveyard's eastern rim, his name and dates at the bottom of his first wife's grave like a footnote. His death is fresh, just three years before, but Gabriella does not trouble us with any of that.

"Now—who's 'it'?" she asks.

All our hands shoot up. "No," she shouts gleefully. "*I* am!" And as we squeal, we all dash behind stones to hide, and we continue our game, hiding and seeking amid the shadows, racing about, laughing. Dancing on the bones of our ancestors.

IN FORTY-ODD YEARS since, I've been back to the graveyard countless times. Gabriella is buried there now, and so are my father and mother. As with all Sedgwick funerals, their ashes were brought here by a horse-drawn carriage from the Episcopal church in town. Because of some over-crowding in this section of the graveyard, Dad lies beside Babbo's grave, rather than in another ring beyond; his gravestone is topped with the Sedgwick crest, a lion trotting above a cross that is dotted with bells. My mother is beside him; her death is fresh, and her grave still needs a marker. My manic half sister Fan is buried where Dad's feet would be if he hadn't been cremated, which can't please him. She is on him in death as she was in life. May has died, too, but she is buried with her husband's family in New York. The circles are expanding, row upon row, reaching out to me.

It's often said that the circular design is so that on Judgment Day the

Sedgwicks will rise up and see no one but Sedgwicks, but that was a joke told on the Sedgwicks, not an explanation by them. It's like the way that the words "Sedgwick Pie" have caught on as the name for the graveyard. The Sedgwicks didn't think that up, but have found it a convenient short-hand for an unusual arrangement, and it gives a nod to the obvious eccentricity, as well.

The circles upon circles play on my imagination as if they make up some mysterious Stonehenge of their own. Stone by stone, I can trace my lineage: Judge Theodore, then three generations of Henry Dwight Sedgwicks—first Harry, then the son he called Hal, then Babbo—to my father. The ancient, mottled stones, the tufts of grass licking at their bases—the graveyard sometimes seems like a pond whose waves are rippling outward toward some unimaginable shore.

The upthrusting obelisk and the receptive urn there at the center, then the rows of graves beyond, many of them paired in imitation of those first parents: even as the family expands outward in time, we all remain in the grip of those first ancestors, toward whom we all point our toes in homage, as if they are both our source and our destination. We are headed out *and in.*

It's an odd arrangement, no question. Perhaps even un-American. Isn't ours a nation of pioneers, our eyes scanning the horizon, or the heavens, for the next frontier? And aren't families supposed to fling their members forth—to Kansas City, to L.A.—to find better lives elsewhere? But we Sedgwicks are eternally drawn back to our beginnings here in this shady little hollow of the Berkshire hills, as if, following T. S. Eliot, our end is indeed our beginning.

I examine the inscriptions on those early tombstones, greened with moss, the letters blurred by age and acid rain. To see more exactly the lives that they built for themselves, to see what of them is in me, to see what of them has *become* me. I run my hands along the rough stones, as if in touching them I might somehow grasp myself.

YET IN WANDERING almost morbidly about the Pie, I've realized that the answers are not to be found in a family boneyard. With so many writers among them, their stories are not buried with them. They were left behind in countless letters, memoirs, journals, biographies, autobiographies, notes, and other jottings, many of them unpublished, many of them never read by anyone except their maker, and myself when I found them. They are now bursting out of my filing cabinets, filling my bookshelves, and spilling out over most of the surfaces of my office. They come from basements, from attics, from the back rooms of private libraries, from the vaults of Harvard's Houghton Library, the rare manuscripts department of the New York Public Library, the little locked closet downstairs in the Stockbridge Library.

By far the largest single repository is the splendid Massachusetts Historical Society in the Fenway section of Boston, whose now-teeming archives were begun shortly after the Revolution, and where the vast collection of Sedgwick papers run second in volume only to the Adams papers, which a team of researchers has now spent over forty years rendering for publication, and is only half done. For my own forays into such an immense holding, I spent many weeks sifting through the brittle letters, many of them still bearing chunks of the original sealing wax. The documentation of this family has been so thorough that days from as long ago as the Revolutionary War can be reconstructed almost hourly. All of this has allowed me to peer more deeply into the past than I had ever thought possible, and is unimaginable for most ancestral lines.

For the most recent periods of my chronicle, I myself was personal witness, and where I wasn't, I have found others—many in the family, many not—who have been able to extend my gaze. I think of this as a family memoir, and I have written it as such. Familiarly. On occasion, I have filled in gaps in the historical record with my best guesses as to what happened, based on what I do know. A family member's prerogative, let's call it. After all, these are my people, ones I have come to think of not as some distant ancestors but as fathers and mothers, brothers and sisters, from another time, and, like all close relations, versions of my own particular self. In writing about them, though, I felt no obligation to adore them all strictly out of kinship, much as it would have relieved me of certain Oedipal anxieties to do so. I also recognize that my primary interest in my Sedgwick forebears—to discover the source and consequences of our family disease—is bound to put many of them in an unflattering light. But the facts have led me to this interpretation, not the other way around. Besides, to look at the family from this angle does not diminish its many accomplishments; rather, it enhances them, as it places them in the context of the trying emotional circumstances in which they occurred.

The big house that Judge Theodore built for himself in 1785 has remained in the family ever since. It is older than the Constitution, and nearly as old as the republic. If the graveyard held my ancestors in death, the house revealed them in life. If I sought to know myself by knowing my father, and my father's father, and my father's father's father, and so on back through the generations, I knew where I had to start.

PART TWO

THE LEGACY

THREE

A MAN OF PROPERTY

A vision.

OUT OF THE void, a man on horseback, riding through the low-slung Berkshire hills along the well-pounded "meadow road" that is just hardening up after the autumn mud season. It is late in the evening of the twenty-eighth of November, 1783, and he is coming to Stockbridge from Sheffield, some ten miles to the south, to secure the prize piece of land he's picked out for a new house. The son of a farmer who died young, Theodore Sedgwick—yes, it is he, the wintry chill biting his hands as they clench the reins—has become a lawyer, one of just five in Berkshire County, and Sheffield's lone representative in the new General Court of Massachusetts, besides. A rich man in a harsh region that has few riches to bestow, Theodore has styled himself a country squire, and, fresh from a full day of lawyering, he is decked out now in a ruffled shirt and stiff-collared frock coat that befits his station, his long brown hair tied stylishly in a ponytail "queue" that streams behind him as he thunders ahead.

Only thirty-seven, Theodore has the imposing force of a much older man: fierce, scowling, black-eyed, resolute. And he is tall, a good six-two, with a thick frame that makes his favorite mare, Jenny Gray, fully aware of her rider as she flies past Great Barrington—the last settled outpost before Stockbridge, its dim little houses lining the road—and then cuts through the narrow pass that separated the long, hulking Three Mile Hill from towering Monument Mountain. There it opened up onto the Great Hollow, an eerie, moon-silvered swamp, studded with rotting trees, that bred frightening stories among the English settlers about an Indian woman who'd murdered her baby and tossed its remains into the mire.

Other ambitious Sheffielders have picked up stakes to head farther west, ready to try the more open country of New York's Chenango County or Michigan's Monroe County. Theodore has bought lands in New York's Genesee Valley, but only to profit from these emigrants' folly. He doesn't want to move *out*, but *up*. His small frame house in tiny Sheffield is too tight for a man of his aspirations. Overly dependent on the mill at Ashley Falls, Sheffielders are, in Theodore's view, also excessively deferential to the high-steepled Congregational Church, which stands directly across from his own house and smack in the middle of the road at the center of town, forcing horses and carriages to veer past it on either side. And the house holds bad memories besides. He bought it for his first wife, Eliza Mason, but she died there in agony from smallpox. His new wife, the refined Pamela Dwight, has borne him five children, but they've lost two in infancy, and the deaths weigh on her, especially now that she is pregnant again.

Tucked into the southwest corner of Massachusetts, hard by the New York border—a disputed line that Theodore himself will soon scramble over rocky hillsides to officially determine—the Berkshires are no longer a frontier in 1783, but more a kind of hollow between the actual frontier, which by then has pushed out clear to Ohio, and the coastal commercial centers of Salem, Boston, and New York, which lie several days of hard travel to the east and south. The region feels more northerly than it actually is, as if the glaciers have never quite receded. An icy cold drifts down from the rumpled hills, a spur of New York's Taconic Range, or rises up from the lazy, wide-swinging river, the Housatonic, that wanders along the road, its clear waters to harden into thick black ice all too soon. And the chill carries deep into the spring, settling into the bones of these hardy settlers, altering their outlook.

* * *

In the distance, finally, a few specks of light. Compared to humble Sheffield, Stockbridge is almost a booming metropolis, and Theodore can see a flicker of candlelight in some of the windows as he approaches, and a few gay shouts from revelers. Unlike Sheffield, Stockbridge doesn't just watch the mail coach and other traffic pass north-south between Montreal and New York. It is a stopping point for those travelers, and also for the ones passing east-west between Albany and Boston. It has the Red Lion, a thriving inn run by a cranky old bird, the widow Bingham, to serve them a hot meal and get them liquored up. The town is also growing commercially, with a pair of prosperous sawmills and an ironworks within its borders. A new school is being planned, and even a subscription library.

And Theodore's wife's family, the Dwights, live there. Although born in Dedham, well to the east, the late Brigadier General Joseph Dwight was a member of one of the Berkshires' first families; the Dwights' influence and prestige, in fact, spread so widely across the region that they were considered one of the reigning "River Gods" of the entire Connecticut River Valley. And his mother-in-law, the redoubtable Abigail Williams Sergeant Dwight, was even more of a potentate. Born a Williams—of the newly founded Williams College and Williamstown—she was a god to whom the other River Gods paid homage. She owns the most impressive house in town, one that stands atop the hill that marks the pinnacle of social ascension. The whole house, complete with an almost ridiculously extravagant frieze over the door, was delivered by ox-drawn wagon all the way from Connecticut.

With its gentry and farmers, Stockbridge might be any rural New England town on the rise. The only serious incongruity lies well to the west of town, on the plain just up from the Housatonic River, into which the wintry sun has long since sunk. There, in the darkening gloom, stand a few lonesome "hut-wigwams"—half birch-bark tents, and half more permanent structures, an assemblage that reflects the provisional, betwixt-and-between nature of the Indians' position in town. This dwindling enclave is the last remnant of the peaceful Indians who originally populated Stockbridge, and a memorial to the failure of the "noble experiment" to which the town was originally dedicated. For Stockbridge was founded in 1739 as a "praying town" for the moral, intellectual, and religious education of a migrant tribe of Mohicans who had crossed over

from New York State in search of land that would be indisputably theirs.
A reservation, really. Several dozen families of "Stockbridge Indians"—as
they quickly came to be called—moved into town alongside four "model"
families of the enterprising Englishmen who were supposed to show
them the path to commercial prosperity, if not spiritual enlightenment.
Instead, they turned their financial acumen against their trusting wards.

A Yale-trained missionary, John Sergeant, was recruited to attend to
the Indians' souls, and he arrived on horseback, as he wrote, "thro' a most
doleful Wilderness, and the worst Rode, perhaps, that ever was rid." But
he soon caught sight of the glittering Abigail Williams, and when he married her at seventeen, he considered this "most ingenious woman" to be
proof of God's providence. But Sergeant died a decade later of a fever,
leaving Abigail a widow at twenty-seven. Abigail eventually remarried,
to Brigadier General Dwight, and the town fathers entrusted the missionary duties to the fiery Puritan Jonathan Edwards, who did his best to
attend to his Indian charges while devoting himself to such theological
tasks as his expansive essay *The Freedom of the Will,* detailing his moral
philosophy. Meanwhile, the Dwights, Williamses, and other English families, eager to stretch out over the virgin landscape, grew more quickly
than the Indians, and, seeing the commercial promise of the area, other
white settlers wanted in. Worse for them, the Indians found it hard to
get the hang of the crafty legal ways of their mentors; in addition, the "old
Jamaica spirits" served up in town proved dangerously enticing. Increasingly, the Indians had to sell off their land to pay for their liquor—and for
the many errors the liquor led them to.

By now, as Theodore is trotting into town on his snorting Jenny Gray,
the Indian town has become nearly a completely English one, which is
why he's come.

It is to see Timothy Edwards—Jonathan Edwards's son, and a rare,
honest broker between the Indians and the English—that Theodore has
come this evening, as the moon shadows stretch across the wide, rutted
street. The Edwards house is a short ways down from the main intersection; on a still evening, the family can hear the rum-soaked whooping, the
songs, from the widow's tavern. The house itself was a humble one for an
outsized personality like Timothy's father. Lined on the inside with bare

brick, it stood behind a prim white-picket fence to keep out stray pigs and any other errant livestock wandering by.

The Edwards place is directly across from the land that Theodore has come to purchase. A single wigwam stands on the property, the very last Indian dwelling anywhere near the center of town. A gorgeous spot—not the hill, but centrally located, a key consideration for a working lawyer. It looks out from a kind of bluff over the Housatonic toward hulking Monument Mountain and the other craggy hills beyond. To its present Indian owner, the view of the land where the great spirit dwelled is humbling. At its base there is a mound of flint stones, nearly eight feet across and only a little less high, which give the mountain its name. Whenever an Indian passed, he added a stone to the heap, as to an altar, each one a tribute to the great spirit the Indians worshiped. But what use has Theodore for spirits?

Theodore lashes his horse to the hitching post and raps on the door knocker. The two men are of equivalent station, in a world that is extremely conscious of social position, so Theodore greets Edwards warmly with a handshake, an American greeting that befuddles European visitors, who are inclined to bow, nearly as much as their scratchy, backcountry accents. He comes quickly inside to the book-filled parlor where a fire struggles to evict the penetrating Berkshire cold. His seller is already there, waiting. She's only come from across the street, but even to a relatively cosmopolitan man like Theodore, she seems a visitor from another country. Her name is Elizabeth "Pewauwyausquauh alias Wanwianyrequenot," as the deed records it. A heavy, burdened-looking figure of middling age, she wears a solid-colored linen shirt over a wraparound skirt and woolen leggings. With her feet clad in moccasins, she treads silently about the room. A braid of sweetgrass hangs from her hair, lightly perfuming her person with a haylike scent. She greets Theodore with a bow and a few grunted words, heavy on gutturals, that strike him as utter gibberish until Edwards translates haltingly. She's brought along with her an Indian comrade, a mysterious figure named Jehoiakim Naunuhptonk, to even out the numbers.

It is a crossing of two cultures, and two histories. A crossing, and then a passing, neither party ever to encounter the other again. In fact, Elizabeth's ancestors were prominent figures in the Mohican nation. Her

father-in-law translated for Timothy Edwards's father. But Theodore
knows none of this, nor cares to. To him, this is purely a commercial
transaction, a brisk bit of business, nothing more.

As the only lawyer present, Theodore himself prepares the deed, and
he signs it in his usual confident hand. Edwards adds his signature as a
witness. And then the document is passed to Jehoiakim to look over, and
finally to Elizabeth to sign. Edwards explains it to her, in her language.
Although the town was created expressly to educate the Indians of both
genders, Elizabeth is unable to read the deed that legally exchanges the
last of her land for "Thirty Pounds Lawful money." Unable to write,
either, she signs the document that ends her relations with Stockbridge
only with a frail and wispy *X*. Its tentativeness contrasts with Theodore's
assertive scrawl, the ink flowing thick across the page. As I look at it now,
the *X* looks like the faint scar left after a growth has been excised. It
shows not so much where Elizabeth was, but from where she has been removed.

And with that, Theodore tucks the deed into his pocket and bids the
fragrant Indians adieu. When the arrangements are complete, Edwards
brings out a rare bit of bar chocolate to celebrate, and brews from it a pot
of hot chocolate. And then Theodore is off, mounting Jenny Gray once
more to return the way he came.

IN THE NEXT two years, Theodore will buy up three adjoining parcels,
two from an Indian widow named Elizabeth Whwumen, and the last
from Johoiakim Mtochsin, the son of an original Stockbridge settler, and
a captain in the Revolutionary Army, who fought with the Minutemen in
the siege of Boston. By then, virtually all the Stockbridge Indians have
agreed to relocate to a "New Stockbridge" in the pine woods by Oneida
Lake, in Oneida, New York. Wearing traditional blankets over their colonial linen shirts, they make a sad, slow progression out of town, the few
horses they own well packed with whatever possessions they can carry,
the men walking ahead, the women behind, their babies in papooses. It
takes them weeks to get there, but they do not last long in that New
Stockbridge, either. They are eventually forced to move again, in 1818,
this time to the shore of the White River in what will become Indiana,
and then to a small reservation on the banks of the Fox River at modern-
day Kaukauna in Wisconsin, where remnants of the original Stockbridges

merge with the Munsee tribe to be called the Stockbridge-Munsee band of Mohican Indians. After a brief stint by Lake Winnebago, they finally settle, in the 1830s, in the swampy pine forest of a reservation that overspreads the townships of Red Springs and Bartelme in Wisconsin's Shawano County, about fifty miles west of Green Bay.*

---

*While the Mohicans' time in Stockbridge added nothing to their prosperity, it did leave its cultural residue. A traveler to Green Bay for a council of Indians wrote a letter to the *New York Observer* in 1830 in which he took special note of the Stockbridges. While members of the other tribes lay about, scraggly-haired and half naked, the Stockbridge Indians' "dress, manners, countenance, and whole appearance exhibited all the decencies of common civilized life," the author noted approvingly. "I have found it a refuge, and a luxury to fall into the society of the Chiefs and principal men of the New York Indians"— as he refers to the Stockbridges. "Among them, I could be sure of exemption from anything vulgar, profane, indecent, or intemperate."

FOUR

MR. SEDGWICK BUILDS HIS DREAM HOUSE

On November 22, 1785, Theodore sold his house in Sheffield for £550, loaded his furniture, law books, and trunks of clothes onto wagons, and moved his young family and several house servants, including a formidable former slave named Mumbet, to the Edwards house in Stockbridge.

The previous month, Theodore had contracted with a pair of local builders to build a rather grand house for his family. The work did not go well. After an unproductive winter, Theodore summoned the two into his office in the Edwards house, echoing with the slow and relentless tick of the jeweled mantel clock, to press on them a binding "Building Agreement." Almost merciless in its specificity, the new document possibly reveals more about Theodore than it does, even, about the grand and spacious "mansion or dwelling place" that was to be built "in an elegant, good, faithfull, stylish, and workman-like manner."

In a town whose best houses were still fairly cramped and gloomy, Theodore had something quite ambitious in mind. His mother-in-law's house, for example, measured thirty-six by twenty-two, with ceilings low enough that a man of Theodore's height could easily press his palms up against them. *His* house was to run a full fifty feet wide and forty feet deep, with the ceilings on each of the two floors unreachably high—

eleven feet on the first floor, ten on the second. While most of the houses in town opened, as Abigail's did, into an anteroom that closed off a full view of the interior—and made the house seem, if anything, even tighter than it was—Theodore's would open by a "good, handsome well-formed and strong" door onto a broad "entry or space way" that led directly to the long, gracious stairs (he specified eighteen steps), and clear through to the far side of the house, affording an immediate, mind-swelling appreciation of the mansion's full dimensions.

A large, square room, twenty feet across, would occupy each of the four corners, making manifest the clear, four-square order of his household. The boards would run "full length" across the downstairs floors, not pieced together in the usual slapdash fashion. Eight-inch *white* pine in the "best Parlour" to the right of the door as visitors came in; ten-inch *yellow* pine in the other parlor to the left. And, to avoid any unnerving give to floorboards, he specified that the underlying four-by-four joists should come at tight two-foot intervals, with the interstices stuffed with lime to deaden any clanking of boots. His floors would be solid as lead.

And on and on the document went, spinning the brains of his poor backcountry builders. The roof was to be hipped, not gabled, and stained "Spanish brown," with fine dentil cornices all around to conceal the functioning gutters within. The clapboard walls were to be painted a glossy white. The downstairs walls would have "Dado" moldings, the very latest style of chair rail. The kitchen would be at the back of the house, with a vast seven-foot fireplace, one of eight in all. And most impressively of all, the light would pour in. Instead of the usual six-by-eight-inch panes on the standard "six over six" windows, his glass would be a lavish eleven by fifteen, all of it shipped at terrific expense from England.

IT'S EASY TO see why the builders had such trouble getting going that first time, and of keeping to the stated deadline of the following December—missing it wildly, in fact, by almost a full year, to the mounting exasperation of the house's expectant owner. For Theodore was proposing an entirely new kind of house for this part of the world, one intended not merely for shelter but for display. It's now termed the Federal style, in deference to the emerging Federalist politics that guided the country after the ratification of the new constitution in 1787, and the creation of the powerful, centralized government that would instill—if not impose—that very sense of order that Theodore was all about. Indeed, Theodore would, as a congressman, senator, and, ultimately, Speaker of the House, be very much part of this political movement, guiding much of Alexander Hamilton's Federalist agenda through the first six sessions of the U.S. Congress.

But the Federal architectural style that he took for his house was, in fact, English. It had its roots in the balanced Georgian style that had flourished in England since the time of King George I earlier in the century; with the addition of the Platonic geometry of the circle and the square to the configuration of rooms, it had become the height of London fashion, the quintessence of urbane Englishness.

As such, it was as suited to the hardscrabble Berkshires as Buckingham Palace. A world away from London, the Stockbridge house's square rooms and finery had a dazzling formal purity—but also a potentially irritating aura of cultivation and superiority. Not to mention, for a small country town, an intimidating sense of scale: just as the land stood over the spreading Housatonic Valley below, so the house rose up above all the other houses around it.

This posed endless problems for the builders. An appended last paragraph recorded the sorry particulars: how the work blew past the original deadline, modifications were necessary, and certain blunders simply had to be accepted. The workmen, for instance, had installed the downstairs shutters on the inside of the windows, not the outside, as specified. And a spherical roof over the front portico that Theodore had requested was just too difficult: Theodore would have to tolerate a more conventional pitched roof over his front door instead.

Still, in the summer of 1786, just as farmers all across the western part of the state were starting to militate against their government, the house was sufficiently finished for the family finally to cross the street, push open that massive front door, and feel, at last, their spirits rise as they came inside. Outwardly imposing, inwardly gracious, Theodore's house told the world that the owner had arrived and intended to stay. This would make him, and his house, a target soon enough.

FIVE

A FRIEND OF ORDER

If there is irony—no, *folly*—in an aspiring American politician turning to aristocratic England for architectural inspiration so shortly after winning a war to liberate his country from England's oppressions, Theodore Sedgwick was not the man to see it. The psychologically astute portraitist Gilbert Stuart captured—or is it mocked?—him in a head-and-shoulders portrait of 1808, depicting a proud if not imperious Yankee gentleman with lengthy sideburns, forelocks curled vainly over his forehead, and a contemptuous glint in his penetrating eyes. His nose, however, is perhaps a shade too bulbous for the perfect image of genteel refinement Theodore sought, and there is, in the swelling expanse around his shoulders, the slightly bloated effulgence of a fifth-generation English aristocrat. However he comes across, the smugly superior element, the self-assurance of it, is striking for an up-from-nothing country gentleman of the era, and it is one that, almost as if it has become a genetic trait, mutates into any number of variants in the portraits of his descendants as well: an airy insouciance with this one, a lustrous confidence with another, and a nearly surreal detachment with a third. In the Stuart portrait, Theodore is the mountain from which they have mined these qualities. His nose is lifted a little, raising his gaze away, one imagines, from the daily struggles of the workaday world—struggles that he, as a frantically busy lawyer and energetic politician, had known only too well. Sturdy to a fault, Theodore always subscribed to the aristocratic ideal of equipoise, of never showing strain, difficult as that would be to maintain.

A social reactionary while a political centrist, Theodore was usually under some pressure to keep these internal contradictions from tearing him apart, but never more so than during the ideological turbulence of the war for independence that had just concluded when he undertook his house in 1783, as he had paraded as a staunch revolutionary here, and stuck up for the propertied classes there.

In the Berkshires as elsewhere, class rank continued to be publicly de-noted by such titles as *yeoman, esquire, mister,* and *gentleman,* which were everywhere affixed. On that 1786 building agreement, for instance, Theo-dore proudly designated himself "Esquire," a word that still carried its original knightly connotations, even as it started to become an honorific reserved for professional lawyers. Theodore always made a social distinc-tion between those exalted personages who might be allowed to enter by the front door and commoners who would be required to use only the kitchen door on the eastern side of the house. "I have seen his brow lower when a free-and-easy mechanic came to the *front* doorway," one witness later recalled, "and upon one occasion I remember his turning off the 'east steps' (I am *sure* not kicking, but the demonstration was unequivocal) a grown-up lad who kept his hat on after being told to take it off."

THEODORE DID NOT invent the social ladder. He merely climbed on. But in doing that, he acquired a status awareness—and, with it, an anxiety—that would govern his descendants for generations.

He did not build his house as much as *re*build it from the memory of the old order that had gone before. One of Puritanism's most salient features—far more than its supposed sexual repression—was its class-boundedness. In creating their own New England, Theodore's Puritan forebears gleefully imposed class stratifications that were even more strin-gent and explicit than in the old. As part of the sweeping blue laws that seemingly regulated every aspect of behavior, they established sumptuary regulations that declared what could be worn by whom, as if one's dress were, like a military uniform, a public declaration of rank. In 1641, the General Court of Massachusetts declared that only those whose estates were worth over two hundred pounds "shall weare any gold or silver lace or gold and silver buttons, or any bone lace above two shillings per yard, or silk hoods or scarfs." Anyone else caught wearing such finery was liable to a fine of ten shillings.

THIS WAS THE social order of Theodore's great-grandfather, Robert Sedgwick, who arrived in Charlestown in 1635. Robert Sedgwick's tale is instructively Shakespearean in its rise and fall, although it is unclear how much of the full truth of the matter Theodore knew, and how much of it he preferred to overlook. Robert had come from a line of Yorkshiremen, natives of the rugged moors far north of London that had produced a disproportionate share of rebellious Puritans. A Puritan himself, he sailed to America with his wife, Joanna, aboard a ship called the *True-Love* that left London on September 19, 1635. They settled in Charlestown, just across the mouth of the Charles River from the slim peninsula on which Boston stood. Robert moved quickly to establish himself as an industrial dynamo. Within a year, he'd built an iron foundry; a brewery that is said to have been the first in America; a wine import business; a construction company that, in turn, built ships, warehouses, wharves, and a hoe mill; and a fine house for himself where the Old Corner Bookstore now stands in downtown Boston.* By 1652, he'd risen socially, too, from a mere "Inhabitant" of Charlestown to major-general of the colony. In 1654, the Puritan leader Oliver Cromwell, ruling England after his Parliamentary allies executed Charles I in 1649, recruited Robert to reclaim from the opportunistic French a pair of forts in the Bay of Fundy, between New Brunswick and Nova Scotia, and another in Penobscot Bay along what is now the Maine coast. When Robert smoothly accomplished these tasks, Cromwell sent him to the Caribbean with a fleet to reinforce the British troops holding the recently captured Jamaica. Retaliating for being repulsed from the larger island of Hispaniola (which Haiti and the Dominican Republic now share), British soldiers had set upon the local Jamaican populace with a vengeance, slaughtering 20,000 of them and destroying crops to starve the rest. But these natives would not be exterminated, and the survivors had, Robert told Cromwell, turned "wild" and much harder to kill. Frightened and disorganized, some of the British soldiers had

---

* A friend bestowed the name of Robert's native Woburn on the town hacked out of the woods just north of Charlestown.

tried to mutiny, which the local military commander, a man named Richard Fortescue, had put down by executing the leaders. Many of the rest were dying painful deaths of famine and disease. When Robert arrived, he found hundreds of pale, stinking corpses strewn along the roadsides, or under trees where the dying had dropped, seeking shade. Robert was sure the survivors would soon join them. "You cannot conceive us so sad as we are," he wrote Cromwell.

When Fortescue died in the contagion, Robert pleaded desperately for Cromwell to bring him back to England, or failing that, to send him provisions and reinforcements, or failing either of those, to relieve him of any responsibility for the outcome. The lord protector refused, sending instead something Robert absolutely did not want: "sole and complete command" of the British detachment in Fortescue's place as its major-general. He had complete responsibility without any hope of success. The fate of his troops was on his neck; and he was helpless to save them. The garrison's food supply was dwindling, and his men wasting away into "living skeletons" and then dying in agony around him. That official commission came late in May of 1656. Within days, Robert was dead. As his secretary—a young man named Aylesbury—reported to Cromwell's secretary of state, Robert had not been "of any visible great distemper, only a little feaverish." His illness, Aylesbury thought, was "inward." Ever since he received his commission, Aylesbury noticed, Robert had not been himself. "It will break my heart," Robert had worried, over and over, referring to his new title. And, the loyal secretary sorrowfully concluded, "I verily believe it did."

When she heard of her husband's death, Joanna Sedgwick returned to England. Their children stayed on in the colonies but displayed none of their father's commercial talent, and his fortune was soon dissipated. The Sedgwicks here barely held together for three generations. Theodore's grandfather started a farm in northern Massachusetts but, frightened off by the Indian raids there, fled to western Connecticut. When he died in 1674, he left behind only a son, Samuel, who, in 1716, sired Theodore's father, Benjamin, as the last of eleven children.

Born in 1746, Theodore, the third youngest, was just ten when Benjamin died suddenly of an apoplectic seizure. The eldest, John, was only four years older, but he took over for his father, and soon exceeded him. He extended the farmland Benjamin had been clearing in Cornwall

Hollow well past the few hundred acres his father had envisioned; it would eventually cover several square miles. Through John's exertions, he was able to send Theodore—who had already started to impress the family with his verbal faculties—to Yale at fifteen, the first Sedgwick to attend college. Caught up in one of the many student rebellions that periodically convulsed the tightly regimented college, Theodore, however, was thrown out in his senior year before he secured his degree. That year, Yale's president, Thomas Clap, moved up commencement by six weeks to speed the exit of Theodore's class, which seems to have been particularly unruly; in response, the night before the ceremony, a mob of students surrounded his house, threw rocks through his windows, and made off with his iron gates.

Theodore had initially intended to join the ministry. But a stint attempting to learn theology from the crusty Reverend Joseph Bellamy in Bethlehem, not far from Cornwall Hollow, disabused him of that. And so he turned to the other career open to an educated young man interested in attainment—the law.

The Berkshires were growing faster in population and in commerce than in lawyers, who would be needed to handle the increasing number of financial transactions. In the winter of 1765–66, Theodore moved to Great Barrington and apprenticed himself as a clerk in the law office of Mark Hopkins, one of the Berkshires' precious few legal lights.

Theodore quickly took to the law—its interior logic, its detail, its basis in precedent, and its need for forceful expression—and he was admitted to the Massachusetts bar just a few months later. After a false start in Great Barrington (no clients came), he moved south to Sheffield and bought that tiny house directly across from the church. It had a separate entrance for the front room—a pleasant parlor, which Theodore used as his office—and, just a few steps to the rear, another identical entrance that led into the kitchen, dominated by a massive fireplace.

The cases were small, but his practice thrived. Petitions, suits of trespass, attachments, flew from his desk in a vigorous, forward-slanting script. He ingratiated himself with local merchants sufficiently to serve as their business agent, pressing for payment from their sluggish debtors, arranging for the sale of parcels of land, purchasing cattle for them. In his elegant knee-breeches and buckled shoes (he always dressed *up*), Theodore was a frequent and rather conspicuous presence in court. He made a point of dealing only with men of "integrity," meaning ones who were certain to pay up, and of receiving payment in hard cash, not IOUs, or, if cash was temporarily unavailable, in land, which he knew full well would prove more valuable than any of the many inflation-prone currencies of the era.

As his activities increased, he improved his clientele to serve many of the River God families—the Williamses, the Dwights—whose ranks he would eventually join. Within four years, he'd added to his little Sheffield estate a half-share in a thirteen-acre plot of farmland, twenty-three acres of pasture and woodlot, several horses, a house servant, a cook, and, among his many fine clothes, a dressing gown that cost three shillings and a pair of English shoes that cost even more. And best of all, by virtue of his holdings he'd secured for himself the official title of Gentleman.

He'd also acquired a wife, Elizabeth Mason, known as Eliza. The daughter of a deacon in nearby Franklin, she was at twenty a delicate, tender, fair-haired beauty, earnest but soft-spoken, and Theodore found her utterly captivating. Then, tragedy. A year into the marriage, Theodore caught smallpox in one of the contagions that routinely swept through the colonies. (One ran almost contemporaneously with the Revolutionary War and killed more American soldiers than the British did.) To contain outbreaks, Sheffield quarantined the victims in one of the shantylike "pock-houses" on the marshy meadow on the outskirts of town. Theodore was lucky; the attack was light, and the disease passed quickly. Frustrated by the isolation, miserable with idleness, and worried about Eliza, who was eight months pregnant with their first child, Theodore pleaded with the town physician, Dr. Hillyer, to release him. When Hillyer finally relented, Theodore returned home to the little Sheffield house. He resumed his law practice there and, the greater pleasure, lay at night with Eliza in their rickety "field bedstead," so called because a beam ran across the footboard, in their bedroom upstairs. But the virus had not passed, and he gave it to his wife. (Years later, he told his daughter Catharine that Eliza contracted it from him when she tended his long brown hair with her ivory comb.) It lodged in Eliza's respiratory tract, then spread to her lymph nodes, producing flulike symptoms—burning fever, pounding headaches. She tried to downplay the signs, but when they were followed by a rash of painful sores all down her throat and deep into her nasal passages, there was no more pretending. As was his duty, Theodore summoned Dr. Hillyer, who, cursing himself and Theodore, sadly sent her to the pox house. It was there, while she lay sweltering in the airless shed, that the disease burst into the hideous and agonizing bubbles all over her skin that made the diagnosis unmistakable. Immune now that he had already had the disease, Theodore stayed with her, and did his best to soothe her with a cool damp cloth that he ran lightly over her blistered skin. But when Dr. Hillyer, exhausted from his rounds, returned to check on her, he could see there was little hope: the pustules were beginning to bleed into each other, making it all but certain that the disease had turned lethal.

Theodore did not leave her side. Torture everywhere on the body, the pox is excruciating on tender parts like the palms of the hands and the soles of the feet, where the relentless itch can drive a sufferer mad. But the skin everywhere—Eliza's beautiful face, her once-smooth legs, her slender arms, the round belly where their child grew within—all of it gradually dissolved into oozing pulp, cracking as it dried, and then, as she writhed, moaning, in her narrow bed, peeled off in long sheets that smeared the bedclothes with blood and filled the air with the stench of rot. So died Eliza, and the baby within her.

For the first and only time in his life, Theodore was convulsed with grief. In his dreams, he sought Eliza out again and again, well into his next marriage, well into his next family. He dreamed of Eliza coming to him in their wide field bed. He would rise and reach for her, grasping only air. The dream's repetition suggests lacerating guilt as much as it does longing. But, perversely or not, this recurring vision of his late wife comforted and sustained him, and, when it came to him later in life, it would leave him radiant in the morning, and later still, he would whisper conspiratorially to his daughter Catharine, always his closest confidant, "I have had my dream."

SIX

AMONG THE RIVER GODS

Theodore had had his eye on Pamela Dwight since she was twelve. Mark Hopkins, the legal mentor who had first drawn Theodore to Great Barrington, was married to her much-older half sister Electa Sergeant, and little Pamma, as she was called then, often came to stay. She was a delicate child, prone to the febrile condition the colonists termed "ague," and painfully shy. Her mother, the fearsome Abigail, complained that when she attended a Harvard commencement as a young teenager, she'd hardly spoken to a soul. Then again, Pamela had reason to be timid, for by the time Theodore came along, she'd been exposed to any number of terrors. When her parents first brought her to Stockbridge as an infant, a hideous cry suddenly went up in the dead of night: "The Indians are coming!" In a panic, Abigail handed the little Pamma to a servant to hide her in the hills. He took off, never to return, and ditched the squalling baby at the first opportunity. A small band of Mohawks did indeed come, burst into a house just down the road, and with their tomahawks, hacked to death a servant, a small boy, and an infant still in its cradle. Once the rampage had subsided, another, more loyal servant soon found the whimpering Pamma under a copse of trees by the roadside. Afterward, Pamela was always fearful that the Stockbridge Indians were plotting something (which, in fact, they sometimes were: the more militaristic Mohawks tried to turn them against their English "protectors"). Pamma's youngest brother, Wollaston, had died in infancy; and the year she met Theodore, her father, the brigadier general, suddenly collapsed and died, too.

Pamela had attended the local elementary school for a few years. After her father's death, Abigail had hoped to get her into a boarding school in Boston, but that never came to pass. In contrast to Theodore's invariably formal and precise letters, Pamela's early missives are a tangle of country-girl bad spelling and irregular grammar, and in her more despairing moods later on, they are full of apologies for any "imbecillity" such clumsiness reflects. Still, her writing always conveys the brisk, pizzicato movements of what is obviously a lively and sensitive mind.

An engraving from her later teens shows the Pamela that attracted Theodore: a young gentlewoman in a fashionable bonnet and prim dress with full skirts, her hands crossed virginally over a damask rose on her lap. She has limpid eyes, a nose that dominates her face with its verticality, and a small, delicate mouth from which few words came, and no loud ones. A later portrait done in 1795 by Joseph Steward, a minister turned itinerant painter, is possibly more revealing, as it shows Pamela as a mother, with six-year-old Catharine standing by her side, their hands lovingly crossed together, in that fine house that Theodore built for them. Pamela's dress is draped with fine black lace this time, indicating prosperity as only elaborate textiles could in those days. Significantly, Steward has included a view of the splendid Mission House off her left shoulder. High up on its hill, a mile away, it was not, in fact, visible from the Sedgwick house. But it did represent Pamela's lineage as the product of not just one River God, but two. With Pamela, her background was always in the picture.

In Pamela, Theodore saw a future Abigail, one who could be immensely useful in advancing his interests. An impressionable young New Yorker, a Miss Susan Morton (who went on to marry Harvard president Josiah Quincy), visited her the year Theodore arrived in the Berkshires, and in a memoir offers a portrait that conveys the spell—partly of wealth, partly of her own innate hauteur—Abigail cast over him. She was "tall and erect, precise in manner, yet benevolent and pleasing," Morton writes. "Her dress, of rich silk, a high-crowned cap, with plaited border, and a watch, then so seldom worn as to be a distinction, all marked the gentlewoman and inspired respect." When she went to church, young Susan herself rode double on a horse, others went by "waggon," but "Madam Dwight" always rode alone in her own chaise.

LIKE WILLIAMSES ELSEWHERE, Abigail had added substantially to her luster through her marriage to Brigadier General Joseph Dwight, a former Speaker of the House of Representatives in the Massachusetts General Court, who conducted himself with a soldier's tautness and strut. He'd come to Stockbridge to help resolve a controversy involving some shameless self-dealing in the Williamses' management of the Stockbridge schools for the Indians. Jonathan Edwards, just arriving as the new missionary, was counting on Dwight to put the Williamses in their place. Dwight seemed to be inclined to do so—until he met Abigail and was bewitched. Within a year he was her husband, and, as a professional merchant, had become the official steward of the Indian schools, selling them their supplies at inflated prices from his own stores.

With such a heritage, Pamela would have caught the eye of any number of suitors, but, when they drew close, the more discerning might have detected something else about her to put them off. Delicate, sensitive, and alert as she may have been, she also had a neediness, a sense of oppression, that is manifest in that sad "worthlessness" that comes through so many of her letters, especially the later ones. Even in her reduced circumstances as a widowed mother, Abigail overwhelmed her Pamma, and there is more than a hint of resignation, of defeat even, in the limpness of Pamela's pose in that early engraving. It becomes even more marked in the grim solemnity of the later painting by Joseph Steward. To Theodore, though, the frailty may have been part of her appeal. It had to have been pleasant to be needed by a woman of standing.

Is that why Abigail was dead set against the match? Was she onto him? Or was she just being naturally overprotective of her vulnerable teenage daughter? Either way, it was risky to cross a man so obviously on the rise, so, instead, she picked away at an aspect of Theodore's character that was more openly questionable: his apparent atheism.

Now that his ministerial ambitions had been put aside, Theodore was no regular churchgoer, and, radically for the time, probably not a believer. At Abigail's urging, Pamela dutifully interrogated her suitor on his religious views in a long letter in June of 1773, when their courtship first turned serious. How Pamela labored over it, writing and rewriting. From the first silky *S* of "Stockbridge," her penmanship, often slapdash, is almost seductively curvaceous as it salutes "the most Generous of men" and then subjects him to a kind of catechism of the faith. Does he subscribe to the essential Calvinist tenets, she asks, and can he instruct her in them, as good Puritan husbands were expected to do? Significantly, she hints she needs guidance in "the Paths of duty" as well, for she is "always wandering." In pleading, desperate tones, she wishes "to heaven that there was no Essential difference in our sentiments" before concluding with a fervid declaration of her eagerness: "O what solid satisfaction what exquisite Pleasure would you give me Could you convince me that you are walking with the Lord."

Theodore's reply has been lost, but it evidently was not reassuring.

After she read his letter, Abigail refused to let Pamela see him, but for once, Pamela defied her. When her mother found out several months later that she was, as Abigail peevishly put it, receiving "Mr Such-a-Ones visits," she gave her daughter such a slap that Pamela squealed for forgiveness, even as she remained adamant in her longing for Theodore:

> *I have often Longed to open my heart to you Madam but the fear of your displeasure has hitherto Deterred me from It. And will can the best of Parents forgive me If I confess to her, with tears and Blushes I Confess to her that I have a very grate and tender affection for a Person that I believe is worthy of my regard. I need not say who. perhaps Madam you may be astonished to think I have not done with all thoughts of this affair some time agoe and may think that I am wholly Led away by Idle misplaced affection. but Pardon me If I say I think I have had a sincere regard to Duty in this affair If I know what a regard to Duty is ... I know that I am submiting my self to the Judgement of one of the kindest of Parents and to one whose own Delicate tender feallings will teach what are the fealings of her Child.*
>
> *Pardon Pitty and Love still madm you weak your worthless Child and Let her know what is her Duty.*

Abigail let her know what her duty was—to give him up. But Pamela refused. In frustration, Abigail consulted the good Reverend Stephen West, minister of the town's new Congregational church, who recommended she meet this suitor of her daughter's, and see for herself what to make of him. So, one chilly March afternoon, she summoned Theodore to tea in the Mission House parlor to look him over privately.

As soon as Theodore appeared in her doorway, Abigail knew that he was a force to reckon with. He blocked out the light as he stood before her. Handsome, tall, forceful, obviously intelligent, capable—he was hard to dismiss out of hand. And then, as they sat together before the sputtering fire in her parlor, a memory bloomed in her mind, of the scintillating Ezra Stiles, her own true love, who had wooed her assiduously in the early years of her widowhood, and then slipped away—later to become president of Yale, no less—over stupid, doctrinal issues. She sensed a warmth in him that made her a doting girl again. She nearly kissed him when he left. She went straight to her desk to issue a glowing report. "He Opened himself with so much wisdom sincerity piety & most tender love and affection to you, that it melted my heart," she wrote. "He expressed no Anger at you, but pity & pure love & anxious concern for you,...all full evidence of his real sincere affection of you—Oh! My dear Child may you be wife & happy."

How Pamela's own heart must have blossomed to read that! And so the union was sealed. With Abigail's blessing, Pamela and Theodore were wed just a month later, on April 17, 1774.

SEVEN

THE WAR WITHIN THE WAR

That spring, rebellious sentiment was mounting throughout the colonies. Common ancestry had obscured the diverging interests on either side of the Atlantic until the crown, strapped for cash to pay for its expensive wars against the French, started boosting taxes on the colonists. Back in 1765, the Stamp Act had first aroused resentment up and down the coast for imposing onerous taxes on the publications and legal papers that the colonists required for their news and daily business. All the colonies railed against it, but in hot-tempered Boston, mobs took to the streets to man-handle British tax collectors. The act was repealed the next year, but the damage was done.

Far to the west, in Sheffield, most of the leading citizens had been loath to throw over a British political order that had served them fairly well. To Theodore, eager to gain power now that he had wealth, that left openings in the leadership on the revolutionary side, and on January 12, 1773, he joined a dozen other Sheffielders at the home of the sixty-six-year-old grandee Colonel John Ashley to take a more assertive political stance.

For Theodore, Ashley was the best possible ally—the most important man in Sheffield, owner of its ironworks and much of its desirable land, and the local region's sole delegate to the state House of Representatives, besides. Unusually for him, he was now a man in need, for he'd voted to disavow a letter sent out by the House opposing the Parliament's latest tax plan, the Townshend Acts, which had levied heavy duties on much-needed British imports like lead and glass. By a vote of 92 to 17, the House refused the crown's demand to take the letter back. In a fury, the British closed the House down altogether, and then shuttered virtually all the other colonial assemblies down the coast as well, angering the colonists all the more. Ashley, however, had been one of the "infamous seventeen," as the Massachusetts House's conciliators were quickly dubbed, not the "glorious ninety-two" who were in open rebellion. The citizens of Great Barrington had voted to deem Ashley's conduct "repugnant."

To make amends, Ashley drew Sheffield's leading men to his front parlor to craft what became known as the Sheffield Resolves, one of many such documents formulated throughout the colonies, to express a measure of defiance to the young King George III, to whom most Berkshiremen still instinctively deferred. Because of his reputation for eloquence, Theodore was chosen by Ashley for the delicate job of finding the precise language. The result was a tentative statement of discontent that Jefferson would put far more forcefully—and memorably—three years later. In his most decorous penmanship, Theodore began by sweetly professing "the most amiable regard and attachment to our precious sovereign" and offering "that deference and respect due to the country on which we are and always hope to be dependent." But then he touched on some of the themes the Declaration of Independence would later ring out, noting that the purpose of government was to provide its citizens something more "than was possible in a state of nature," namely the right to "undisturbed Enjoyment of their lives, their Liberty and"—tellingly—"Property." To Theodore, property *was* happiness.

Provocative as Theodore might have considered them, the Sheffield Resolves soon seemed very tame. Insensitive to the delicacy of relations with the colonies, the British government awarded a monopoly on the colonists' tea to the East India Company, drastically boosting the price on a favorite beverage. A clutch of Boston men dressed up as Mohawk Indians and, whooping, dumped the East India tea into Boston Harbor on December 16, 1773. The British retaliated by closing the port of Boston, canceling most of the self-governing provisions of the charter of Massachusetts, and replacing the colonists' judges, sheriffs, and other officials with appointees of the crown.

Clearly, it was time for bolder measures. On July 6, 1774, three months into his marriage to Pamela Dwight, Theodore hurried to Stockbridge's Red Lion Inn to join a boisterous crowd of sixty other Berkshiremen pushing to put the county on a war footing. Still not ready for open rebellion, he politicked for sending food to the "distressed inhabitants" of Charlestown and Boston who were suffering from the British shutdown of Boston's port; to boycott all British goods, substituting rough local leather for fine imported textiles (a personal sacrifice for Theodore, who fancied British linen); to build up economic independence by raising local sheep for their wool; to avoid the self-defeating "Mobs and Riots" that had broken out elsewhere in the Commonwealth—a particular concern of Theodore's; and to treat with "Neglect"—that is, to ostracize—any citizens who did not go along.

Such recommendations were hardly a declaration of war, but, along with the many other less temperate responses, they heightened tensions between the colony and the crown. The First Continental Congress was convened in Philadelphia's Carpenters' Hall, but Boston remained the frenzied heart of the resistance. British troops poured into its slim peninsula to tame it; the shore was ringed with masts of Royal Navy ships. But the heavy military presence proved only a greater irritation. While the colony girded itself for war, General Thomas Gage, the prickly commander of all the British forces in North America, was made royal governor of Massachusetts as well, making it all the more evident that the British government likewise was preparing for battle. Unnerved by word from his spies in April that the colonists were assembling vast arsenals in Lexington and Concord, to the west of Boston, enough to support 15,000 troops, Gage decided to strike before the colonists could use these munitions against him. But word of his plans got out; the silversmith Paul Revere made his fabled midnight ride to Lexington; and when the British troops arrived at dawn, the Minutemen were already assembled on the village greens. The British demanded they disperse, but the Minutemen held their ground. Grapeshot was fired, a few soldiers fell, and the American War of Independence was on.

The electrifying news, fanning out through the colonies by post riders, reached Sheffield two days later. The town's Minutemen were, in fact, drilling on the green by Theodore's house when a post rider came thundering in on the morning of April 21. By noon, twenty men were on their way to aid the besieged Bostonians.

Theodore Sedgwick, however, was not among them, for Pamela had just delivered him his first child, and he was needed at home. It wasn't until a year later that he signed on as Major General John Thomas's military secretary in an ill-fated campaign, led by the tightly wound Benedict Arnold, to seize heavily fortified Quebec in June 1776 as part of an imprudent scheme to take all of Canada. It proved a fiasco. Among other calamities, smallpox broke out among the American expeditionary forces. Theodore's previous exposure saved him, but General Thomas succumbed; the disorganized invading forces scattered into the woods; and Theodore was left frothing about the "Follies of Villainy with which I am surrounded."

The following fall, Theodore joined his old legal tutor Mark Hopkins, now a revolutionary colonel, in the battle of White Plains, New York, in which Washington's ragtag Continentals, retreating north from New York City, were chased down by General Howe's redcoats. When Hopkins fell wounded during the fracas, Theodore ignored the whistling musket shot to find a litter and men to carry him off to safety, but his old mentor died of his wounds. When Washington's army continued to fall back, as it would repeatedly until the depths of winter, Theodore himself returned home once more and shifted over to providing military service of a sort for which he may have been better suited. He became the commissioner of supply for the northern department of the Continental forces, producing through private contracts everything, from meat to nails, that would be needed for the military campaign in western Massachusetts and northeastern New York.

THERE WAS THE war, and then there was the war within the war. While Theodore threw himself into one, he got caught up by the other. Unlike the hot war that pitted American patriots against their British overlords, the cold inner war set American against American. Ostensibly, it was the separatists against the loyalists, but actually, the division was at least as socioeconomic as it was political, as it most frequently pitted the commoners against the untitled but very real aristocracy. A class war, in other words. In the early days of the gathering revolution, it was fairly benign. In Sheffield, the revolutionary zealots erected a liberty pole, which was promptly cut down by a group of Anglophiles led by Dan Raymond, the chief merchant in town, whose imposing brick house stood across the road from Theodore's. When the British sympathizers were caught, they were forced to walk a genteel gauntlet, bowing their heads and abjectly begging the pardon of everyone in town. Only the man who had actually cut down the tree was dealt with severely. He was tarred and feathered, which was no joke: boiling tar was slathered on his bare skin. Finally he was mounted on a broken-down horse to go about seeking everyone's pardon.

But that was 1774. By 1776, Samuel Adams's Committees of Safety, which had been started to prosecute the war, had expanded their agenda, turning radical and, to Theodore's mind, insidious in their determination to control the revolution's politics along with its military objectives. The committees had sprung up all over the commonwealth, subjecting professors and ministers and others to ideological review, and taking for themselves the role of dispensing justice, refusing to allow proper courts to sit within the county. Theodore found it appalling, and not just because it threatened his own livelihood once the war was over. A single judge, operating without a jury, could brand a dissenter a Tory, have him tarred and feathered, seize his property, intern him in a remote detention camp, or banish him from the colonies altogether. But—who was to say who was a Tory? Who could look within himself and not find some lingering fondness for Mother England? And, when property was to be had for calling someone a Tory, how could anyone be sure that the real "crime" was not simply owning that property? After all, the Committees of Safety almost never went after a Tory who was not a rich one.

To Theodore, this was the work of a mob, and he would not stand for it. When in 1778 his good friend, and his wife's uncle, Colonel Elijah Williams, was arrested for treasonably aiding the enemy, Theodore tried to protect some of Williams's assets by buying his cattle in Theodore's capacity as supply officer. When the West Stockbridge Committee of Safety learned of that move, it turned its wrath on Theodore, seizing the cattle and then punishing him by taking some of his horses besides. Theodore was incensed. "The cause in which we are now embarked has professedly for its object the preservation of our property," he lectured the committeemen, glowering. "Think you that a lawless violation of property can tend to support such a cause?" How dare they seize Williams's cows, and Theodore's horses, without a trial? When the men told him to curb his tongue, he fired back: "This may seem free language, it is so. But remem-

ber that Freedom of Speech is one of the Liberties for which we are con-
tending." Was this to be a nation of laws, or of individual men wielding
power for their own ends? "If for the latter," he concluded, "I would as
soon submit to British as American tyrants."

Theodore scrambled to do what he could for other friends who found
themselves on the wrong side of the new sentiments, as a large number of
them did. The following winter, a gang of thirty Tory hunters came to
Theodore's house to "call me to account," as he wrote a friend, for such
attitudes. They came at night with torches, banging on Theodore's
kitchen door. When he did not immediately appear, they burst the door
in, and charged up the stairs. Leaving Pamela in the bedroom looking
after their second child, just six months old, Theodore met them in his
nightshirt in the hall. The gang was all liquored up for this encounter,
which did not serve them well. There were angry words on both sides,
but the men's threats were put down by a piercing look of indignation on
Theodore's part. "Their numbers so thin that they dare not attempt the
encounter," Theodore wrote, rather gallantly, since he in fact had been severely outnumbered. "This is however just a Beginning—what the End
will be can only be guessed at."

IN THE FIELD, the Continental forces commanded by General George
Washington, nearly decimated at Valley Forge in the winter of 1776, had
advanced smartly on the British in the fall of 1777, when Patriots out-
maneuvered overextended British troops in two battles at Saratoga, New
York, capturing six thousand men. Those victories brought the French,
always looking to press any advantage against their British nemesis, to the
American side, and tipped the war decisively in the Americans' favor.

By 1778, the time had come to look to a new, independent future. By
then, all the colonies but Massachusetts had written new constitutions removing the crown from their political affairs. In the Berkshires, the radical element, determined to take literally the Declaration of Independence's
stated ideals of equality, threatened to force the county to secede from
Massachusetts and join with New York or Connecticut if its state constitution did not include provisions that were more favorable to the common
man. Theodore did his bit to squelch these "mobbish, ungovernable, re-
fractory people," and their secessary movement was put down.

When, in 1780, the state constitution was finally written—by John

Adams, with a clear tilt toward the propertied classes—Theodore put himself forward to serve as Sheffield's representative to the lower house. On an election day in Sheffield made festive with gingerbread and home-made beer, the town's legal voters pushed inside the meetinghouse across from Theodore's little house, and, by voice vote, making everyone accountable for his choice, they elected Theodore the town's first representative to the new General Court of Massachusetts. It appeared that a Sheffield majority was favorably disposed toward what Theodore liked to term "the friends of order." For now.

EIGHT

WILLIAMS FAMILY SECRETS

In marrying into the Williams family, Theodore was availing himself of an impressive social network that would put him in good stead politically as well as financially. But a family is, of course, more than its connections. In the case of the Williams family, it was also a shadowy emotional underworld for which Theodore, as a starry-eyed rationalist, was woefully unprepared. There were wicked secrets in the Williams family—the betrayal of the Stockbridge Indians primarily.

Among other injustices perpetrated against the Indians, Ephraim Williams Sr. peremptorily built a mill for himself on Indian land. When the Indians objected, he turned to an English court on which sat many friends, and it agreed that it would be "inconvenient" for Ephraim to be deprived of the mill. He later sold it to his son Josiah for £160. He sold Josiah a second parcel he was not entitled to, as well. Shortly afterward, in the fall of 1752, the sixty-one-year-old Ephraim started behaving strangely. He rose before dawn one morning and wandered all about Stockbridge, flushing the residents of virtually every house on the Plain out of their beds to offer them lavish sums for their property—as much as twice their own purchase price—to be paid in the cash he'd stuffed into his pockets. It was a kind of deranged parody of his land dealings: after taking the Indian property for nothing, he was offering the English ridiculously too much for theirs. And not just in Stockbridge. He rode all day through a thunderstorm to Wethersfield, Connecticut, then the next morning on to New Haven, then to Newington, then back to Deerfield, and finally clear to Boston, offering huge sums for houses nearly everywhere he went. Horrified, Abigail pleaded with her brother, Ephraim Jr., to stop their father before he undermined "our piblick affairs & intirely ruin[s] us, for he is...by no means...In one quarter of his notions. I beg you Do all in your Power to get him ye mind of Coming Home as Soon as may Be, if you have any love for him or us."

But her father did not come home. He returned to Deerfield, and he stayed there. He sold all his Stockbridge land, some 1,505 acres in all, to his son Ephraim Jr., whose estate would later go to the founding of Williams College, and included "my negro servant Moni, my negro boy London, also my Negro Girl Chloe." He requested that his son Elijah send the "Red Jackit & blue millatary Britchis...also the thing I put over my head to keep my Ears warm which I button under my chin." He couldn't remember the word *hat*. Dementia had set in. By 1754, Ephraim Williams Sr. was dead.

By then, of course, he had passed on this seed to his descendants. What, exactly, did it consist of? Moderns would locate it on the chromosomes, as if such specificity lent clarity. They would cite the *Diagnostic and Statistical Manual of Psychiatric Disorders,* trying to determine if this cluster of symptoms was sufficient to drop Ephraim's behavior into this category or that. There are elements of what is currently termed hypomania, a low-grade version of the full-blown mania, in Ephraim's deluded buying spree. But useful as such diagnoses might be in the present, they are dubious in the past, especially the distant past, when different concerns bred different disorders. In any case, I don't think of this particular dark seed as anything so gothic as a propensity for evil. Rather, I view it as a certain psychological vulnerability, and I seize on it in Williams's case because it marks the first occasion in the historical record that any sort of madness manifests itself among the Sedgwicks or their forebears, and it defines a theme that, with variations, would reappear at regular intervals throughout the generations since, and eventually come down to me, like a soup-spoon wrapped in cloth that is all that is left of the family silver. This vulnerability manifested itself in Ephraim Williams as a kind of vestigial innocence, a soulful fragility that crumpled under moral pressure. And, for most of the affected Sedgwicks since, it has likewise been curiously passive, a too-great openness to feeling; it's as if a window has been thrown open in us, where for others it has been raised just a crack. In Williams's case, the pressure came from guilt, from the recognition that his life was built on a deception. Under the guise of generosity toward the Indians, he had stolen from them. As he aged, the belief that he would shortly face an all-knowing Calvinist God must have added a dreadful anxiety to his shame: the flames of hell were very real to all the Williamses. In his delusion, he tried to undo his crimes by frantically buying up these properties, as if by wildly overpaying for them, he could somehow cancel this psychological debt and remove its stain. No, the dark seed did not lead him into evil, but into a full awareness of it. He became a prisoner of his guilt.

The seed passed to Pamela, where it lay waiting to be awakened once more, in yet more explosive forms, in her and in her offspring.

NINE

THE HOUSEHOLD
DID NOT RUN OF ITSELF

Naked lust is usually left out of eighteenth-century epistles, only in part because the letters were likely to be passed around and read aloud. In Theodore's case, his letters to his new wife could be counted on to be read by her mother, Abigail, who became, as Pamela declared, her closest "Female Friend." Nevertheless, Theodore's eagerness was sometimes too much for him, and his itch is frankly expressed. For all his aristocratic trappings and aspirations, his broadcloth and silk, Theodore had his passions, fierce ones even—for friendship, for the finer things, for hatred (he was a terrific hater), for his family, for glory, and, not least, for sex. Well into the marriage, when his brood already numbered four, he wrote to Pamela from New York—where he was staying with the Massachusetts delegation—with startling directness for the time: "I wish my dearest love you would spend not only an *evening*, but a *night* at my hotel." You can almost see the tender Pamela flush at that. But she returned his eagerness, if more delicately. "You will suffer me to be a Little Angree with Congress," she wrote him when his congressional work dragged on, "for so long detaining my Husband from my Arms."

Once the second Mrs. Sedgwick set up housekeeping with her husband in his little Sheffield house, the couple reserved for themselves the bedroom at the top of the stairs. There Pamela lifted her linen petticoats for him, and settled him down between her thighs to rock together on that field bedstead, taking seriously the divine injunction to be fruitful and multiply.

The first baby, a girl, did not come for nearly two years—on April 30, 1775. Pamela delivered her in the roomy kitchen downstairs, while standing nearly upright astride two chairs. A Yale graduate, Dr. Samuel Barnard, had taken over from Dr. Hillyer as the most-trusted area physician, although he was based in Deerfield. For reasons of modesty, he did not attend births. A midwife did the honors, assisted by a crowd of local women, Abigail being one of them.

The baby's name itself is revealing, for Theodore insisted on naming the child after his first wife, and Pamela couldn't think to do anything but go along. And so she was Elizabeth Mason Sedgwick, known as Eliza.

She would soon grow into an eye-catching little girl, "perfectly symmetrical in her form, with pretty, dark eyes and hair, and a very gentle, modest retiring manner," as Catharine described her, adding that she was the daughter who most resembled their mother.

Two years after Eliza's birth, in March of 1777, an unnamed child lived only a day, a victim of the spring mud season that turned the roads into deep oozing corridors of slop, and made it impossible to summon Dr. Barnard from Deerfield to provide emergency help. Theodore, at least, was home from the war by now, buried in the details of his work as supply agent. Pamela bore the loss, but not well. All the same, another child, Frances Pamela—with her "fair skin, and blooming cheeks" according to Catharine—came the following spring of 1778, just as Theodore was beginning to antagonize the more radical elements of the separatist movement.

By standing up to this mob, Theodore reinforced Pamela's impression that her husband was a god, but the incident also reminded her of how much she had to fear. As the war advanced, Pamela's anxieties mounted, especially when Theodore was away, and her health suffered. It was the beginning of her own psychological woes, ones that would only deepen with time. There is a great divide in modern psychiatry over the question of how much of such psychological distress is organic, and therefore inevitable; how much of it can be brought on by genuine, debilitating suffering—isolation, hardship, pain; and how much, as seems most likely, results from a complicated interplay between the two. Would Pamela have descended if she had been free of that oversensitivity that her grandfather demonstrated? It is hard to know, but I doubt it. Plenty of lives were harder than hers. But then again, hers was by no means pain-free. In 1777 she acknowledged suffering "frequent turns of the headache." Three years later, in 1780, she admitted to her Boston friend Eliza Mayhew she'd been experiencing "but little health for this two years past." That was the year that Theodore won election to serve in the General Court in Boston, beginning a political career that would take him away from her for ever longer and ever more debilitating stretches. A small thing at first, her ill health, but it grew in time.

As SHE GOT older, little Frances—or Frannie, as she was known— became a devoted reader of romances; she was "excitable, irritable, enthusiastic, imaginative," in her younger sister Catharine's appraisal, characteristics that would lead her into romantic trouble later on. Two years on came the sturdy and dutiful namesake, Theodore Jr., in December 1780, who, inevitably, gloried in his father's attention, and on the strength of it became so bossy toward the younger children that they dubbed him "the Major." Pamela delivered a daughter named Catherine early the following summer, but, to Pamela's distress, she died the next spring. She was followed by a son, Henry Dwight, the spring after that, and he did not last a year.

Of Pamela's first six pregnancies, half the babies had died before age one. Normally in Sheffield, only one out of ten babies died (although the death rate was higher elsewhere), and it made Pamela, already sensitive, feel that the fragility that she'd always imagined was hers alone was now being extended to the children she bore. And, worse, that the bubble of safety that protected her from the many hazards of the world was dangerously thin. It wasn't hard to think that way in a remote farming village, with its penetrating cold from fall to spring, contagions, Indian uprisings, political upheaval, and now, with the revolution, open warfare. It was a kind of curse.

After each death, Pamela valiantly washed the little body on the big kitchen table, just as she had seen her mother do with her tiny brother Wollaston, and covered it in a linen shroud that tied below the feet, and then placed it in a small pine coffin to be covered in a pall and taken by

Theodore and other "underbearers" from town to the Sheffield burial ground while the massive bells of the meetinghouse tolled. The ordeal brought on more headaches, then a fretful sleeplessness, and finally a general lethargy that proved hard to dispel.

PAMELA GAVE BIRTH one more time, in September of 1785, to another Henry Dwight, called Harry, before Theodore decided that he had finally had enough of Sheffield. The new house in Stockbridge was not completed; work in fact had hardly started. But Pamela was desperate for a change of scene, so Theodore moved the family into the Timothy Dwight house while his own was constructed across Plain Street. After those months of exasperating delay, in the summer of 1786, the house was sufficiently enclosed that the wagons could be loaded up once more and the Sedgwicks' furniture carried across Plain Street and through the big, wide front door of the grand house that Pappa had built for them all.

The petite Eliza, already starting to remind people of her mother, was eleven that summer; her little sister Frances, a dreamy child of eight; Theodore Jr. a stalwart five. Little Harry was not yet one, and as a sickly, nervous baby a constant worry to his mother, who feared he wouldn't live out the year.

Still, to cross that threshold must have seemed to the older children as if they were pushing into a dream. Who had ever seen ceilings so high? A staircase so long? Such tall windows? And so many bedrooms! Not one apiece, quite, but infinitely more space than in Sheffield, and a lot more than in the Edwardses'. For Pamela, it would be a bustle just to fill all the rooms with furniture. Much of it would have to come from Boston or New York, if not London. Local shops could never provide the sort of dining table the new dining room required, or suitable couches for the two front parlors, especially that "best" one to the right of the door as visitors came in. And they'd need pictures for the walls, and bedding for the bedchambers, and silver and plate for formal dining. So many things, and seemingly all at once.

SUCH A BIG household did not run itself, though. There was cooking, washing, baking, cleaning, polishing, sheep-shearing, spinning, sewing— a whole bustling private economy to keep a large family going from one day to the next. Just heating the house, with its high ceilings and all that glass, was backbreaking labor, especially when you consider the depth and extent of the brutal Berkshire winters. A house of this sort would burn nearly fifty cords of wood in a typical winter, enough logs to fill the two massive front parlors of the Sedgwick house up to head height. It's hard now to get a feel for the cold, the constant need for heavy clothes, for insular drapes around a bed, for two (or more)-in-a-bed "bundling" that raises the eyebrows of moderns, and for steadily tended fires. Even so, in those days before even the slightest insulation, and before efficient fire-places, or "Franklin" stoves, the temperature difference between indoors and out was not always very great. It was not uncommon for a day room's temperature, when taken halfway between fireplace and outer wall, to fall below fifty degrees, and a bedroom without a fireplace would rou-tinely plunge below freezing overnight, leaving any bedside water frozen solid in the morning. One settler records his astonishment at looking into the fireplace one frigid night and see the log burning in the middle, but the sap frozen solid where it had oozed out at the end.

No wonder one of Catharine's sharpest memories, from when she came along a few years later, is of the roaring fireplace. "Oh, that blazing fire!" she nearly shouts in a memoir. "There may be such in Western homes, but they will never again be seen on this side the Alleghenies. As the short winter day closed in, a chain was attached to a log, and that drawn by a horse to the door-step, and then rolled into the fireplace, shak-ing the house at every turn. Then came the magnificent 'fore-stick,' then piles on piles of wood—and round the crackling fire what images appear!"

But the logs did not bring themselves into the Sedgwick house. They were hauled by a flock of servants, retainers, apprentices, and, in a couple of cases, runaway slaves. There were always at least half a dozen servants working in the house, although none of them lived in it. They were obliged to sleep outside in the crude outbuildings to which Theodore, in his building contract, gave no regard, any more than he would have wasted his time on the "necessary," as the outdoor privy was termed. The-odore was not one for mixing. Much as Catharine worshiped her father, she couldn't help noting that his "house was one of the few where the do-mestics were restricted to the kitchen table," rather than dining with the family, as was more customary. This created, in the household, a world that was split into sunshine and shadow. His family, of course, existed in

the full sun. Indeed, one might almost say that Theodore was himself the
sun that radiated upon them. In the shadows were the domestics—a word
just coming into favor—who slipped silently about, many of them almost
completely unnoticed.

Yet there was a ladder for them no less than for Theodore, and as for
him, ascension for them was a product of talent, hard work, pluck, and
connections. The head man, Agrippa Hull, was the son of freed black
slaves in Northampton farther to the north. He'd come to Stockbridge on
his own at age six to "apprentice" as a servant, then caught on with the
household staff of a local colonel in 1777 before winning the more presti-
gious role of personal servant to the heroic Polish engineer and general
Thaddeus Kosciusko, who'd come to advance the revolutionary cause.
Hull was a bit of a card, and Kosciusko once caught him dressing up in
his general's uniform, even blackening his bare legs and feet to look like
he was wearing shiny boots, for the entertainment of his fellow domestics.
Or was he mocking the dandified ways of his strutting master? Turning
his servant's show against him, the general dragged him—by the ear, one
imagines—into the tents of several officers, to pass him off, Catharine
says indignantly, "as an African prince," and then dismissed him with a
kick.

After the war, Grip (as the Sedgwicks always called him) returned
to Stockbridge, where he became Theodore's manservant, often accompa-
nying him to Congress, and assisting him at home. When Grip heard
that a slave named Jane Darby had fled her master, a Mr. Ingersoll, in
nearby Lenox, Grip nervily persuaded Theodore, as a lawyer, to arrange
for her "discharge," whereupon Grip married her, and she joined the
Sedgwick household. In his later years, Grip himself became something
of a man-about-town in Stockbridge, always ready with a quip or a
rhyme, and was a regular figure at local weddings. Still, the Stockbridge
townspeople never overlooked his race. In his prayers, according to one
near-contemporary account, they smugly claimed he "gave thanks for the
kind notice of his 'white neighbors to a poor black nigger.' "

There was a mean-tempered black woman dubbed Tip-Top, whom
Catharine regarded as "the Baberlunzie, the just and terror of my childhood!" More appealing were the runaway slave named Sampson Derby,
who did the cooking in that vast kitchen fireplace, and the "impish lady
Prime, also known as Betty or 'little Bet.' " A young black man named

Moses Orcutt apprenticed himself to the Sedgwicks in exchange for being taught how to read and compute basic sums and then given, at twenty-one, such help in acquiring future employment elsewhere "as indifferent men shall determine he ought." The only true disaster was Cato, who, for "good food & Clothing; 2 suits of Clothes & a bible," signed with the family as an indentured servant. Shortly thereafter, Cato was convicted of "assaulting and ravishing" a nine-year-old girl and spent the rest of his life in jail.

But the undisputable queen of the domain was Mumbet. If Grip was the shadow father, Mumbet was the shadow mother and more. She'd come into the family as a paid servant back in Sheffield, but her work was so grueling that it must have been hard for Mumbet to distinguish it from the slavery she'd left: spinning wool on a clattering wheel; weaving cloth on a loom; churning butter with a long wooden oar; baking bread in the "reducing" flue oven that was hollowed out of the brickwork beside the kitchen fireplace; and the endless, exhausting effort—through frigid winters no less than hellish summers—of laundering, drying, and ironing the family clothes, from Theodore's long silk stockings to Pamela's linen undergarments to the newborns' gowns.

Beyond her skill in conventional housework, Mumbet was also an accomplished midwife. Even as a slave, she was always being sent for by the neighbors at odd hours of the night, and she'd acquired all the folk medical knowledge that went with the trade. Herbs, roots, Indian remedies, and the like were all part of the household armamentarium. She assisted at all the births, and at those early burials, too. And as Pamela declined in her grief, Mumbet ascended. Catharine, for one, placed her at the moral center of the family. As she writes, Mumbet

> was never servile. Her judgment and will were never subordinated by mere authority.... [She] had a clear and nice perception of justice, and a stern love of it, an uncompromising honesty in word and deed, and a conduct of high intelligence, that made her the unconscious moral teacher of the children she tenderly nursed.... I do not believe that any amount of temptation could have induced Mumbet to swerve from truth. She knew nothing of the compromises of timidity, or the overwrought conscientiousness of courage and loyalty. In my childhood I clung to her with instinctive love

and faith, and the more I know and observe of human nature, the higher does she rise above the others, whatever may have been their instruction or accomplishment. In her the image of her Maker was cast in material so hard and pure that circumstances could not alter its outline or cloud its lustre....I well remember that during her last sickness, when I daily visited her in her little hut—her then independent home—I said then, and my sober after judgment ratified it, that I felt awed as if I had entered the presence of Washington.

TEN

ALL MEN ARE BORN FREE AND EQUAL

Illiterate, Mumbet was unable to record her own story, so it, like her for much of her life, became largely the property of others. She stands in time like Monument Mountain, as it is seen from the Old House—towering and majestic, but also somewhat cool and remote. Of the relatively few evocations of Mumbet from the time (many more would come later, when her true significance became known), two stand out to bring before us the full, sweaty reality of her.

The first is a small watercolor painted by a daughter-in-law of Theodore and Pamela years later, in 1811. It's so small, just two inches by three, it bears the look of a votary, but nonetheless it shows Mumbet as she must have been: plump and rather crotchety-looking, especially as she got along in age. It's the only likeness that exists of her (and one of the few of any African-Americans from this period), and in it Mumbet stares out rather suspiciously at her portraitist. The folds of her plain blue dress pull tightly on her, revealing a full bosom and meaty arms. A maid's white bonnet that ties around her face makes her black skin blacker still; its modest frills contrast starkly with her earthy physicality. And she wears a slender gold necklace. Whether she got that from the Sedgwicks, or to keep up with them, the small vanity offers an intriguing hint of vulnerability—or pride—in a homely figure that otherwise betrays none. But the skepticism of Mumbet's face is arresting. Her head is turned dismissively, her thick lips turned down in a harrumphing attitude. There is power in her disdain.

The second is an incident that brings this image to life. Catharine set it down in an article called "Slavery in New England" that she wrote decades later, in 1853, for the London magazine *Bentley's Miscellany*. Essentially an account of Mumbet's extraordinary campaign for her own freedom, the article details Mumbet's years as a slave in the Sheffield household of Theodore's sometime political ally Colonel John Ashley and his crabby, prudish Dutch wife, Aadije, better known as Hannah. Catharine seems to have memorized the tale, right down to the smallest details, as if she had heard it often as a girl.

"It was in May," Mumbet began, "just at the time of the apple blossoms; I was wetting the bleaching linen, when a smallish girl came in to the gate, and up the lane, and straight to me, and said, without raising her eyes, 'where is your master? I must speak with him.'"

Mumbet knew right away what the trouble was. "*Gals* in trouble were often coming to master," meaning Colonel Ashley, who doubled as a justice of the peace in the small town. White gals, she means. You can see the raised eyebrow when she says that. It's the look of the portrait. "In trouble" had the same meaning then as now, but there was more to it this time than usual, Mumbet immediately sensed. "I never saw one look like this," she told Catharine. "The blood seemed to have stopped in her veins; her face and neck were all in blotches of red and white. She had bitten her lip through; her voice was hoarse and husky, and her eyelids seemed to settle down as if she could never raise them again." Mumbet brought the girl— her name was Tamor Graham—into the house, and shut her into the bedroom by the kitchen so her mistress wouldn't find her. "She had a partic'lar hatred of gals that had met with a misfortin'," Mumbet explained. But Hannah Ashley did notice the girl, and immediately ordered her out of the house.

Mumbet coolly told the girl to stay where she was.

Her fury rising, Hannah insisted she go, but Mumbet told her no. "If the gal has a complaint to make, she has a right to see the judge; tha's lawful, and stands to reason beside."

Mumbet arranged for the girl to eat with her in the kitchen while they waited for Colonel Ashley to appear and then offered the girl food from her own portion of dinner, since Mumbet believed she had no right to "madam's food." But the girl was too agitated to eat.

Finally, Ashley arrived, and Tamor's story tumbled out. She had been raped by her father, just as Mumbet suspected. She had come to see him prosecuted, even though he'd be hanged for the crime if convicted. Ashley gave charge of the girl to Mumbet, while he sent word to the authorities. Shortly afterward, Tamor received a message to meet her mother. Unsuspecting, she went alone—only to be kidnapped by a couple of her father's henchmen, and locked in a hut deep in the woods to keep her from testifying. The militia was called out, and she was safely brought to testify. The father was found guilty, and duly hanged. Tamor left for a "distant province," never to be heard from again.

It is quite a story. Unfortunately, it may not be true. While there was a family of Grahams in the area, no account of an incestuous rape involving any of them exists in any surviving historical records. What does exist is a similar tale from 1806—the sensational case of Ephraim Wheeler's conviction and execution for the crime of raping his thirteen-year-old daughter, Betsy. (It has recently been retold in the book *The Hanging of Ephraim Wheeler,* by Irene Quenzler Brown and Richard D. Brown.) That story spread, via newspaper accounts, as far south as Virginia. And Catharine surely knew of it, because her father was the presiding judge in the case, and she wrote a fictionalized version of the story in the 1830s. Did Catharine take that case and transpose it back a half century, and place Mumbet at the center of it? It is a definite possibility, and it poses a moral conundrum, for Mumbet—or at least the Mumbet that Catharine evokes—would never have countenanced such a radical departure from the literal truth, and would have been embarrassed to be the beneficiary of such a fiction, all the more so in an essay hailing her moral discernment. All I can think is that, as a novelist, Catharine was not above stretching the facts in the pursuit of a larger truth about Mumbet's rectitude. And that point is hammered home here, as Mumbet refuses to submit to any injustice, whether perpetrated by her own mistress (in trying to throw Tamor out), by herself (by offering her food that is not hers to give), by a poor white girl's scurrilous father, or by society (in denying Mumbet her liberty). She divines the shocking truth about Tamor before the girl has to reveal it, and she makes sure that Tamor speaks to her master so that justice is done. Even though Ashley is the putative justice of the peace, Mumbet is the true moral judge.

MUMBET HAD BELONGED to a Dutch trader named Peter Hogeboom of Claverack Landing, a small trading town along the Hudson about twenty-five miles south of Albany. Hogeboom owned a store there along the wharves. Through the Dutch West India Company, the Dutch had been importing Negro slaves into what they called New Netherland since 1626, and Mumbet may have been descended from some of them. More likely, her forebears had been brought to America from West Africa as part of the triangle trade that flowed mostly through Newport and other smaller harbors in Rhode Island. That trade had evolved from the practice of exchanging the incorrigibly warlike Indians captured during the Pequot war of 1637 for more docile slaves from the West Indies. The slaves proved particularly useful in the Narragansett region of Rhode Island, which needed a steady supply of labor to run the large plantations there. By the end of the century, Rhode Island slavers were acquiring nearly all the slaves that came to the New World to run the New England economy. Typically, they rounded them up from the slaving grounds that extended from modern-day Senegal down the Gold Coast to what is now Angola, paying for their purchases with the potent "Guinea Rum" that was Rhode Island's chief export. (More delectable to local palates than the French brandy that was its primary competition, the Rhode Island rum became the local currency in European fort towns.) The slaves were then transported back in three-masted schooners across the Atlantic, shackled naked belowdecks in the cramped holds that had been filled with the rum that was "spent" to acquire them.

But slaves were widely for sale in Massachusetts, as well. A Boston merchant named Hugh Hall offered slaves along with rum, sugar, and "sundry European goods," and a few tavern keepers in the city made their pubs available for the display of slaves for purchase.

The success of the trade bred more success still, as it fueled a broad range of industries, from distilleries to shipbuilding. By 1755, one Rhode Island adult out of ten was a black slave, and a year later, slaves made up 14 percent of the population of New York. Although Massachusetts had many fewer African blacks, its total increased every year up to the Revolution, with 5,249 in the census of 1776. Most of them were concentrated in the area around Boston. "There were household slaves in Boston who drove the coaches and cooked the dinners and shared the luxuries of rich houses," Catharine noted in that essay on slavery, "and a few were distributed among the most wealthy of the rural population." Slaves were taxed like property, and they were subjected to corporal punishments, chiefly flogging, for even minor crimes like being caught in the street after dark or breaking a streetlamp.

A relative rarity in western Massachusetts, blacks were considerably more common in New York State in the 1740s when John Ashley journeyed to Claverack Landing to do business with the Dutch trader Peter Hogeboom, who owned slaves and sold them along with other commodities. Ashley fell in love with Hogeboom's blond, blue-eyed daughter, the shrewish Hannah, married her, and brought her back with him to Sheffield, where he built that fine house for her in Ashley Falls along the Housatonic. When the elder Hogeboom died in 1746, he passed all of his slaves on to his children, awarding to Hannah the teenage Mumbet and her infant daughter, Lizzy. When she arrived in Sheffield by sleigh that winter, with Lizzy bundled up in the straw at her feet, the only possession that Mumbet brought with her was a treasured "short gown" of her own mother, who, if she still lived, had evidently been kept behind or sent elsewhere.

In the Ashley household, Mumbet helped raise the Ashleys' four children while she did her best to raise her own. But Hannah was always jealous of any attention that Mumbet gave her own daughter, and the tension flared up late one afternoon when Hannah discovered little Lizzy making for herself a small "wheaten cake" from some scrapings left in the bowl after Mumbet had prepared cakes for the family dinner. Enraged, Hannah grabbed the iron shovel, red hot from the oven, and started to bring it crashing down on the girl, but Mumbet raised an arm to protect her daughter, and took the blow herself. It burned her terribly, leaving a horrid scar that she—always alert for moral advantage—wore like a badge for the rest of her life. "I never covered the wound," she told Catharine, "and when people said to me, before Madam—'Why Betty! What ails your arm?' I only answered—'Ask Madam.'"

It must have been galling for Mumbet to see the colonists speak hotly of *their* rights to "liberty" in the very house where she and her daughter were slaves. It explains why Mumbet no doubt listened carefully when

Colonel Ashley brought together the leading gentlemen of Sheffield, Theodore included, to craft the Sheffield Resolves. And why she would squeeze inside the Sheffield meetinghouse there in the center of town to hear a reading of the new state constitution in the spring of 1780, when it was first submitted to towns across the state for ratification. As a slave, she must have been stunned by the opening proclamation that Adams drew from Jefferson's audacious Declaration of Independence:

> All men are born free and equal, and have certain natural, essential, and unalienable rights; among which may be reckoned the right of enjoying and defending their lives and liberties; that of acquiring, possessing, and protecting property; in fine, that of seeking and obtaining their safety and happiness.

She often told Catharine, with a shake of the head, "Any time, any time while I was a slave, if one minute's freedom had been offered to me, and I had been told I must die at the end of that minute, I would have taken it—just to stand one minute on god's *airth* a free woman—I would." This, at last, was her chance.

Mumbet assumed that Theodore, as Sheffield's representative to the Massachusetts General Court and a prominent lawyer besides, was the man to see about securing her rightful freedom under the constitution's edicts. She went to him at his front-parlor office in May 1781 to do just that. In all his finery, with his imperious ways, Theodore could be intimidating—and he must have been stunned by Mumbet's audacity. He was certainly not inclined to view an uneducated black slavewoman as his equal, so he must have been flabbergasted when she proceeded to tell *him* the law. Aside from the broad social and political issues Mumbet's case would raise, there were some special problems. She belonged to the wealthy and powerful Colonel Ashley, who had once been a judge in the courts where Theodore had presented cases, and might well be again. Besides, even if Theodore were disposed to take her case, which he was not, was he supposed to provide his legal services to her for free?

But there were other considerations. It might be useful for a man with Theodore's political ambitions to put his stamp on such a prominent portion of the state's constitution, and, by extension, on the Declaration of Independence itself. And it was possible—Theodore had to concede this—that Mumbet had a point. If the high-flown phrase about everyone being born "free and equal" meant anything, surely it meant no one could be enslaved from birth simply because of the color of his or her skin.

And one thing more. As Mumbet stood before Theodore in his cramped office, he may have seen something of himself in this proud, assertive black woman who had, to an even greater degree than he had, come from nothing. *She did not even possess her own self.* And yet the determination, that glint in her eye—he took notice of that. She was taking a huge risk. If the lawyer sent her away, and Hannah Ashley ever got wind of what she had tried, Mumbet might face terrible retribution. If she were to press the case and lose, she might well return to a condition of slavery far worse than the one she'd left, exposing her backside to the snap of the whip. But Theodore knew what brass looked like. He saw it every time he looked in the mirror. He could not tell her no.

Once Theodore committed to the case, it became the talk of the town, if not the county. *Sedgwick vs. Ashley.* Both sides enlisted more lawyers, compelled by ego as much as practicality. From Connecticut, Theodore brought in Tapping Reeve, the tremendous legal talent who would go on to start that first law school in his native Litchfield, training a generation of political leaders from Aaron Burr to John C. Calhoun. Ashley responded by hiring two legal cannons of his own—David Noble, who would later serve as a judge of the Court of Common Pleas, and John Canfield. The two adversaries were very competitive men, and neither could afford to lose.

Theodore and Tapping Reeve swung into action by immediately issuing a "writ of replevin," demanding Ashley release Mumbet and another slave named Brom, who also signed on to Mumbet's petition, for the duration of the trial. But when the county sheriff arrived at Ashley's door to enforce the writ, Colonel Ashley stoutly refused to give them up, insisting they were his "Servants for life." That is, they were *his*. This, of course, was the crux of the case: could a human being legitimately be the personal property of another? Stymied by Ashley's refusal, the sheriff went away empty-handed.

Several times, the lawyers sent the poor sheriff back to enforce the writ, but each time Ashley held firm, effectively imprisoning Mumbet and Brom in his house until the trial finally started at the Berkshire County Court of Common Pleas in Great Barrington, in the last blast of

sweltering summer heat on August 21. By then, Theodore was no longer
Sheffield's representative. To his distress, he had been voted out in May, a
victim of the rising frustrations of the hardscrabble farmers, angered by
their lack of a political voice. Sensitive as he was to his own political interests, this may have disposed him to make the most of Mumbet's suit,
viewing it as an opportunity to make a name for himself on such a prominent case.

The Great Barrington courthouse fronted that "meadow road" where
it ran through the center of town, and the local citizenry jammed the
public benches for the finest legal show in some time. The rhetoric flew;
furious motions and countermotions were made.

Speaking for John Ashley, Noble and Canfield demanded an immediate dismissal of the matter, insisting that, as the court records put it, "the
said Brom and Bett, are...the legal Negro Servants of the said John
Ashley during their lives." In presenting his case, Theodore took Mumbet's own line as he stood before the jurymen, all of them poorly educated
farmers, sweeping out his long arms and fixing them with those black
eyes of his: Slavery simply could not be allowed under a state constitution
that provided freedom and liberty for all people. There could be no ex-
ceptions. All people meant all people, not just people who happened to
possess white skin. There was more to be said, and Theodore doubtless
said much more, but it was as simple as that. When he was finally done,
the jurymen all filed out to perform their deliberations. They were not
gone long, and when they returned, the foreman stood up to say that the
jury found for the plaintiffs: "Brom & Bett," he declared, "were not...the
legal Negro Servants of him the said John Ashley during [their] life." The
throng of onlookers started a murmur that quickly turned into a roar. For
on top of their verdict, the jurors awarded Brom and Mumbet thirty shillings in damages (at a time when a little over a single shilling would pro-
cure ten pounds of lamb) and, in a final insult, charged Ashley with nearly
six pounds of court costs, a prodigious sum. With that, the court set
Mumbet and Brom free. Thus a handful of Berkshire farmers helped set
the state, and eventually the nation, on a fateful course to animate the otherwise abstract ideals of the country's founders.

Ashley was devastated, although he was too proud a man to show it.
He could not bring himself to look the victorious Mumbet in the eye. She
scarcely glanced at him, but swung around to glower at Hannah, who

was watching, ashen-faced, from the gallery. Through his lawyers, Ashley immediately appealed the ruling to the state's Supreme Judicial Court, and then, his lawyers in tow, he pushed through the crowd that was milling around outside the courthouse, and, curtly summoning Hannah, climbed into his carriage and, with a flick of the whip, departed for Ashley Falls.

By October, though, when the appeal was scheduled to be heard, Colonel Ashley had reversed course. He dropped the case and assented to the judgment of the lower court. Why? By then, the legal winds in Massachusetts on the slavery issue had shifted markedly. When Mumbet began her suit, few imagined that the state constitution had any particular bearing on slavery. Now, judges were starting to declare that yes indeed, liberty and equality can only mean liberty and equality. In particular, there had been a case in Worcester County involving a slave named Quok Walker, who was owned by a farmer, Nathaniel Jennison. That judgment, finding for Walker, came down in September 1781, a month after the jury's verdict in the Mumbet case, but a month before the appeal would be heard. By then, Colonel Ashley may well have figured a legal turning point had been reached.

In fact, though, his capitulation itself *was* the turning point. For the Walker decision was still sufficiently blurry that the state Supreme Judicial Court felt obliged to provide a clarification of its ruling two years later, when Chief Justice William Cushing declared, "The idea of slavery is inconsistent with our own conduct and Constitution; and there can be no such thing as perpetual servitude of a rational creature." That sentence officially abolished slavery in Massachusetts. But it was the very argument that Mumbet herself had made when she first approached Theodore. And it caused the number of slaves in Massachusetts to drop from the 5,249 of 1776 to zero by the end of the decade, and forever afterward.

Humbled, Ashley asked Mumbet if she might kindly consider rejoining his household as a paid servant. She turned him down flat. But when Theodore made her the same offer, she accepted. She would be glad to work for the Sedgwicks.

ELEVEN

THE PROPER OBJECT
OF GIBBETS & RACKS

Mumbet and Theodore. Each owed so much to the other. Yet beyond lawyer and client, and then master and servant, there was scarcely any relationship at all, certainly nothing of the fascinating and complex dimensions of the one between Thomas Jefferson and his slave, lover, and common-law wife, Sally Hemings. Theodore scarcely mentioned Mumbet in his letters, and Mumbet, of course, wrote none. An unspoken admiration ran between them, and a silent antagonism, too. Theodore would have found it vaguely irritating that an illiterate black woman had indeed been able to further his legal education. And I suspect Mumbet did not enjoy relying on a pompous barrister to set her free.

But each came to depend on the other. Once she came into his service, Mumbet, rather than the weak-willed Pamela, covered for Theodore while he expanded his political and business careers, which were always his primary attachments. She provided a measured sense of order in a noisy household, filled with the squalling children that Pamela always complained about, to say nothing of the other servants who were often little more than children themselves. For a time, when they moved into the illustrious house on Plain Street in Stockbridge, she took the place of the master himself.

Theodore was not there for the move. After being painfully voted out by Sheffielders, he'd won back a seat in the Massachusetts General Court, in the Senate this time. And he'd held it until June 1785, when, by a joint ballot, the House and Senate voted to have him represent the state in the Continental Congress in New York City the next winter, where he addressed himself to the appalling state of national finance, among other pressing concerns.

Theodore could see he should be home. "Believe me, my sweet prattler," he wrote ten-year-old Eliza after he first moved down to New York, "that you can not, more than I do, regret our separation. Should it so happen that my duty will permit, I will fly on the wings of Love to see and embrace my lovely, sweet children." He was particularly worried about Pamela, and he enlisted Eliza to give her the help he could not: "Be kind to your mamma. She is good. She deserves all your attention. Remember that you are the eldest child, and that you can reward your parents' care by a good example."

The family's need for a father grew all the more acute the following summer as terrifying news started to come in from surrounding Berkshire towns, some by letter, some by sketchy accounts in the extremely primitive newspapers, and much of it just frightening gossip among townsmen. The fragments were hard to grasp, and seemed almost impossible to credit, but it appeared that hundreds, if not thousands, of common citizens—farmers, most of them, but by no means all—were starting to rise up against the state government.

The underlying truth was, for all the new state constitution's handsome rhetoric about liberty, it was reviled in the backcountry, where it seemed only to tilt power further toward the mercantile elite of the more prosperous east. Among its many clauses, the constitution established steep property requirements for voters, and steeper ones still for officeholders, making the government into an instrument almost exclusively of the rich. Meanwhile, there was a huge debt to pay off after the war, about £1.25 million. Eastern speculators held most of the notes, yet the state legislature in faraway Boston had raised taxes on everyone to pay the interest.

Writing to Pamela from New York, Theodore was quite unsympathetic to the plight of the dispossessed. While he acknowledged the "very general disposition to uneasiness" among the citizenry owing to the "almost intolerable burden of taxes," he was instinctively indignant, fulminating that anyone who whipped up the common people over such an issue were "the proper objects of gibbets, & racks."

Later that summer, the state legislature voted him off the Massachusetts delegation as a sop to the rising anti-lawyer sentiment across the state. To Pamela, Theodore airily claimed he wouldn't miss public life a jot, but he was obviously seething. "You know the pride of my heart," he confided. But Pamela was ecstatic at the prospect of having her husband back. Unnerved by all the talk of a mounting insurrectionist sentiment, she'd started to feel toward the local people the way she had once felt toward the Indians. She wasn't sure, when they looked at her, whether they saw a friend or an enemy. When she returned to the Sedgwick house, she was relieved to swing the stout door shut behind her.

But the house still wasn't finished! There were workmen about all day long, and an endless din of sawing and hammering, the noise setting off poor little Harry into rounds of whimpering. She longed for the day when the house would be safely enclosed, impermeable.

At Pamela's insistence, Theodore left for home promptly on the last day of the session, Tuesday, August 22, but the roads were dreadful, and he still hadn't arrived a week later, August 29, 1786.

That day, word filtered back to Stockbridge that the unthinkable had happened. The rebels had struck the Northampton courthouse, twenty miles to the northeast, just as it was to open for the fall session. Several hundred farmers, many of them veterans of the Revolutionary War, paraded into Northampton before dawn, in time to ambush the three startled presiding justices in their wigs and robes before they could enter the building. Thus they seized control of the rule of law that had only scorned them—and commanded attention for their own grievances, chiefly involving the high taxes imposed to pay off the war debt. Until these issues were addressed, the rebels would simply not allow any other legal matters to be attended to.

The uprising would soon become known as Shays' Rebellion after Daniel Shays, an embittered farmer from the hill country mid-state. A former captain in Washington's army, Shays had been active in the town's committee of safety, one of the vigilante organizations whose extremism so alarmed Theodore, but he was following behind the rebellion as much as he was leading it. To term it a rebellion, actually, is to suggest that it was itself a discrete incident, like a riot. In fact, it was far broader and more portentous, the violent expression of the class hatreds that had been building since the earliest days of the Revolution. Now, to Pamela's horror, the class war burst forth, with America's home-grown aristocracy as the new redcoats.

Theodore arrived a few days later, as details of the Northampton insurrection spread across the state. Shocked as he was, he did his best to assure Pamela that no such thing could possibly happen in Stockbridge, which was blessedly free of such rabble.

But then, early in September, word came that another hundred men with bayonets had blocked the door to the courthouse in Worcester.

And then—it was too shocking—news arrived that the rebels had taken the courthouse in Great Barrington, a scant eight miles from Stockbridge! A thousand militiamen had been in position, ready for them, but when the rebels pulled in, the rascals insisted, with enormous cheek, that the soldiers be allowed to decide for themselves which side they were on, and incredibly, eight hundred of them crossed over to the ranks of the malcontents, surrendering the courthouse. The rabble also took over the debtor's prison and set all the inmates free.

Theodore could scarcely believe it. Pamela was terrified.

To Theodore, the closing of the courthouses was a personal affront. The rebels were overturning the very state government that, along with many others (including many of the malcontents themselves), he himself had faced down British muskets to create. And there was something more disturbing. The courthouses were the aspect of state government that most involved *him*.

Theodore was not the only one to recognize that. The rebels did, too. Despite his political career, he was still known best as a lawyer. It was not just his livelihood that the Shaysites were attacking, in the very courthouse where he practiced. It was himself. The rebels were gunning for him, Theodore Sedgwick, by name.

So said none other than William Whiting, the chief justice at the Great Barrington court, whom—all the more galling—Theodore had personally recommended for the post. Himself a wealthy man who Theodore had had every reason to believe was reliably conservative, Whiting had astounded him by openly siding with the rebels, assailing the Massachusetts aristocracy as "overgrown Plunderers." In so saying, he had Theodore Sedgwick uppermost in mind, as he'd widely groused to colleagues that Squire Sedgwick had taken unconscionable advantage of the state's exorbitant fee schedule—one that Theodore himself, as a legislator, had helped set—to rake in over £1,000 a year from his legal services, enough to build *two* Sedgwick mansions per annum.

In essence, the knock on Theodore was simply that he had been successful. That was the nature of a class war, after all. But it was the *way* that he had succeeded that was particularly insulting: by gaining a college education (still a rarity in Berkshire County and, as now, a dividing line in society), and then lording it over people with that pompously erudite tone of his—that style, that set of references, that grammar—all of which said, *I am a better man than you.*

On October 2, Theodore defiantly rode to Great Barrington to size up the situation for himself, and on behalf of Massachusetts governor James Bowdoin. This took a bit of daring, to canter right up into the teeth of the enemy that had vowed to kill him. Covered in a common overcoat to disguise himself, and with his tricorn hat drawn low over his face, Theodore trotted by the courthouse swiftly, careful not to seem too curious about the goings-on. Still, the sight was disturbing. The courthouse was occupied by renegades who stood about in some disarray, to be sure. But the numbers were off-putting—at least thirty of them, and more arriving every minute. Theodore rode on to the outskirts of town, sneaking back once or twice more for further updates for the governor, and for his own information. By nightfall, the mob had grown to over two hundred men.

Theodore took lodging with an old friend (whom he scrupulously does not name in the dispatch to the governor), who let him know that William Whiting had directed the insurgents to bring "destruction" to Theodore Sedgwick's "life and property." In his letter to the governor, Theodore doubted "that any *sufficient* body of men could be found who could be induced to adopt such deadly intentions." But then, late that night, a defecting Shaysite pounded on Theodore's host's door, and after some earnest interrogation, he was brought to Theodore in his bedchamber, where he pleaded with Theodore to be off, immediately, or he feared the consequences. As Theodore told Bowdoin, "the last violence was resolved against my person." At that, Theodore saddled up Jenny Gray and galloped safely home to Stockbridge by back roads in the darkness, arriving well before dawn.

And not a moment too soon. In Great Barrington, the Shays men went on a rampage that very night, unhindered by a drenching rainstorm that had swept through the mountains. One renegade burst into the house of the upstanding Captain Walter Pynchon, pointed his firelock at his chest, and pulled the trigger. Fortunately, the gunpowder was too sodden to explode. After Pynchon's neighbor Ezra Kellogg was heard to heap scorn on the insurgents, other dissidents discharged a "volley of arms" into his house, then burst inside to seize him. Kellogg had already jumped out a back window and fled into the woods—leaving behind his wife, whom the Shaysites found cowering inside. One of the ruffians pressed the sharp blade of a bayonet to her breast and threatened to slice her open unless she told them where her husband was. She pleaded with them that she didn't know, and they left her convulsed with tears, but unharmed.

Alarmed by reports like Theodore's, Governor Bowdoin convened the legislature to take action against the insurgents. Even though old revolutionaries like Sam Adams had themselves had a hand in closing the courthouses during the war for independence, Adams took a hard line now, demanding the hooligans be rounded up and slaughtered wholesale. After considerable debate, the state legislature decided on a more cautious course, repealing some of the taxes that had so inflamed their backcountry constituents and offering pardons to any of the rebels who were willing to renounce their seditious ways. But, to firm up the defenders of order, the legislature also voted to indemnify any sheriffs who killed rioters, and directed any convicted rebels to be given "39 stripes on the naked back"—and thirty-nine more every three months they remained in jail.

The rebels were not intimidated. Three hundred Shaysites swarmed the courthouse in Springfield to shut it down at the end of December.

Bowdoin turned to harsher measures. He assembled a proper 4,000-man army, and placed it under the command of a Revolutionary War hero, General Benjamin Lincoln, with the order to hunt down the insurgents and exterminate them.

By then, marauding bands of Shaysites had fanned out through the snowy Berkshires, spreading terror through the region by filling the air with a peculiar sort of rasping music that became their revolutionary anthem. Local believers drew boards across sawhorses that had been smeared with rosin; the vibration produced a weird shrieking whine that echoed eerily around the town. Then they swept down to force on the citizenry the hemlock sprigs they took as their revolutionary emblem. In Stockbridge, some of Shays's men grabbed Theodore's young law clerk, a lad named Henry Hopkins, and rode him on a rail up and down Plain Street until he was willing to replace the white paper cockade in his tricorn hat with a bit of the insurgents' hemlock.

THE NEWS SENT Pamela into a panic. Over and over, Theodore had assured her that the Shays men would never come to Stockbridge. They wanted only to shut down the courthouses they viewed as the symbols of political oppression, he'd told her; and Stockbridge had no courthouse. He had not mentioned to her anything of Whiting's threats. He hadn't been entirely sure what to make of them himself, although subsequent events in Great Barrington were hardly reassuring. But now there was no denying it; Shaysites had indeed come to Stockbridge, and tormented Theodore's own clerk. Was anyone safe now? Was he? His family?

Realizing he could not wait for Lincoln's army to defend him, Theodore took charge and organized a force of volunteer soldiers to turn Stockbridge into an armed garrison, with himself the general in charge. He dispatched spies to conduct local reconnaissance, and report to him in code. He had some of the more reliable boys in town keep watch from upstairs windows, ready to sound the alarm if they spotted any rebels sneaking into town. And he armed them with muskets to pick off the Shaysites if they dared enter. When seven-year-old Theodore Jr. heard about this, he volunteered for the job, offering to take a position at the window of the little bedroom at the top of the stairs. But Pamela gave him a smack and sent him to Mumbet.

Late in January, Theodore's spies picked up a rumor that Daniel Shays himself was taking refuge nearby. A large group of his supporters had already gathered in West Stockbridge under a blustery Shays lieutenant named Paul Hubbard, and they were set to join forces with the rebel commander as soon as he arrived. Theodore had a war council at the house. Determined to lance this vile cancer before it spread, Theodore assembled five hundred volunteers from surrounding towns at the center of Stockbridge and prepared them to march down Plain Street toward West Stockbridge. His plan was to attack Hubbard's men from three sides at first light, but first he needed to determine their precise location.

Well after midnight, Theodore saddled up Jenny Gray and, as stealthily as he could, led a small advance party of three dozen foot soldiers and six other men on horseback down Plain Street to scout out Hubbard's position. In the dark, however, a couple of Theodore's horsemen happened upon Hubbard's sentries, who opened fire, quickly bringing the full brunt of the rebel force upon Theodore's men.

Instead of galloping off to safety, Theodore settled Jenny Gray where they were, took the measure of the opposing troops, and then coolly advanced toward them, speaking to them like a father to obstreperous children. Calling out to some of the local lads by name, he commanded them not to fire, but to be sensible and lay down their arms. Confused by this fatherly bravado, the malcontents froze. Then, as Theodore continued to advance toward them, they broke ranks and fled into the woods. Theodore had his small cavalry give chase, and they quickly caught up to the frightened rebels, capturing the snarling Hubbard and eighty-three of his men.

Theodore had his troops tie the hands of the prisoners, then marched them back up Plain Street, where they endured the catcalls of the loyal townspeople, and past Theodore's fine house, to lock them all in a makeshift jail behind the Red Lion Inn.

THE OTHER SHAYSITES were no more soldierly, and usually crumbled at the first sign of serious opposition. As the winter wore on, they were also increasingly bedraggled and dispirited, and they stood little chance against General Lincoln's well-organized, well-paid professional soldiers. Guessing that the Shays forces would make a final, go-for-broke attempt to capture the arsenal in Springfield, Lincoln's major general, William Shepard, arrived there first, to hide several hundred troops inside. When the Shays troops did indeed come for the arsenal, Shepard first dispatched artillery fire over their heads to paralyze them, then mowed them down with cannons aimed at "waistband height." Four men were killed, many more wounded. Terrified, the rest of the Shays forces bolted.

Daniel Shays managed to regroup some of his men deep in the snowy woods outside Petersham, some thirty miles to the east. But Lincoln braved a snowstorm to surprise the bivouacked Shaysites there. Shays and some of the other rebel leaders melted into the woods, and eventually slipped away into Vermont. But Lincoln captured 150 of the men. That took most of the fight out of most of the rest—but not all.

Later that month, thinking the insurrection had been securely put down, Theodore had—over Pamela's frantic protests—boarded a coach-sleigh for the long, three-day journey into Boston. He had some business there, and a bit of politics, too. A number of political intrigues were developing. Despite Theodore's claims that he was fed up with government work, he missed the glory of high office, and there would be another election that spring. The victorious General Lincoln would be running against his putative commander, Governor Bowdoin. And the wealthy landowner and gallant John Hancock, Theodore's erstwhile colleague from the Continental Congress, had thrown his hat into the race as well.

Plus, a convention had been organized in Philadelphia to beef up the Articles of Confederation and its too-feeble Congress. Theodore hoped that the Shaysite threat might bring new power to a federal government—and to any legislators in a newly organized Congress. Theodore wanted to remind members of the General Court that he would be delighted to serve.

Pamela begged him not to go. After the battle with Hubbard's men, she was terribly afraid the Shaysites were spoiling for revenge against him. She feared for herself and her children, too. If they couldn't find him, they could easily find his family. She pleaded with him to take them, too, if he had to go. And not just to Boston. Farther! Across the ocean to England, to join his Loyalist friends there. They could be happy in England, she was sure of it. They could start fresh. She was done with Stockbridge and its many terrors. She wanted to live in a place where she could at last feel secure.

But Theodore would hear no talk of leaving. He'd built his house, and he planned to live in it. The rebels were all in captivity, he assured her, or they'd fled to Vermont, if not to Canada, never to return. He did take the precaution of depositing his legal papers with her cousin Rev. Stephen West and, to assuage her, agreed to remove his family from Stockbridge, boarding them all with friends in another town, until he returned.

Of all the household, he left only one person behind to protect the house.

It was Mumbet, of course. By then, through some odd psychological alchemy, Mumbet had come to think of the house as an extension of herself. But she knew perfectly well why everyone else was leaving.

Mumbet lived in an outbuilding herself, but while the family was away she would spend the day, and much of the night, too, inside the mansion. She would keep the kitchen fire going, for defensive purposes as well as for cooking.

Now she set a kettle of water boiling, ready to scald any interlopers. She collected up the family valuables, the silver plate they'd acquired, along with Pamela's jewelry, and hid them amid her own dresses, which she'd tucked into a large oak chest at the top of the attic stairs. Thus prepared, she settled herself down to wait for whatever, or whomever, might come. It was terribly cold. With no other fires burning in the house, the rest of the mansion was freezing, with a frosting of rime lining the walls and the inside of the windows. And when the sun set, it was pitch-dark, except where the moonlight angled in across those fine floorboards.

Perez Hamlin, another grizzled revolutionary soldier who'd thrown in with Shays's rebels, had been lurking in neighboring New York while General Lincoln scoured the Massachusetts countryside for the last of the insurgents. Shortly after Theodore left for Boston, Hamlin gathered up a small posse and swept across Massachusetts' western border. He snuck his men up Plain Street in the darkness. The local citizens, confident—just as Theodore was—that the days of rebellion in the Berkshires were over, had slid comfortably into their beds. Hamlin commandeered a back room of the Red Lion Inn as a kind of field headquarters, and from there he sent out marauders in all directions.

A few went farther to the east down Plain Street, where they burst into the Woodbridge house. The widower Jahleel Woodbridge was asleep in his big bed with his young son Timothy. The renegades hauled out Jahleel and sent him barefoot in his nightclothes out into the snow while they gathered up his guns and munitions. At Deacon Ingersoll's nearby, the deacon's wife thought more quickly. When the Shays men burst into her house, she reached into the cupboard for the couple's best brandy and won lenience in exchange for a bottle, which was quickly downed to dispel the cold of that raw night. Another group of diehards surged down the Great Barrington road to the south. The farmer Ira Seymour heard them coming, and he dashed out of the house and into the woods before the men could spot him. Yet another party charged up the hill, break-

ing into Captain Jones's place to steal his muskets before they pushed on
to the fine Mission House at the crest of the hill. Abigail was asleep there.
The Shays men let the acrid old woman be, but they hustled off her son—
Erastus Sergeant, the village doctor—along with a couple of medical students who were boarding there, and locked up all three in the inn.

One last hunting party, however, headed up Plain Street the other way,
toward the biggest house in town, dark except for a few flickering candles' dim glow in the kitchen at the back, their footsteps muffled by the
soft snow that was heaped up in drifts along the road and hung off
the hemlocks lining the street. The scent of wood smoke rising from the
chimney raised hopes that they might possibly find the villainous Squire
Sedgwick sleeping inside. They climbed the snow-covered granite steps
to the portico held up by those proud columns, and they brought heavy
fists down on the thick oak door.

Mumbet, alone in the kitchen, roused herself from a fitful sleep at the
sound echoing through the house. Arming herself with a kitchen shovel,
she advanced down the hall, carrying a candle to light the way.

Through the sidelights, she could see the gathered men. Their fists
beat the oak again. Who's there? she demanded. We've come for the
squire, the men called out, giving no names. When she told them he
wasn't at home, they replied that they'd determine that for themselves,
thank you.

The house was no fortress. Those ostentatious panes of English glass
bordered the door, and the men threatened to smash them in if Mumbet
refused to unbolt the door. She had no choice but to oblige, finding a half-
dozen men, shiftless, dirt-poor, massed there. In the candlelight, she recognized the leader. It was the broom peddler Sam Cooper. He'd come
through town pushing his wares just a week or two before; out of kindness, she'd bought a broom or two from him for the family, but found
them not much good. Seeing him now, she spat out a few words of deri-
sion. Cooper snapped a few insults back, calling her a nigger besides, and
then pushed past her into the wide entrance hall, tracking snow onto the
beeswaxed floorboards.

The men swept through the front parlors, the dining room, the
kitchen, in search of Theodore. They climbed the handsome staircase to
the bedrooms above and, with Mumbet looking on, thrust their bayonets
under beds and into clothes closets.

Then to the cellar. Spotting the place where Theodore had laid in his liquor, Cooper cracked open a bottle of Theodore's best brown stout, but then spat it out and wiped his sleeve across his lips with disgust. "Is there nothing better here?" he demanded, according to Catharine's account, which may well have been embroidered to put Mumbet in an even more heroic light.

"*Gentlemen* want nothing better," Mumbet curtly replied.

The men glared at her, then climbed the narrow steps once more, ascending all the way to the attic, where their eyes lit on Mumbet's chest, locked with a padlock. The men demanded the key.

She turned indignant. "You and your fellows are no better than I thought you," she sneered. "You call me 'wench' and 'nigger' and you are not above rummaging my chest. You will have to break it open to do it!"

Embarrassed, the men left the trunk that hoarded all the Sedgwick jewels and silver, and turned their attention elsewhere.

When the Shaysites didn't find Theodore or anything worth stealing, they went around to his adjoining law office, where they discovered his two law clerks, the young Henry Hopkins and Ephraim Williams, curled in their beds, having slept through the break-in. Cooper took his wrath out on them, dragging them out of bed and then shoving them outside into the snow. Then they rampaged around in the office and, finding some frilled shirts and fancy underclothes, tossed them out into the snow, too. In the barn, some of the men found Jenny Gray, and they slapped a saddle on her to ride her away, but after Sam Cooper swung a leg over her, the mare pitched Cooper onto the icy ground and bolted across the field and into the woods.

The men hung on to Hopkins and Williams as hostages to safeguard their exit from town. They returned with their catch to the Red Lion Inn, where they joined up with Hamlin and the other men and their prisoners and then made their way south along the Great Barrington road. Foolishly, Hamlin halted at a public house for the night, giving Colonel John Ashley time to bring up eighty men from Sheffield to the south. They set upon Hamlin's men as they came onto Egremont, where they fought the last battle of the rebellion. Like most of the others, it went badly for the renegades. Ashley's men blazed at them from the woods, killing thirty, including Hamlin himself. None of Ashley's forces were killed, but one of the hostages—a teacher named Gleason—died in the crossfire. Hopkins and Williams, still in little more than their nightclothes, were returned safely to Stockbridge.

Much of the rebellion's leadership fled to other states, just as Shays had. Of the followers, most were let off with a small fine if they were willing to take a loyalty oath. But several did come to trial, some because they'd been sued for damages in civil cases, and others facing criminal charges for the injuries they had caused. As a gesture of forgiveness, Theodore himself defended some of the accused in Berkshire County. Despite his efforts, six of the men, including a lieutenant of Perez Hamlin's named William Manning, were sentenced to be hanged. Theodore wrote Governor Bowdoin pleading for clemency, claiming the six were just following orders. Bowdoin then pardoned all but two, a pair of common laborers named John Bly and Charles Rose. They were taken to church in Lenox, just to the north of Stockbridge, where the minister preached a sermon against them, and then, trailed by makeshift coffins, they went to the gallows outside of town and were hanged.

As for Theodore's nemesis William Whiting, he was stripped of his judgeship, fined £100, and sentenced to seven months in jail. Another government traitor, Moses Harvey, was obliged to stand for an hour at the Northampton gallows with a hangman's rope around his neck, awaiting a drop that never came.

THE MOST ENDURING consequence of Shays' Rebellion came after it was over, as the national leadership surveyed the damage to the tender psyche of the young nation. From Philadelphia, John Jay sounded the alarm to Thomas Jefferson in Paris, where he was serving as the nation's envoy to France. "In short, my dear Sir," Jay concluded, "we are in a very unpleasant Situation. Changes are necessary, but what they Ought to be, what they will be, and how and when to be produced are arduous questions." Abigail Adams, writing to Jefferson from London—where her husband was ambassador—was even more shrill in her anxiety about the country's future in the face of "ignorant, wrestless desperadoes." Jefferson responded with ethereal nonchalance. "I like a little rebellion now and then," he wrote her. "It is like a storm in the atmosphere." He embroidered the theme later in the year in his famous formulation: "The tree of liberty must be refreshed from time to time with the blood of patriots and tyrants. It is its natural manure."

Normally imperturbable in his retirement, George Washington, however, grew considerably less sanguine as he read the dire reports from western Massachusetts in his cozy study at Mount Vernon, where he'd returned after the war as a gentleman farmer. He feared that Shays was just the first among many other "combustibles" in other states, ready to explode. If Massachusetts could scarcely contain an eruption of some vengeful backcountry farmers, how would the country as a whole respond to a wider conflagration? To Washington, the Shays affair was frightening proof that the country, chopped up into thirteen feeble sovereignties, was far too loosely confederated to long survive, and if he didn't rouse himself to back the effort to create a stronger national government, the American Revolution—*his* revolution—would come to naught. Shays' Rebellion did what his own conscience could not. It brought him to sweltering Philadelphia, where he placed his considerable reputation behind the Constitutional Convention by serving as its president.

TWELVE

BOTTLED LIGHTNING

With the stately Washington there to watch over the proceedings, the conventioneers summoned the courage to scrap the ineffectual Articles of Confederation altogether, and to replace it with a radically new concept of national government. The powerful chief executive, the two branches of the legislature, the independent judiciary—the arrangement seems to us now as inevitable as the solar system, but to the former colonists it was at least as astounding as Franklin's bottling of lightning in a Leyden jar.

To his disappointment, Theodore had no role in the writing of the Constitution. But he did get himself elected as Stockbridge's representative to the Massachusetts state convention at the State House in Boston to decide whether to ratify the new constitution, as three-fourths of the states needed to do for it to take effect. He spoke most memorably in favor of the lack of property requirements for federal legislators. Showing new political adroitness, he neatly pirouetted to profess astonishment that anyone might wish to exclude from service "a *good* man, because he was not a *rich* one." After much debate, the delegates voted narrowly to ratify the Constitution, making Massachusetts the first of the heavily populated states (Virginia and New York were the others) to get behind the new government.

When the time came to elect a new Congress the following December of 1788, Theodore put himself forward to represent Massachusetts' western counties with all the feigned reluctance that General Washington, in his own campaign for president, made fashionable for ambitious public men. Unlike Washington, however, Theodore was not chosen by universal acclamation. Far from it. He was one of four congressional candidates, including his archenemy William Whiting, by now released from prison. Being the best known of the four, Theodore took the lead in the first vote that December but, to his frustration, fell short of the necessary majority for election. The three other candidates then united in attacking him in vituperative, anonymous newspaper articles that drew off a good deal of Theodore's support in the next round of balloting in January, and still more in the one after that, in March, dropping him into second place, a hundred votes behind Samuel Lyman, the presumed author of an especially vicious (and unsigned) article Theodore angrily termed the "Lyman Insult." No quitter, Theodore rallied the faithful in the April balloting by spreading malicious insinuations of his own, and the fifth election, in mid-May, proved decisive, delivering Theodore Sedgwick a scant majority of just 11 votes out of the 3,946 cast. As exultant as he was exhausted, Theodore left home a fortnight later by the stagecoach that stopped by the Red Lion Inn, and traveled south by way of Poughkeepsie to the federal capital of New York City. Several supporters rode along with him past Great Barrington nearly to the Connecticut line before they left him to his future.

THEODORE TOOK ROOMS at Mrs. Dunscomb's somewhat threadbare lodging house on Dock Street in lower Manhattan, with a few other members of the Massachusetts delegation.

A bustling city of 3,300, New York had none of Boston's seaside charm, or Philadelphia's mannered elegance. To Theodore, it was a boisterous, rowdy place filled with smiling hucksters, and he devoted most of his first letters home to venting his outrage at the prices charged by these "villains." By then, the government was well settled into its official quarters in the former city hall, where the Continental Congress had sat. As befit its new powers, the building had been thoroughly refurbished by Pierre Charles L'Enfant as "Federal Hall" in an ornate Georgian style. Washington— wearing a plain brown suit of "superfine American Broad Cloth" that he hoped would set an example for patriotic gentlemen everywhere—had been sworn in there as the country's first president on April 30. Afterward, he'd delivered his inaugural address from the second-floor balcony, then watched that night as the city exploded into joyful pandemonium: bands played, ships at harbor blazed with lanterns, and the night sky was lit up with two hours of fireworks. The debate raged in Congress about how to address so august a personage without monarchical taint—"His Excellency"? "His Elective Highness"?—before finally settling on Mr. President. Still, a certain regal grandeur did accrue to the office. The presidential cream-colored coach, drawn by six horses, was a frequent sight about the city.

Despite the president's sartorial example, when Theodore presented his credentials to the House as Massachusetts' fifth and final representative, he was attired in conspicuously European fashion: black satin, white silk stockings, and a stylish broadloom coat. And although he had professed to his wife astonishment "that I ever dared to put myself or permit myself to be placed in a station so responsible," he jumped right into the debate then raging in the lower chamber about the power of the president in selecting members of his cabinet. Theodore, always eager to give greater authority to one's social superiors, thought the president should have a free hand. But James Madison, who had himself crafted much of the Constitution, preferred to subject the president's wishes to Senate approval, and that position won the day. Still, Theodore's efforts to promote the powers of the chief executive did not go unnoticed at the presidential residence on Cherry Street, and to his delight, Theodore soon became a regular visitor there, often dining at the head of the family table, where Washington himself, out of modesty, never sat. At one gathering, Washington even led the attendees in drinking Theodore's health. Theodore confided to Pamela that the president's approving gaze on him "excited a pleasing sensation of gratitude difficult to describe."

Theodore added a few strokes to the filling-in of the constitutional outline, most of them to enhance the power of the executive branch, that quasi-royal element that, to him, was the pulsing heart of Federalism. His most memorable touches, though, were all rhetorical, adding a note of lusty vituperation to his strong opinions that made them stick in the mind. For instance, he branded the few New England representatives who had strayed to the anti-Federalist cause "pittiful, low and malignant" turncoats, a "desperate faction" that had plainly been "seduced" by "insidious," "profligate," and "incorrigible" devils.

Through it all, Theodore found himself paired with Washington's dazzling young Treasury secretary, Alexander Hamilton, who had insinuated himself into the creative center of the new administration as the Federalists' Federalist. To Theodore, as to Hamilton, the essence of Federalism was economic—only a powerful national government could assure the interests of the propertied class, on which, they heartily agreed, the future of the nascent country depended. The two had known each other in New York during the waning days of the old Articles of Confederation, Theodore with the Massachusetts delegation, and Hamilton as a rising and politically astute Manhattan lawyer who had served on Washington's military staff. Each had come to recognize the need for a strong federal government, and now that Washington had installed Hamilton as his Treasury secretary, Theodore was of a mind to help him push his economic bills—for the federal assumption of the war debt, for the creation of a national bank—through the House.

Bound by a common philosophy, the two were close allies, but not particular friends. A look at them tells why: a slim, dashing five-seven, Hamilton had a dancer's body and, intellectually, a quick, lithe grace, while Theodore was, at an ever-thickening six-two, a considerably stodgier personality, demonstrating the more plodding virtues of constancy and conviction. If Theodore's industry was shaped by a lifetime of grim New England winters, Hamilton, born in the bright, enchanting West Indian island of Saint Croix, had the brisk energy of a child of the sun. Proud descendant of Puritans, Theodore put great stock in lineage, while Hamilton was practically self-generated. He was born out of wedlock—the "bastard brat of a Scottish peddler," in John Adams's memorable phrase—and, coupling with another man's wife, would shortly sire an illegitimate son of his own. But the two of them were alike in believing their success their own, and so they both shared the recognition that whatever they—and the many accomplished men like them—had been able to make for themselves should be recognized and preserved by any government worth the name.

Mindful as he was that the debt issue had precipitated Shays' Rebellion, Theodore still regarded it as a matter of gentlemanly honor for the new national government to make good on its colossal war debt of $76 million, no matter how burdensome to repay it. As a matter of policy, this was definitely a virtuous position, for the country needed to foster the confidence of the wealthy, but one doesn't have to scratch too deep below the surface of Theodore's philosophy to see its self-serving qualities, as such an attitude was destined to win the appreciation of the bondholders, those men of entitlement whose good opinion Theodore always sought, in hopes that he would join their number. In this regard, he remained very much the squire of Stockbridge, eager to take, and keep, his place at the pinnacle of society, no matter how much this would estrange him from the common people he professed to represent.

Formidable opponents like James Madison were willing to make whole the original debt holders, but Madison had no patience for the speculators who'd since bought up the notes for pennies on the dollar in hopes of a tidy profit. Theodore, however, like Hamilton, believed in an unfettered free market, which by its very nature would always attract opportunists. Greed, to Theodore, was not an entirely unattractive quality. (While his political opponents charged that this was because Theodore was a speculator himself, records show he held only some war bonds he had purchased in his military days.) Still, the disagreement between him and James Madison grew so heated that Theodore started referring to Madison as his "once friend," and the two stopped speaking to each other. "He is an apostate from all his former principles," Theodore fumed. By April, when the bill to square the debt came up short on its first vote, Theodore was so distraught, he pronounced a lugubrious "funeral oration" over it and, when he was done, actually broke down and wept with frustration—much to the amusement of his many enemies in the press, who teased him for his "puritanic gravity."

It took Hamilton to rescue the situation through a brilliant piece of political horse-trading. He won over the critical votes of various northern delegations, leery of southern influence, by offering to move Congress to Philadelphia for a decade while the long-planned permanent home for the new nation's capital was erected farther south on the banks of the Potomac in Washington. The northerners, hoping that once in Philadelphia, the capital would stay there, found that prospect enough of an inducement to seal the deal on the assumption bill.

THIRTEEN

THE STATE OF WIDOWHOOD

Back in Stockbridge, Pamela had assumed the mantle of leadership in the little country that was *her* domain (an analogy she would make repeatedly). But she was not at all pleased about it. Winter was never Pamela's season anyway, and she'd had an especially hard time that first winter when Theodore had embarked on his endless campaign for a congressional seat. She had not wanted him to leave, certainly not on an errand that would only increase the chances he would be away a longer time in the new Congress.

I see her sitting at her desk in the parlor, where frost blurred the glass of the large windowpanes. A shawl was draped over her shoulder; the fire flickering in the hearth did little to dispel the icy dampness in the air. The three-year-old Harry and infant Robert, born the spring before, were fussing loudly in another room, a din she wearily does her best to close out, the better to concentrate on the bitterness that was consuming her heart.

When Theodore had left, she'd snapped at him, using words she won't repeat, but would obviously like to have back. It gave her "verry Sencible pain" to have spoken that way, she wrote him, considering that he had gone only to perform his "duty," just as she remained to perform hers. Now, she wanted him to know she'd tried to conceal her feelings "least theay might give Pain To my love whose Happiness is ever near my heart."

Yet it seemed her pain never reached him somehow. It remained entirely with her, eating away at her self-respect, her strength. Not the pain solely, but the inequity of the pain. *She* suffered for his duty, while *he* did not. He was greeted with huzzahs wherever he went, and she? She returned to a lonelier house than ever. She was nothing to him at all. Utterly worthless. A kind of savage.

> *When I reflect on my own Imbecillity & weekness I think with the Poet*
> *that you have lost your mate and been Joind by one from a Barbarous*
> *Land—*

There was a loud cry. Her children needed her:

> *I am Interrupted. Adieu my dear If you mate again I pray It may be on*
> *more Equal Terms Yours affectionate*
> >   *Pamela*

She held the letter for a few days, debating whether to send it. A week passed, and then another, and finally she added a postscript:

> *If this letter is neither common sence or Tollerable English you will*
> *escues it as I have not wrote a singgle line without twenty interruptions.*

Then she handed it off to a friend who was headed Theodore's way.

THEODORE'S REPLY HAS been lost, but he must have tried to be reassuring, for she admitted in her letter back to him that his sentiments were "flattering to my Vainety and Soothing to my heart." And yet Pamela could not escape the idea that Theodore did not care for her, for he ignored her pleas and, instead of returning straight home from New York City, took a leisurely side trip to visit his old Tory friend Peter Van Schaack, just across the New York border from Massachusetts, in Kinderhook, whose estate Theodore had helped restore after the war. (Few men in public life were as openly sympathetic to old Loyalists as he.) Van Schaack had not gone blind, as his friends had once feared, but his eyesight was decidedly poor, and his wife had died of tuberculosis. More than Pamela, Van Schaack deserved a little cheering up.

That spring, when Theodore went off to join the rest of the Massachusetts congressional delegation in New York, Pamela had tried to put the best face on it. "You are too wise my dear not to make the Best of your Lot in life," she told him. "It is the Earnest wish of my heart that you may succeed In your generous Porpose and be made a Blessing to your country." But that was early June, and life for her always was easier in the warm weather. In July, after the children had spent a delightful day berrying, then passed the evening bathing by moonlight in the Housatonic, and were at last tucked into bed, Pamela seemed for once actually to enjoy the gentle longing she felt for her husband, who'd sent word that he'd just celebrated the fourth of July by taking a "ramble in long island" with some friends. She wrote back:

> *I have just been regaling myself with a little Turn in the Back porch of our House—the Sky Clear the full Moon rissen uppon us in soft Majesty—all silent around me—nothing could then have added to the Pleaseure of my Sensations but the company of my Beloved That I might have leand uppon his Bosom—and softly Stroked his Cheek——The same moon will shine oppon you may you enjoy It I am sure you doe— Good night to you my dearest*

And so it had gone for Pamela, hard winters followed by somewhat easier summers, year after year.

IN A RARE stroke of luck, Theodore was there the next winter to help Pamela through the delivery of Catharine Maria Sedgwick in December of 1789—not that he was allowed in the birthing room. Mumbet ruled there, with her pans of boiling water, her strong hands. Catharine had been two months premature, and Pamela's milk did not come in. Distressed by her lack of breast milk, and fatigued besides, Pamela took to her bed, leaving the baby to her eldest, Eliza, to sleep with through the "cold watches of the night," as Catharine later wrote in her memoir, and to feed her cow's milk from a cloth.

Theodore temporarily took charge of the household, but come January, as the ground froze and then whitened with windblown snow, he was gone for a new round of legislative battles in New York City. How the big, high-ceilinged house hollowed out for Pamela then. As the time dragged on and Pamela's irritations mounted, her distress became visible on the page. At her best, her handwriting was tightly controlled, with charming little flourishes atop her *d*'s. But in May of 1790, when she learned that Congress, which had already held her husband two months longer than expected, would not adjourn until the end of the summer at the earliest, her words started to sprawl and blur into an angry miasma across the page. "I am Tired of Liveing a Widdow and being at the Same Time a Nurse," she raged. "I Sicken at the thought of your being Absent for so long a Time. Tho I have very long been deprived of the pleasure of your company yet I cannot Possibly reconcile my feelings or make my self Happy in the State of Widdowhood. It grows more & more disagreeable To me every day." She added that Harry, nearly five, was getting impatient, too. He "says we have killd the Fated calf for Mr Sedgwick and he will not come To Eat him." As the months rolled on, all the children were beginning to sound a chorus. "Is Pappa Come is Pappa come was Ecchod Twenty times evry one of those days from evry Part of the House. Think then my love what must have been my disappointment."

SUMMER EASED THE tension somewhat, but as the next winter came on, Pamela railed at her pathetic dependency, and begged her husband to return in virtually every letter to him. She was a "mourning dove without a mate." Only his letters cheered her in the "gloomy season." Could he at least have a likeness done, one "as Large as the Life," so she would have something of him close by in his absence? Adding to her other woes, her aging mother had grown "verry week" that January of 1791 and was suffering frightening nosebleeds as well. "Life can hardly be thought worth haveing after we have out Lived all our enjoyments," Pamela moaned. Two weeks later, when Abigail was still "languishing," Pamela stayed overnight with her up the hill at the Mission House. The stay brought out the bleak Calvinism in her soul: all was "Vainity and Vecsation of spirit." Pamela gathered the children around her mother's bedside for a solemn leavetaking. When Abigail had "prayd for and blessd" them all, she "Lifted up her dying Eyes and then Sunk into a more Stupid frame," but somehow survived another week before expiring at last.

Theodore couldn't bring himself to leave for home until Congress recessed on March 3, 1791, and then it took ten days travel over "execrable" roads to reach Stockbridge. He arrived cross and exhausted. To raise his wife's spirits, he took her with him on a business trip to Boston that May. When he returned again from Congress that summer, Pamela broached the topic of his resigning, but Theodore would not hear of it. He stood for a second term as congressman that fall, and this time won handily on the first ballot.

By then, Pamela was in the last difficult months of yet another pregnancy, her tenth, grieving over her mother's death; as the winter darkness gathered, her sorrows mounted to a new and more frightening height. Normally oblivious, Frances sent her father an alarming report about her mother's distress, and his loyal legal assistant, Ephraim Williams, not one to intrude on his domestic affairs, echoed Frances's anxieties. Theodore found the situation annoying. "If...you wish me to return home tell me," he wrote Pamela from Philadelphia, adding with more than a touch of anger: "I know I need not assure you that I should not think it any trouble to travel thousands of miles to administer to your comfort." He vowed to quit his seat if that was what she wanted, or to bring her down to live with him in Philadelphia. But she knew he would never do either.

Still, he was alarmed by the reports, and he dispatched Peter Van Schaack's brother Henry to come down from Pittsfield to size up her condition. When he arrived by sleigh late that November, Pamela was descending the front steps in her winter finery, off to pay a call on her brother Henry Dwight, who lived in town. Van Schaack thought she looked fine, resplendent even, the chilly air bringing the color to her cheeks, and told Theodore so. "She is much more chearful than she has been."

But she was not at all well. It was one thing to muster a bit of spirit for a sudden visitor, and quite another to face the daily grind of running a sprawling household, albeit with a half-dozen servants to assist her, a hundred matters daily crying out for her attention, when she was exhausted, defeated, and ballooned out with yet another child she did not want. A week later, Pamela sat down before the fire to deliver a report of her own. She began with deceptive politesse as she demurely thanked "the kindest the Tenderest of Husbands" for inquiring, as he routinely did these days, about her health. Then, in one long barely punctuated sentence that, in its breathlessness, perfectly tracked the anxious flitting of her mind, she let him know—feel—*precisely* how she was:

> *friends would perswade me I am not well but this I have no reason to*
> *Believe but shall I tell can I tell you that I have lost my understanding*
> *what is my Shame what is my pain what is my condition to think of this*
> *what Evils wait my Poor Family without a guide without a head For*
> *their sakes I wish you at home—for your sake I wish you not to come*
> *you must not come it would only make us both more wretched let me*
> *beg you my dear to make yourself easy dont be anxcious about a*
> *Creature utterly worthless to herself and to you—It may Please god to*
> *give me reason and disposition to improve that Reason to some good*
> *purpose.*
>
> > *If it should you will then here again from your once affectionate and*
> *ever obliged*
>
> > > > *Pamela Sedgwick*

When he ran his eyes down the page, Theodore must have gasped. It wasn't the terms themselves. *Shame, pain, evils, wretched, worthless*—such words were, if anything, overly familiar. No, it was the frightful incoherence with which they were strung together. His wife had lost her reason.

"O my dearest Love my heart melts with tenderness inexplicable, I cannot be much longer separated from you," Theodore immediately wrote back. "I will administer comfort to your dear distressed heart." But not that day, December 14, or the next. As it turned out, with Mumbet's assistance, Pamela delivered her baby that day, an always harrowing event that Theodore had intended to be home for. It proved to be an agonizing birth. Baby Charles was so scrawny that Pamela pushed the child away, sure that he would die, an unbearable prospect, and thrust the bedraggled little thing into Mumbet's arms. If Mumbet could keep the boy alive, fine. Pamela was done with the business either way.

Ten days' travel away in Philadelphia, Theodore was oblivious to the crisis of Charles's birth, and he did not return. As chairman of the House Judiciary Committee, he was busy trying to bolster the president's administrative powers in the legal arena. "My friends here," he wrote Pamela, "have been so importunate as to induce me to put off the Journey for a few days." Besides, Philadelphia offered him everything that Stockbridge did not: clarity, purpose, order, progress. What's more, the Congress gave him significant responsibility for a far larger and far more needy entity than a family of six. (He wasn't even aware that his brood had reached seven.) He didn't return until nearly the end of December. As always, the children all dashed to the door to greet him when his coach-sleigh pulled up. All except for the newborn, Charles, who was in Mumbet's arms.

And for Pamela, who remained alone in their wide bed upstairs.

FOURTEEN

A DISORDER OF THE BLOOD

This time, when Theodore climbed the long staircase and turned to her corner bedroom, he could tell by a single look that he would need to remain home longer than he had planned. Pamela was a terrifying sight—her once-beautiful hair in scraggles around her pale, drawn face, she lay wasted, limp, tiny, in the great marriage bed. Her sunken eyes displayed none of their earlier zest as she gazed up at him; she was scarcely able even to find the voice to reply when he asked her how she was. To her, the question itself was an affront. He should not have needed to inquire. If he had done what she asked and been there with her for her ordeal, he would know.

For the first bewildering days, Theodore tried to tend to Pamela himself, largely by secluding her in that big, square bedroom. It was exclusively "mamma's room" now, a phrase spoken by the children with a degree of dread for what they might find there.

At her worst, all that winter, Pamela raged at her imbecility and worthlessness in the most vile (and, by Theodore, unrepeatable) terms, tore at her clothes, refused to eat, flooded herself with tears, went without sleep for nights at a time. At her best, as she was for Theodore's first appearance, she was deathly somber, gazing expressionless out the frost-ringed window at the snow-laden hemlocks, or at the flickering embers of the fire.

In the morning, when he'd enter her bedchamber to open the heavy draperies to the low-angling winter sun, he tried to bring out the dimpled smile he remembered from the first stirring days of their courtship. But although he lived by words, and trusted in his command of them, he couldn't find any to cheer her. The sight of her, so distraught and troubled, scrambled his brains. And so, time after time, he would withdraw far sooner than he intended, an iciness all over him, and feel nothing but relief when he closed the door behind him.

In fairness to Theodore, Pamela's clinical depression—to call her condition by its modern name—was more than anyone could handle in this distant county, where psychological distress was still thought to be more a province of religion than of medical science. A matter for earnest prayer, that is, if not actual exorcism to release her from her demons.

Pamela's half brother, the town doctor Erastus Sergeant, had done his valiant best to bring the latest medical knowledge to Stockbridge. In January of 1787, during a lull in Shays' Rebellion, he and a local medical herbalist gathered with other local medical emissaries at the widow Bingham's tavern. There, they formed a western outpost of the Boston-based Massachusetts Medical Society, for the purpose of exchanging information on "the art of physick."

That art did not much extend to the treatment of emotional complaints. But, in his desperation this January, Theodore summoned Uncle Doctor, as the children called Sergeant. Theodore took from him his hat and heavy coat and led him up the long staircase to the second floor, while the younger children gathered about Mumbet's skirts to watch nervously from the front hall.

Tall and so gaunt as to seem brittle, Dr. Sergeant wasn't known for bigheartedness. He betrayed no particular unease at seeing Pamela with her nightgown nearly undone, her hair spilling raggedly down over her face. He spoke calmly to her in low tones, but his words didn't quite catch hold. To him, Pamela seemed as blank as a pane of glass at nighttime.

He was inclined to view Pamela's distress as stemming from a disorder of the blood, and, if pressed, he would have recommended leeching. But her obvious frailty stopped him; he didn't dare sap whatever scarce vitality she could still muster. He made mental notes on her condition, but for now, told Theodore only that he thought rest and seclusion were indeed for the best. With that, he descended the stairs once more, clapped his brother-in-law on the back, and retired from the house.

Only Mumbet was able to reach Pamela in the dank place her soul occupied. There was something about Mumbet's earthy somberness that touched her. Catharine was only an infant during this first episode, but over time she came to see that Mumbet treated Pamela exactly the same, whether she was lucid or not. This took fortitude on Mumbet's part; it was not easy to respond reasonably to the frequently outrageous behavior of a woman so distraught. But her years in the Ashley house as a slave to a cruel mistress had taught her the necessity of forbearance even under the harshest conditions, and this time the forbearance was tempered by sympathy. Unlike Theodore, Mumbet knew all about misery.

To take over Pamela's place in the household, Theodore had turned to his oldest daughter, Eliza. Short, brown-eyed, with her long, thick hair, she reminded Theodore acutely of the young Pamela, especially now that age and responsibility had tamed the high spirits of her earlier years. She remained in full charge of little Catharine, who trundled after her wherever she went, bawling if she disappeared even for a moment; and she was supposed to keep her eye on Harry and Robert, too. Frances and Theodore Jr. were old enough to look after themselves. Theodore Jr., with his tousled hair and sleepy eyes, was buried in his schoolbooks; that winter he was preparing to follow in his father's footsteps to Yale. Frances, an eager, flighty youngster with chestnut-colored hair and a voluptuous mouth, was secretly seeking out some real-life romance to match what she had read in her blood-stirring novels. But the "little boys," as the troublesome Harry and Robert were still called, remained a problem. Six-year-old Harry, with the flashing eyes in which his father saw nothing but trouble, was always prodding the more genial Robert into tiny bits of little-boy mayhem. In exasperation, Theodore employed Pamela's widowed cousin, Mrs. Gray, to look after them. And finally there was tiny Charles, just weeks old. His pitiful, mewling cries provoked Pamela only to greater fits of weeping. Pretty much given up for dead by Theodore, too, the scrawny little fellow remained Mumbet's to restore, or not.

Theodore was as strict as ever about his dress and his manner as he went about town, but old friends like Peter Van Schaack's brother Henry, who came around from Pittsfield to cheer him up over a glass of stout, could tell that he was clearly tried by his ordeal, even though he never spoke of it. Instead, Theodore filled the air with lamentations for the fledgling national government, which, as he angrily wrote fellow representative Benjamin Goodhue in early January, in his absence "was hastening towards disgrace and impotency."

The new congressional session was due to start in mid-January, but even Theodore could see he couldn't leave Pamela, and he was forced to write to Governor Hancock for permission to stay home for the first few weeks of the new term; he felt unusual gratitude when he quickly received word that permission was granted.

It wasn't until early February, well over a month after her collapse, that Pamela was at last able to dress herself and come downstairs, where she attempted to take charge of the children—who regarded her warily, all except for tiny Charles. Dandled in Mumbet's heavy arms, he reached for his mother whenever she was near. But Pamela was still in no mood to take him. She wasn't fully better, as all the other children could sense, and indeed she never would be. Under Mumbet's care, Charles was at last starting to fill out into a plump, rosy-cheeked baby. In gratitude, Theodore had reached into his change purse and given his loyal servant a silver crown as an earnest expression of his thanks. Mumbet never spent the coin, and kept it until she died.

THE FRAILTY THAT was always part of Pamela's character thinned out whatever membrane it is that keeps a person's essential self safe from the adversity that life inevitably throws one's way. Amid the bleak uncertainties of backcountry life, Pamela had so many things to fear: each new baby only reminded her of the agony of losing three of her previous ones. Everyone's health was so precarious, it seemed that even the halest among them could drop dead at any moment. Shays' Rebellion showed how quickly politics could turn deadly, too; the rabble—as her husband had inclined her to think of the poor—were all around her, ready to attack at any moment. The very things her husband did for their security— making a name for himself, creating their marvelous prosperity—only increased their vulnerability, their exposure, and her fears.

But none of these fears would have eaten at her the way they did if Theodore had stayed home, as virtually every other Stockbridge husband did. It only compounded her misery to sense the many prim Calvinists in town clucking over her breakdown, which to them betokened a loss of her Lord. But in a sense, the neighbors were right. Her ever-certain husband had become a kind of god to her, as every groveling, worshipful letter to her "Lord and Master" attests. It was as if she'd made a bargain with the devil: her husband's majesty came at her own expense.

And, worse—much, much worse—she sensed he'd realized that. He'd become impatient with her, and finally disgusted, turning his back on her when she needed him most. *Because* she needed him most. It confirmed her deepest fear: that she was, in fact, unworthy. He who knew her best knew that.

———— ◆ ————

FIFTEEN

AMONG THE MANIACS

———— ◆ ————

Three weeks into February, Pamela seemed sufficiently recovered in Theodore's eyes for him to take leave of her again, returning to Philadelphia by way of Princeton, New Jersey, where he left his sleigh and horses, and continued on by coach. He arrived just in time to celebrate President Washington's birthday at a large ball, to which all the members of Congress were invited.

It was a presidential election year, the first to involve a sitting president. As the very embodiment of the young republic, Washington must, of necessity, be reelected. His vice president, John Adams, was another matter. While some of the more militant Federalists would have "his rotundity" replaced, Theodore stood firm for his fellow Bay Stater. In the next few months, Theodore often rode out for breakfast with the Adamses at their fine house on Bush Hill, and was particularly taken by Abigail, whom he described to Pamela as "one of the most excellent of women," a comment that cut his wife to the quick.

After hours, Theodore had long been in the habit of taking in Philadelphia's cultural offerings, like the artist and impresario Charles Willson Peale's celebrated American Museum with its displays of "skins and beast, & birds, minerals, fossils, coins, shells, insects, moss & dirt," and the lending library that Benjamin Franklin had organized, with the largest collection of books in America, some ten thousand volumes. In his rambles,

however, there was one local attraction that Theodore had avoided: a view
of the deranged inmates in the psychiatric ward in the damp cellar of the
new Pennsylvania Hospital, a sprawling, otherwise proudly modern facil-
ity that occupied an entire city block downtown. It was just a few minutes'
walk from Dr. Jackson's residence, and, like London's Bedlam, it had long
been one of Philadelphia's preeminent attractions. Founded in 1751, it was
the very first general hospital in all the colonies, and in the early 1790s was
still the only one to take in mental patients as well as sufferers of the more
conventional somatic complaints. It was Benjamin Franklin who first
made the case for treating the insane, arguing that, with the general rise of
the population, their numbers were threatening to overwhelm the jails
and almshouses where they would otherwise end up if their families could
not keep them. For the first decades, though, the hospital served as little
more than jail; the "Maniacs," as the deranged were termed, were often
chained within cells about ten feet on a side, with sturdy walls, thick iron
bars, and a stout door with a small portal, through which their food was
passed. Some of the inmates lay about naked, raging and snarling like
beasts, confirming, in the eyes of their watchers, the policy of displaying
them like zoo animals.

In the lingering belief that the insane were being willfully unruly,
some doctors set about to "reform" them by squeezing them into body-
enclosing restraints like the "Madd-shirt" (akin to the straitjacket of more
recent times) or having them flogged by their keepers. With time, leading
physicians began to think that the mentally ill might be suffering from as-
yet-unknown diseases no less than the victims of consumption, the ague,
and other complaints they presumed to understand, and subjected them
to the "heroic" therapies that Dr. Sergeant had hesitated to try on Pamela:
bleeding, blistering, purging, drugging, even starving the disease out of
their bodies.

In 1789, the dazzling polymath Benjamin Rush, who traveled in the
same political circles as Theodore (although that didn't keep Theodore
from later vigorously opposing Rush's nomination as treasurer of the
mint), took over the insane ward at Pennsylvania Hospital, where he'd already served on the staff for several years. Rush had become convinced
that insanity was a disorder of the blood, a system that he had come to
think of in the mechanistic terms that reflected the emerging scientific ra-
tionalism of the era. To reduce the internal pressures on his lunatics, he

drew off blood with a vengeance. In one case, he extracted 470 ounces, or nearly *four gallons*, of blood over forty-seven bloodletting sessions. For those whose pulse was, in his view, dangerously elevated, he invented the "tranquilizer," a chair with straps to immobilize patients more benignly (although no less completely) than the Madd-shirt. For those whose blood circulation he considered too weak, he developed a blood-stirring "gyrator" to whirl them about. Occasionally, he attempted to shock his patients *into* their wits by pouring ice-cold water down their sleeves, or threatening to drown them. While many practitioners assumed that such mental disorders were the special province of the lower classes, Rush believed that members of the upper classes were actually more vulnerable by virtue of the greater delicacy of their circumstances. As evidence, he pointed to the case of King George III, who was himself, in the years after the Revolution, straitjacketed for his frequent episodes of howling madness.

Despite the seeming brutality of his methods, Rush treated his patients with considerable tenderness, insisting on comfortable quarters that were well ventilated in summer and well heated in winter. And his gentleness anticipated the "moral treatment" of the insane that, following the romantic precepts of Jean-Jacques Rousseau, would soon be articulated by the physician Philippe Pinel, and put into practice at Salpêtrière hospital. Yet Rush continued the practice of exhibiting inmates behind bars as if they were exotic beasts. Many visitors found it amusing to pelt the wretches with small stones to rouse them out of their customary stupor and into a raving fury. Pamela's condition must have seemed too perilously near such degradation for Theodore to find such an outing edifying.

Through the winter and into the spring, Theodore sent Pamela assiduous letters pleading with her to attend to her health, and late that summer of 1792, oppressed by his cares, he actually did seriously consider retiring from public life to devote himself to his family. But his fellow Federalists insisted he stay. Fisher Ames was alarmed that he could even contemplate "leaving us to fight without hope and against odds"—fight, that is, the rising forces of the Jeffersonian faction that opposed the Federalist idea of a powerful central government and favored returning authority to the states. When Theodore allowed his name to be entered for the new congressional election once more, a weary Pamela wrote him from Stockbridge that she hadn't really expected otherwise. She knew his "Public Cares" were so "exceedingly fascinating" they'd become "a kind of agreeable Drudgery." His work was his life, no matter what he claimed. For a political animal like him, "its Interests require his constant attention from which he cannot be saperated without fealing a Distressing and painful Void—as a fond Mother does when She is by some misfortune Detached from the care of her Family."

When Theodore returned to Philadelphia for the new congressional session in January 1793, Pamela had hoped to join him there, but, cruelly, to one who found the cold so oppressive to her spirits, mild temperatures turned the roads into rivers of mud and slush. From January through March she gathered almost daily intelligence on the road conditions to the south, but every report was the same: the snow was too patchy and soft for sleighing. "I feal my Self grieveiously disopointed," she told Theodore. "I had set my heart much on going."

SIXTEEN

THE REIGN OF TERROR

As ever, there was much else to preoccupy her husband. A few days after his return to Philadelphia, an awestruck Theodore was among the dignitaries watching with President Washington as a colorful, handpainted balloon rose up from the grounds of the Walnut Street Prison, hoisting the French aeronaut Jean-Pierre Blanchard, accompanied by a small Labrador retriever, two miles into the air and then fifteen miles downwind, where the flying contraption landed gently in a field in Woodbury, New Jersey. It was the first lighter-than-air flight in the United States, and since Blanchard was bearing a letter in his pocket, the country's first air-mail delivery, as well. "It is impossible to describe the majestic sublimity of the objects," Theodore rhapsodized to Pamela. "Never in my life have I had such sensations."

But not all French-inspired developments were so uplifting. The French Revolution, which had begun as a joyous echo of the American fight for independence, turned terrifying with the beheading of Louis XVI by guillotine in a Parisian public square later that January, beginning the orgiastic bloodbath that was the Reign of Terror. The news didn't reach Philadelphia until March, shortly after President Washington's inauguration for his second term. Three thousand citizens would be guillotined in Paris alone, with thousands more slaughtered in the countryside.

To Theodore, this was nothing short of a cataclysmic, nationwide eruption of the Shaysite delusions he had faced down in the Berkshires. It demonstrated once again how essential it was for sensible men like himself to redouble their efforts to maintain public order and secure private property here in the United States, which were under attack from Jacobins of their own.

As Washington closed out his first term, Theodore was intent on building up the Federalist faction in western Massachusetts, both as a bulwark against the rising anti-Federalist sentiment and as a political machine that would return him to his post. To this end, he cultivated Loring Andrews, the impressionable young editor of the new Stockbridge newspaper the *Western Star*, and freely used the paper as a mouthpiece by which to present his staunchly Federalist views to the local electorate. To him, the spreading contagion of the French Revolution demonstrated the hazards that awaited a republic if the "friends of order" acceded to the whim of the Republican rabble. The French, it appeared to him, had gone stark mad.

As if to underscore the dangers posed by a revolution that spun out of control, a number of French noblemen "condemned to cruel exile" (as Theodore put it in a letter to the children) by their country's revolution came through Philadelphia in the months afterward, and Theodore extended a brotherly sort of friendship to several of them. In particular, Theodore became close to the duc de La Rochefoucauld-Liancourt, an adviser to Louis XVI, an agricultural experimenter, a man of generosity and principle, a bit of a philosophe, and someone in whom Theodore saw something of himself. For his part, Liancourt regarded Theodore as "an excellent man," but he could not help noting that "in his politics he is somewhat warm, and not a little intolerant." By that, Liancourt referred delicately to Theodore's growing tendency to fly into a rage over political disputes. All the same, Theodore had Liancourt up to Stockbridge (regardless of Pamela's condition, Theodore was always having friends in; the big house was like a hotel sometimes), and the nobleman greatly enjoyed his rambles about the "beautiful country, fine land, well cultivated, all in meadow, and a most excellent soil."

Shortly after Theodore returned to Philadelphia, a fearful scourge of yellow fever broke out, killing two dozen residents a day well into the fall. Thinking the pestilence was infectious (it is actually spread by mosquito), citizens walked down the middle of the street to avoid contact with any possible carriers, and virtually no one dared shake hands.

Overwhelmed by her fears for her husband and for herself, Pamela's emotional health started to fail her once more, and by fall, she was again scarcely able to function. On Theodore's advice, Pamela remained secluded in her room, with only Mumbet to tend her. In her misery, Pamela berated herself, and took no care about her appearance. Her face had gone deathly pale, and she'd thinned down terribly. Normally unshakeable, even Mumbet found it hard to be in her company for long. The older children sent letters to their father pleading with him to return and look after their mother. Every time a packet of letters from him arrived, they would tear theirs open, hoping to find word that he was on his way back.

When little Catharine found no letter for her, she made one up, claiming he had sent an invitation for her to come with Mrs. Gray to visit her papa in Philadelphia. In this, she echoed everyone's wishes, but it wasn't until late October that Theodore actually braved the muddy roads to head north. When he burst in the front door, creating a near pandemonium of joy from the children, and then rushed up the stairs to see Pamela, he had only to take one look at her—ashen, listless, scarcely able to stir herself to greet him—before he made his decision. Within days, Pamela was headed by carriage to an establishment for the insane run by a Colonel Lovejoy in Andover.

SEVENTEEN

COLONEL LOVEJOY'S METHODS

A prosperous farming town in the northeast corner of Massachusetts, Andover was nearly as far away as New York, a good 150 miles. This was part of its appeal; Theodore needed a clean break, both for Pamela's sake and for the family's. With its bucolic setting, safely removed from the smallpox, diptheria, and, later, tuberculosis epidemics that regularly infested Boston, the town had a reputation for healthy air, and its general restorative qualities were extended into the mental realm largely through the efforts of Dr. Thomas Kittredge, an unusually enlightened professor of anatomy at Harvard College. He began, late in the eighteenth century, to arrange for the insane to be boarded with carefully vetted families headed by "strong, fearless, capable and good natured women" who kept the patients "under little restraint, though subject to constant supervision." *

---

* James Otis Jr. was probably the most famous of the deluded souls who ended up in Andover. The revolutionary Boston firebrand, famous for his early rallying cry, "No taxation without representation," Otis had been struck violently on the head during a quarrel in 1769, and it appeared the blow had turned him into a "mad freak," according to a contemporary diarist. In 1770, after several rounds of increasingly outlandish behavior, he threw rocks through the glass of the Boston Town Hall and shot his pistols out the win-

But private madhouses could also be torture chambers. A fiendish practitioner named Dr. Willard, who operated an establishment down by the Rhode Island border, made a point of breaking his patients' will. With the particularly strong-minded, he'd try to jolt them into sensibility by enclosing them in a coffinlike box that was pierced with holes, then lowering them into a tank of water and leaving them submerged "until bubbles of air ceased to rise," according to a later account. Only then was the patient brought back up to the surface to be "rubbed and revived"—assuming he could be.

Were these Colonel Lovejoy's methods? Years later, when Catharine recounted Pamela's many sufferings in her memoir, she gave out a kind of sob. "But oh! I cannot bear to think—it has been one of the saddest sorrows of my life to think how much aggravated misery my dear, gentle, patient mother must have suffered from the ignorance of the right mode of treating mental diseases which then existed." Still, the specifics of her particular mode of treatment are unknown. Not a single letter alluding to her stay survives, quite likely because none was sent. Pamela was too deranged to write, and considered to be in too delicate a state to receive any letters. She was gone.

PAMELA DID NOT return to Stockbridge until the following spring of 1794. Considerably revived, she was thought to have recovered enough to make the trip to Philadelphia she'd had to put off the previous winter, and to bring Eliza along. Together, they sailed down the Hudson to New York.

It was one of the very few times in her married life that Pamela was truly happy. She and Eliza prevailed upon Theodore to escort them to the theater a few nights after their arrival, and they stayed out past midnight, a wonderful, giddy extravagance. She and Eliza went by carriage around the whole city, taking in the attractions that her husband had raved about

---

dows of his boardinghouse. Temporarily rehabilitated by a stint of country life, he reentered politics, then passed the Revolutionary War as a teacher on Cape Cod, but turned erratic again, several times jumping out of windows. At other times, he would be found at all hours wandering aimlessly about the yard at Harvard College, alarming the students. That is when he ended up with Mr. Henry Osgood in Andover, who kept a large farm for the insane. Otis died there in 1783 when he was caught out in a storm and struck by lightning.

in his letters home—that most peculiar museum of Mr. Peale's, Dr. Franklin's extensive library, the many parks, and best of all, that glittering social life of her husband's that had so long been denied her. The high point of her season was an unstintingly European-style ball for Mrs. Washington, a vast affair with dancing and fine gowns. An engraver captured the scene, and in the crowd that swirled about Mrs. Washington, he caught a beaming Pamela, her face no less radiant for being demurely tipped down, in the company of Mrs. John Jay. Theodore, uncharacteristically, is nowhere to be seen.

The two Sedgwick women stayed on with Theodore only through the end of that congressional session in June, but the excitement carried Pamela through a full year. In Stockbridge again in the fall of 1795, Pamela took yet another plunge. If before, she directed her anger inward as she lamented her many shortcomings in the classic manner of depressives, now she was starting to lash out, sometimes using the most grotesque vulgarities. The children, who had learned to keep their distance when Mamma was in one of her moods, now stayed away most of the time. Even little Charles, just four, knew to be wary.

This time, Theodore did not return to examine his wife himself. His sympathies never did run to the weak, the needy. Besides, Pamela's derangement seemed to be aimed at him. Her unreason was a retort against his rationality. He could not face it anymore. He sent word from Philadelphia, directing Ephraim Williams to remove Pamela from the house, this time to a home in nearby Sheffield. "Mrs. Sedgwick as longer afflicted with her dreadful malady must not be at home," Theodore declared. "To remove her...will to me be very painful, but infinitely less so than to have her in the family where she may give wounds to sensibility which can never be healed."

He wrote a letter to the older girls, Eliza and Frances, to assure them he had done all he could. "Now [their mother's] mind cannot admit the force of arguments, and [until] a gracious providence should restore her mental faculties, the subject itself is of too tender a nature to allow the necessary explanation. But yet we must, without deviation, pursue that line of conduct which our best judgment may dictate."

When Theodore did return home at the close of Congress a few weeks later, he did not see his wife, even though he passed through Sheffield on the way north. He told everyone that he was afraid his presence would only inflame her, but in truth he simply could not bear it. Everything about her spelled defeat to him. Instead, he asked Dr. Sergeant to go in his place, and to give him a full report. When the children learned Uncle Doctor was going to see their mother, Eliza, Frances, and Harry (young Theodore being away now at Yale) all insisted on visiting, too, and joined Dr. Sergeant in his carriage.

The house was small and tight, like so many of the houses in Sheffield, with open grounds outside that were off-limits to the patients, who did not number more than three or four. When Dr. Sergeant and the children were admitted to Pamela's room, his blood froze at the sight of her. When he had first visited her in Stockbridge, she had at least been able to keep still. This time, she was so agitated, he feared the devil had indeed gotten hold of her. She was still "frantic," he wrote Theodore, and used "very profane" language. She demanded to know what had become of Theodore. "I told her you had returned," Sergeant reported, "and would soon visit her." The children could barely bring themselves to enter the room, and they stared at their mother with saucer eyes. "The Children were all affraid of her," Dr. Sergeant wrote.

Still, Theodore did not come. Instead, he returned to Philadelphia to attend to the nation's business. When further reports from Sheffield showed no improvement in his wife, he instructed Ephraim Williams to take the family chaise to Sheffield, remove Pamela, and deliver her to the care of a Dr. Waldo in Richmond, a few miles to the northeast of Stockbridge.

When Williams arrived at Sheffield for this miserable duty, Pamela screamed at him that she wanted only to go *home,* but Ephraim had his orders: "Mrs. Sedgwick must if possible be soothed, and be made to understand that her absence from home is the effect of calculation for her good." Williams gathered up her things, made arrangements with the proprietor, and with Pamela up beside him, made the hour's drive to Richmond, where Dr. Waldo took charge of her, setting her up in a small room with a fireplace.

Unlike the Sheffield establishment, Dr. Waldo did not allow family visits, explaining that they "tend to disconnect her mind and render her more uneasy to stay there." Theodore turned again to Dr. Sergeant, who obediently rode out to Richmond a few days later to examine the patient. Hoping to put Theodore's mind at rest, he declared her "much better."

She no longer tore at her clothes, but kept them "decent," managed to stay in bed through most of the night, and lay down "frequently" during the day. Continuing his clinical appraisal, he noted her "regular stools, and the other evacuations moderate and regular—her appetite about middling for her." She "seldom" wetted the bed, but the fire had to be "removed every Night." But the key point was this: "She appears to be intirely under the control of the Doctor, which he has obtained in an easy and gentle manner, he rides with her every day when he can attend it...."

Clearly, Dr. Waldo's direct, humane interest in her was an improvement over the Sheffield establishment, but the initial gains proved short-lived. Pamela's brother Henry Dwight was allowed in to see her in February, and he wrote Theodore that she'd turned frantic again. "[Even] a very small matter seems to unhinge her." It distressed her especially to think of being seen by her children or her husband in such a pathetic condition. "When she will recover," Dwight concluded, "and whether, are beyond our reach."

Theodore stayed in Philadelphia. As he wrote the children that March, he "dare[d] not desert the post assigned to me." It could be that his vote alone "may save my country from that ruin with which the passions and ambition of wicked men threaten it." He was referring to the congressional ratification of Jay's Treaty, which, for all of Theodore's hyperbole, was probably the most fateful issue of the Federalist era, and certainly the most divisive one politically, as it offered the tantalizing possibility of not just rapprochement with England but with much that England represented, and Theodore hungered for personally—a recognition of an established elite, a certain traditionalism, and, by assuring a trade agreement, an alignment with the mercantile interests. To Theodore, England's major attraction was that it was not France, which is to say (at the risk of oversimplification, but Theodore himself was inclined toward oversimplification), it was not beset by the sort of dangerously excessive passions too amply demonstrated by that bloody and riotous revolution. The treaty proved to be a critical political divide, creating the Federalists as a modern-style party, and establishing their Republican opponents, led by Jefferson and Madison, as their entrenched antagonists.

So it is interesting to think how much of Theodore's own political stance on such a momentous issue was derived from his horror at what had become of his wife. Surely, the depth of his revulsion for revolutionary France owed something to his mounting distress at his wife's emotional instability. In supporting England, he was endorsing reason, order, logic, clarity—all the things he desperately believed in, and all the things that seemed to be under attack in his wife. Thus, at least for Theodore, his foreign policy was based on his domestic one. His fervid support of Jay's Treaty roused him to his greatest height of oratory in March, when he delivered a long, impassioned speech on the house floor promoting it. When he was done, representatives on both sides of the debate rose to applaud his eloquence, although more lustily on the Federalist side. "Your speech for matter, style and delivery exceeds anything I ever heard, and I have heard much good speaking," Vice President Adams wrote him. And a fine lady watching from the gallery made Theodore a present of some "sweetmeat of Orange," she was so overwhelmed. Seeing the measure through, Theodore did not return to Stockbridge until the close of the session in June, whereupon Dr. Waldo declared Pamela fit to join him after all. She always did do better in the summer, especially when Theodore was home.

THAT SUMMER WAS an exciting one, for Eliza had a suitor, Dr. Thaddeus Pomeroy. In his early thirties, the Harvard-educated Pomeroy was nearly a decade older than Eliza, and he was well established as a druggist in Albany. Truth be told, Pomeroy was rather sedate, if not wooden, but Eliza, worn down by running a household, had lost most of her youthful high spirits and scarcely seemed to notice. She'd also become alarmingly receptive—in eight-year-old Catharine's view—to the gloomy Calvinism that had held their mother in its dark thrall. Duty was Eliza's lot; and she may have been eager to escape her many responsibilities in Stockbridge. Catharine was leery of a stultifying match, and not at all drawn to Pomeroy, but when Theodore gave the marriage his blessing, the couple set a wedding date of April 1.

That fall, the Massachusetts legislature selected Theodore to fill out the remaining two years of the Senate term of Caleb Strong, after Strong had done what Theodore could not bring himself to do: resigned to return home at the urging of his ailing wife. Delighted to join this more rarefied political club of the Senate, Theodore served on many committees and, after Washington quit the presidency on completion of his second term, was thrilled to rub shoulders with the new administration of John Adams. It wasn't unusual for him to spend the morning with the secretary of war, then President Adams himself for the afternoon, then the secretary of the Treasury before looking in once more on the president until Theodore returned to his chambers at the Jacksons at nine thirty.

Meanwhile, Pamela distracted herself that winter with the wedding preparations. But even this domestic happiness was not unalloyed. What Pamela looked forward to with such uncharacteristic eagerness, the younger children, especially Catharine, approached with dread. To them, the nuptials marked the loss of the surrogate mother who'd seen them through the many collapses of their actual one. Nevertheless, the wedding was conducted that April before a throng of friends and family. No sooner had the minister launched into his opening homily than Catharine burst into tears at the prospect of her sister moving away to Albany. She made such a scene that her father swept her up into his arms, which may have been her object, to try to soothe her.

When that still didn't stop the tears, Theodore summoned his manservant, Grip, to carry her across the hall into the other parlor, where a scowling Mumbet did her best to hush her, still to no avail. Finally, once the service was over, Pomeroy himself came into the room and whispered to Catharine the words he assumed she longed to hear. "Your sister may stay with you this summer," he told her.

But Catharine was furious. *"May!"* she recalled in her memoir years later. "How my whole being revolted at the word. He had power to bind or loose my sister."

Catharine was not the only sister in turmoil on Eliza's wedding day. At twenty, Frances had been pursued by a suitor Theodore did not favor: Loring Andrews, the darkly handsome young editor of the *Western Star* who'd proved such a valuable confederate to him. In the battle over Jay's Treaty, Andrews had extolled Theodore in iambic pentameter, concluding: "Firm as the unshaken Andes meets the sky, / Thou stand'st the bulwark of thy country's friends." But that earned him nothing from Theodore where his daughter's marriage prospects were concerned. Loring Andrews remained a mere scribbler—and, it was whispered, a drinker besides.

Andrews had first taken an interest in Frances when she was just fourteen; at the time, he was being cultivated by Ephraim Williams on Theodore's behalf. Williams's own infatuation with Eliza led Andrews to pair off with Frances, and the four often took off for long playful rambles about the Stockbridge countryside. Williams's interest in Eliza had been promptly nipped by Theodore, but Andrews's had continued on surreptitiously until the previous summer, when Theodore learned of it. To end the business, Theodore had packed Frances off to boarding school in Philadelphia, telling Andrews that he'd had to take such action because of his "degrading habits." Swearing that Theodore was mistaken, Andrews responded by brazenly asking for Frances's hand in marriage. Theodore was appalled that the upstart should show such cheek. Andrews pleaded for him to reconsider.

> *May I yet indulge the hope that I have not given you offence? To a cooler judgment, and to a more deliberate reflection, it may appear that I have done amiss—but when I listen to the responses of my heart, it does not accuse me of guilt—if that heart, in its indulgences, has forced a passage thro' the bounds of propriety, let its feelings forcibly urge the plea for pardon—if the wish which it dictates could be complied with, I think my share of human happiness would not be small——but you, Sir, must determine—all rests with the decision which your goodness, wisdom and judgment shall prescribe.*

But Theodore's decision was unchanged by these perfumed sentiments. *No.* Miserable, Andrews quit the *Western Star* that spring and arranged to leave Stockbridge. But by the time of Eliza's wedding, he had still not departed, and Frances was done with school, so Theodore thought it prudent to remove the object of Andrews's affection to New York City, where she was to stay with his friends the Penfields.

BACK IN PHILADELPHIA the following spring of 1798, Theodore picked up the cudgel once more against the specter of Jacobin France, this time pressing for the Alien and Sedition Acts to rout out the French spies who were supposedly swarming America's cities, and the scurrilous editors who sympathized with them, and for an army for a show of firmness against the perceived French threat. An intemperate response, certainly, and it would soon prove a self-destructive one, both for his party and for himself. *Paranoid* is the modern word for it. Only three aliens were deported under the act, and for every seditious editor who was prosecuted, a dozen more sprang up in his place to decry the outrageous infringement of liberty. It was as if the Revolution were being refought, this time with the Federalists as the royalists, and the Jeffersonians the common people clamoring for liberty—not a winning position for Theodore's adored party. Far from being cowed, Republican newspapers doubled in number over the next two years, and the Jeffersonians gained with them. In trying to secure his position, that is, Theodore was ruining it.

Now FIFTY-TWO, THEODORE was growing exhausted by all the tumult, and showing some of the wear of a much older man. Seriously overweight from the sumptuous dinners he so enjoyed, he was also beginning to suffer from gout, made all the worse by Philadelphia's long, wet winters. And his eyes, strained by years of late-night reading by candlelight, were troubling him, as well. Once again he considered retiring from public life, and tantalized Pamela by saying he actually planned to bring up the matter with President Adams. But, perhaps inevitably, he could not find the right moment.

Pamela deteriorated again that summer of 1798, and, capping increasingly distressed reports from the children, she herself confessed to Theodore of "a return of my old disorders." By August, her Stockbridge neighbors were saying openly that Mrs. Sedgwick was "deranged."

She held on through the fall, during which Theodore, citing the imagined French threat, let Berkshire County Federalists draft him for his old House seat. When he was elected for yet another two-year term, Pamela drooped again. "My own unworthyness is the cause of depressing my spirits," she wrote him. Theodore tried to persuade her otherwise. "This we both know to be a disease," he insisted. "Would to God we could discover a remedy." Pamela curtly told him she knew the remedy: "I shall be better when I again have the pleasure to see you."

Still he stayed away.

THAT JANUARY OF 1799, Theodore had entered his name as a candidate for Speaker of the House. He'd done this several times before without much hope of success, but by now Jefferson's Republican Party had risen to the point where the Federalists believed they needed a true bulwark to counter him. Who better than the ever-stalwart, if occasionally frothing, Theodore Sedgwick? Having served in every term of all five congresses to date, most recently as senator, Theodore had come not just to voice the Federalist perspective but to embody it. In his expensive finery, his erudite allusions, his grand manner, not to mention his sheer poundage, he was clearly of that aging Royalist old guard that assumed authority in the bejeweled way they asserted it. Although Rufus King, John Jay, and many other early Federalists were still around to guide the party, Theodore styled himself as the last true believer, expatiating endlessly and heatedly about his unwavering opinions, every extra syllable reducing the chances of persuading his opponents and increasing the risk of alienating his allies. Well positioned as he was, he squeaked in by just one vote, the first time the speaker had been elected purely along ideological lines, rather than by a common consensus that transcended party loyalties. The Federalists could no longer assume they spoke for the nation. Certainly, Theodore couldn't.

Rather than temper his attitudes in hopes of finding common cause with his opponents, Theodore, his black eyes burning, went right after the Jeffersonians, summarily banning a pair of reporters for the Jeffersonian paper, the *National Intelligencer,* from their customary seats in the House. They repaid the favor by playing up a story of Theodore's tiff with another observer whom he'd had forcibly removed from the gallery. In retaliation, Theodore angrily had yet another *Intelligencer* man pitched from the gallery. The Jeffersonians in the House moved to have Theodore censured. This was unsuccessful, but Theodore went wild with indignation. The mood was set. The fact was, Theodore's high-handed and quarrelsome manner was starting to annoy even his friends. One of his closest Federalist allies, Gouverneur Morris, was irritated just by the way the plump Theodore lounged in his chair "with his heels on the table."

In his search for enemies, Theodore scrutinized his friends and started reconsidering his loyalty to the blunt, irascible Adams, who, as president, should have been the unquestioned leader of the Federalist party. Their break started with a slight. The previous year, as Theodore looked to the future, he had written an uncharacteristically stammering, painfully obsequious letter asking Adams to consider appointing him to the Supreme Court to fill the place of an associate justice, James Wilson, who'd just died. ("I pray you, Sir, pardon this address which I fear will be thought improper or presumptious...") After Adams dismissed his application, Theodore turned on him, and started to refer to him as "that weak and frantic old man" and "an evil."

A frequent visitor to the Adamses before 1800, afterward, Theodore scarcely even corresponded with his president. On the few times they saw each other, Theodore complained to his friend Henry Van Schaack, "[Adams's] conduct was as cold as his heart."

For his part, Adams thought Theodore a man "without dignity...ever ready to meet in private caucuses and secret intrigues to oppose me." He was right about that. By 1800, Theodore had decided to throw Adams over for South Carolina's Charles Cotesworth Pinckney, Washington's erstwhile ambassador to France, whose chief qualification was that he was not Adams. In this maneuver, he was supported by Alexander Hamilton, who retailed his own loathing for the irascible incumbent in a fifty-four-page diatribe that stopped just short of calling the president mentally unstable. Together, Theodore and Hamilton decided to back Adams and Pinckney equally, in order to throw the election into the House, where Theodore as Speaker could tip the election to Pinckney. Theodore feverishly promoted the scheme to his remaining political allies along the eastern seaboard.

That fall of 1800, as the election began, the nation's capital officially moved south to its new permanent site on former swampland along the Potomac. Annoyed by everything, Theodore found it a terrible comedown from Philadelphia, and filled his letters home with complaints about his "execrable" lodgings, with their "nasty filthiness of every thing from the garret to the cellar," and the sloppy roads through which "it was almost impossible for the Horses to drag us."

But he put all of that aside as word of the electors' activities filtered back to Washington, and it became clear that the South was throwing its support solidly behind Jefferson, who ran with the handsome, accomplished, but perhaps overly ambitious Aaron Burr. New England went for Adams, but New York for Jefferson-Burr, and when the middle states split their vote, Jefferson and Burr were tied at the top with 73 ballots apiece. (While the party considered Jefferson the presidential candidate, and Burr the vice presidential, the electorate did not make the distinction, regarding both men as Republican presidential aspirants.) The election would be decided in Theodore's House after all, only without any Federalist candidates to pick from.

A violent snowstorm whipped Washington that Wednesday, February 11, 1801. To turn back Jefferson, Theodore tried desperately to unite the Federalists behind Burr, whom he thought opportunistic enough to accept Federalist policy in return. He secured all the Federalists but five behind Burr. Because each state's vote was determined by a majority of its representatives, that was enough to keep Jefferson from winning, but it was not enough to tip the House to Burr. The two remained tied, as the first ballot revealed. Despite ferocious politicking on both sides, the vote remained deadlocked for ballot after ballot—nineteen altogether through the afternoon, evening, and deep into the night. As Speaker, Theodore dutifully tabulated each result, and read out the count from the Speaker's chair. On and on it went for four days. Across the nation, anxiety mounted. There was fretful talk of civil war if the Federalists succeeded in denying the Republicans their rightful victory; the governors of Virginia and Pennsylvania braced their states for bloodshed.

Finally, on Sunday, five days into the balloting, Delaware's lone representative, the Federalist James Bayard, decided to spare the country such agony, and, despite Theodore's fierce arm-twisting, abandoned Burr. Bayard still couldn't bring himself actually to vote for the hated Jefferson, but leaving his ballot blank would be enough. When he heard Bayard's decision, Theodore wrote to his son Theodore Jr., "The gigg is up." And so it was. On Tuesday, February 17, Theodore Sedgwick counted the votes of the thirty-sixth ballot, and then he solemnly rose from his Speaker's chair to declare that Thomas Jefferson had won the presidency.

Unable to bring himself to attend the inauguration of his nemesis, Theodore boarded a four A.M. stage from Washington the morning Jefferson was to be sworn in. One of his fellow passengers was the outgoing president, John Adams, who couldn't bear the spectacle, either. Each man blamed the other for the Federalists' sudden, ignominious fall from power, ending the political careers of both. It was a journey of nine days. Neither man spoke a word to the other the entire way.

EIGHTEEN

IT CAN NOT BE TOLD

Stockbridge had come up in the world from the backcountry town that Theodore had left for his career in public life sixteen years before. Loring Andrews's *Western Star* had done much to make the town a node for news, both incoming and outgoing, and its advertisements reflected a new prosperity, as local purveyors offered everything from oysters and carriages to literary wares like Hutchinson's *History of Massachusetts* or handsome editions of Cervantes's *Don Quixote*. For those who could not afford to buy such books, there was now a Berkshire Republican Library, offering 150 titles to its members at $2 per share. When the old schoolhouse burned down in 1793, it was rebuilt, at Theodore's direction, into an "elegant" new structure that all his children attended. And fancy-dress balls were now a regular feature of the Red Lion Inn, although Frances had by now spent enough time in festive New York to think of Stockbridge as a social "Greenland."

When Theodore returned to take his seat at the head of the table for the two-hour repasts that were his standard, his water "tempered" with a splash of brandy, few of the children remained in their places around him. Eliza had moved to Albany with her new husband. Theodore Jr. had graduated from Yale and, after training with his father's old friend Peter Van Schaack in Kinderhook, had also moved to Albany to start a law practice there. Harry and Robert were both sophomores at Williams College.

At twenty, Frances was back from New York City, but she would not stay in Stockbridge long. Theodore had at last found a proper suitor for her: jowly, thickset Ebenezer Watson, a man, in Theodore's confident summary, "of virtue, of unblemished character, and of happy prospects." Frances had first met him in New York when she was at the Penfields under orders to stay away from Andrews. Ebenezer's parents owned the *Connecticut Courant,* the thriving Hartford newspaper, and Ebenezer was trying to get established as a book publisher. For the time being, though, he was marketing rare foodstuffs; in his many letters to Frances in Stockbridge, he often inquired about the price of Berkshire ham and other local farm products.

Initially, Frances would have none of him. "Attachd to him I am not," she took care to assure her father in February of 1799. As she scoffed to her brother Theodore, Ebenezer wooed her with "the baby cant of a whining lover." Meanwhile, her forbidden beau, Loring Andrews, had bowed to Theodore's wishes and decamped from Stockbridge, ending up in Albany. When Pamela and Mumbet journeyed there for the delivery of Eliza's first child that spring, Frances came along—and, still entranced by Andrews, met him on the sly. They continued to see each other in secret, either in Albany or, more brazenly, in Stockbridge, for a full year until he could bear it no longer. Andrews came down once more from Albany, burst into the house, found Frances in the front parlor, and insisted that they elope.

Afraid of doing something so contrary to her father's wishes, Frances sent Andrews away, and he slunk back to Albany, where he soon repaired to the nearest tavern and he was heard drunkenly maligning her. She forgave him this gaucherie and continued her secret rendezvous with Andrews whenever she could. But the affair was hopeless, for her father continued to push Ebenezer on her. "I must like him because my friends do," she pouted, declaring herself ready "to love Watson, if possible." But it was not easy, as Theodore Jr. told his father, "to concenter in an object that to her has no charms."

Essentially, Francis was caught in the same bind that trapped her mother. "I can perform nothing worthy of you," she grieved to her father in February of 1801. "The sense of your excellence rises in my heart—

& the feeling it there raises—send one of those warm suffusions to my face—which speaks the force of an affection and admiration—that no language can copy." Certainly not hers. Much as it had Pamela, the force of Theodore's personality turned the once-buoyant Frances into a wan sniveler: "As for me insignificant as I am—I can hope to cause little happiness any where—but through countless time you will be remunerated for all your goodness to me."

Through all these machinations, Theodore professed to have no power over his daughter. He claimed to be mystified—and not a little disturbed—by Frances's marital choice, even though it was the one he had labored mightily to bring about. "Are not these girls very strange beings all at once to leave their fathers & mothers, and brothers & sisters to go and live with a strange man, who possibly will be very cross and abuse them?" he asked Catharine with stunning disingenuousness. "But so it is, and if Frances will go I had as lief she would go with our friend Eben as with any body."

And so she did. Frances married Ebenezer Watson on April 7, 1801. Once again, the wedding service was conducted in the Sedgwicks' best parlor. Afterward, the couple left for New York City, where Ebenezer proved to be every bit the ogre that Theodore, who chose him, imagined.

THE BIG, DRAFTY house seemed very empty after that, with just Catharine and Charles at home. Theodore had plenty to busy himself with: his law practice in the ell off the side of the house, and the small farm, with its cows and sheep, to run. To Catharine's delight, he allowed her unlimited credit at Mr. Dwight's store in Stockbridge, and she regularly treated her friends to luxurious lunches with "Malaga wine" and raisins, a delicacy, all unwittingly paid for by Papa. And of course he also had Pamela on his hands, frail and somewhat limp now, after waiting so many years for this very moment. Still, her gladness was evident every time she gazed his way, even if her eyes seemed hollow. At fifty-five, Theodore was definitely starting to feel his age, too. His eyes were increasingly bothersome, and often bulging and inflamed. And the gouty ache deep in his bones was all the worse for the Berkshire cold. He'd started to use a cane—a silver-topped one—but on occasion the pain was so severe that he was forced onto crutches, which was deeply humbling.

He had every reason to stay put in Stockbridge, and yet he did not. The following year, he accepted an offer from Massachusetts' governor,

Caleb Strong, whose place in the Senate he had briefly taken, to become a justice on the state's Supreme Judicial Court. Far from a sinecure, it was a roving position, as three of the seven judges were required to preside over each jury trial in every county across the state, and up the raw, damp coast into what is now Maine but was then part of the Commonwealth. That first trip in the spring of 1802, he took some pleasure in his sojourns, which left plenty of time for socializing in taverns and hostelries. But when he found himself up the Maine coast, he admitted to an unusual "homesickness."

In his absence, Pamela wilted once again; and when Eliza's third child, a daughter named Pamela, died unexpectedly the next fall, the elder Pamela was crushed.

Theodore was home with her, his woes compounded by the death of his great Federalist patron Alexander Hamilton the previous summer. Hamilton's final letter before his fateful duel with Aaron Burr at Weehawken was written to his old friend Sedgwick, whom he counted on to circulate to his fellow Federalists his conviction that the northern states, no matter how bitter about Jeffersonianism, must not consider seceding, as some Federalists had proposed, and Theodore himself was considering. "Dismemberment of an Empire will be a dear sacrifice of great position and advantage, without any counterbalancing good," Hamilton lectured Theodore with what proved to be the final strokes of his pen. It was a lesson that Theodore took to heart, using it as his reason not to press too hard for abolition, in the movement that was just forming, lest it cause the slave-holding southern states to secede.

WHEN PAMELA DID not rally, Eliza loyally moved back to Stockbridge with Pomeroy—whose drug business had faltered—and their four surviving children the next winter. In February of 1806, though, when Theodore was again away, Pamela's mood descended further. But when it was time for Theodore to renew his term two months later, he asked to continue. He was determined to capture one last honor—of ascending to chief justice of the Supreme Judicial Court. Francis Dana, a former member of the Continental Congress, had stepped down from the post, leaving Theodore as the senior man. It was customary that the chief justice be drawn from the existing court. So Theodore was thunderstruck when Governor Strong reached outside of it to select a Boston lawyer, Theophilus Parsons, instead.

"I am, indeed, most infamously treated," he griped to his friend Henry Van Schaack. "I will not patiently submit to this unmerited disgrace." After dozens of letters and testimonials from leading members of the Massachusetts political and legal establishment flooded in, imploring Theodore not to resign, he reconsidered, and returned to his judicial post despite his wife's latest plunge. Around this time, his sons stopped referring to him as "Pappa" and started calling him "the Judge." Soon the whole family would. It was not entirely a term of respect.

PAMELA, FIFTY-FOUR THAT year and fairly elderly for the time, struggled through the winter. In March, Catharine stayed on with her into the spring rather than go to Frances in New York, as she'd planned. Harry was off to Litchfield to study law. Only Mumbet remained with her.

The family gathered briefly late that summer, when Ebenezer and Harry hunted pheasant in the meadows, and the others joined in for picnics by the river. Pamela was able to manage only a turn along the bluff overlooking the Housatonic. Theodore looked "desolate and haggard" to Frances. But he was soon off again on his judicial rounds.

He'd left his wife for the last time.

On the morning of September 19, when Mumbet came into Pamela's bedchamber to help her dress, she spotted an ugly bruise over her mistress's eye. Evidently Pamela had banged her head into something—the bedstead, the wall, the windowsill—during the night. Pamela took her seat at the table for the family's customary elaborate lunch in the dining room, but, pleading fatigue, she returned afterward to her bedroom. That evening, Catharine, Harry, Ebenezer, and Frances were all chatting gaily in the front parlor when a friend of Pamela's named Sally Fairman stopped in to see her. Fairman was afraid all the noise in the parlor below might disturb her ailing friend. But Pamela surprised Fairman by saying it pleased her to hear the children "cheerful and happy." They were her last coherent words.

That night, around two, the household was jolted awake by loud groans from Pamela's room. The children came running, as did Mumbet. They found Pamela thrashing about in "paroxysms" on the floor. Evidently she had taken a poison, and probably not for the first time. Pamela might have taken an insufficient dose the night before, leading to a bruising fall. Or she might have tried to kill herself by slamming her head against the wall or flinging herself out the window, but suffered a failure of nerve and decided that *this* time, tonight, she would do the job properly.

Panic-stricken, Mumbet and the children heaved Pamela up onto the bed, where she continued to toss about, frothing, unable to speak. For the next several hours, they tried to calm her, but Pamela's slender body clenched tight in painful, full-body contortions as the poison did its work. Her face horribly distorted, she emitted a terrible gargling sound from deep in her throat as she struggled to draw breath. According to her neighbor Mary Bidwell, she suffered twenty of these wrenching spasms altogether. Overwhelmed by the sight, Catharine fainted repeatedly.

When dawn broke, and it was clear what the final result would be, Mumbet dispatched the servant Cato to bring Theodore back from Boston, and word was sent by messenger to Theodore Jr. in Albany, as well. But this was hardly done before Pamela slumped down onto the bed, and this time she did not stir.

Catharine threw herself on her mother's body and clung to her, wailing.

"We must be quiet," Mumbet reproached her, gently pulling Catharine away. "Don't you think I am grieved? Our hair has grown white together." Still, Catharine fainted several times more that day. Uncle Doctor Sergeant had to attend to her. And she was still in turmoil two days later.

Recalling this time, Catharine swooned: "Beloved mother! The thought of what I suffered when you died thrills my soul!"

WHAT *she* suffered? Yes, that would be Catharine, always raised to a strange ecstasy by her feelings, whatever their source or nature; it is almost as if, in the face of her mother's act of self-negation, she is moved to shriek, *Don't forget me!* It fell to Harry to write his mother's obituary. After summarizing her many virtues of patience and piety, he took up the same theme that Catharine had raised. The dead Pamela was not the victim here, but her family.

What her friends, and, above all, what her husband and children have suffered, must be left to the conception of the reader—it can not be told. But it is hoped that they will try to dismiss all selfish regards, and to rejoice that she is now where the righteous have their reward, and the weary are at rest.

It can not be told. By saying nothing, Harry says everything. He implies that it would not be merely indiscreet to detail the suffering, but impossible: their suffering is boundless. And no doubt it was wrenching for Theodore and his children to witness Pamela's decline from the bright beauty of her youth to the despairing wretchedness of her older age. While that decline tore at the affection of her children, turning their natural sympathies into something closer to horror, it also raised frightening questions about the underpinning that their father had built the family upon, about the Sedgwicks' inherent "elevation," as Catharine once termed it, and, indeed, about the durability of their devotion to each other. In her last act, Pamela exposed that fraud: much as her husband professed to care about his wife and family, he was preoccupied with his own glory. At her direst moment, he wasn't there. His fine Federal-style house was not a monument to his confidence in his country, or a fitting mansion for a quasi-royal family, but a gilded prison for her to die in alone, unloved. *It can not be told.* Just so, for the full implications of what might be said would be devastating. And so Harry restricts himself to the barest hint of what actually transpired in that final night, as he writes: "She was often afflicted with the severest anguish, from an apprehension that her life was useless." In contrast to the flowery encomiums that normally sent off the dead, this was candid enough, shockingly so.

Harry knew what his mother had done, just as they all knew. And they conveyed their knowledge in a manner that moderns would recognize as standard for dysfunctional families, in a code of silence that tries to relegate unpleasant truths to the same state of nonexistence that Pamela took herself to. The truth of Pamela's death was indeed buried with her. No one in that garrulous family ventured a single word about her death in the voluminous family correspondence for four full years. Not in the days and weeks after the death, not at New Year's, when Catharine was accustomed to circulating a family letter summing up the significant events of the past year, not on their mother's birthday, or on the anniversary of her death. Catharine finally broke the silence by alluding obliquely to "that purity which is now sainted in Heaven." In the meantime, hundreds upon hundreds of letters had flown among the various members of this most writerly of families. Her death was that profoundly shocking, a frontal assault on all their delicate sensibilities, their fine feelings. It had no place to lodge.

When Theodore finally managed to make it home two days later, there was a funeral for Pamela Dwight Sedgwick. Her body was laid out in the front parlor, while Rev. Stephen West said a few words. She was placed in the town cemetery down Plain Street, where the Dwights had established a small family plot, returning her to the family from which she had come.

NINETEEN

IN MAMMA'S ROOM

I am in that bedroom in the Old House where Pamela died. It is the corner room upstairs, with unusually large windows (now heavily curtained) that look out to the curved front drive and the sweep of lawn, fringed with maples, beside the house. The fireplace that warmed Pamela through many winter nights is no longer in use; the walls, originally white, are papered a cheerful green. But I can feel her here, if only because it's such a lonely room. At twenty by twenty, it is a little too big—or is that only because I am here alone, while my wife remains at home in Newton looking after our daughters? It seems to take a few too many steps to reach the door; I am conscious of the air, the vacancy. And the Platonic square that Theodore was so enamored of makes the room seem strangely abstract, like the *idea* of a room.

All this only intensifies my feelings for Pamela, passing the last night of her life alone in her bed within these four walls, hearing, as I do now, the trees rustling uneasily in the breeze that sweeps up from the Housatonic, and the creak of the timbers of the big house. Or, frantic with her own distress, did she hear nothing at all that last terrible night except her own panting, her pulse thudding in her ears, as her body heaved about the bed, a body without a mind at all?

I have been thinking about death anyway, a mother's death, in fact, for just before this latest visit to the Old House I moved my own mother into the medical ward of her retirement home in Lexington, not far from our house. Her emphysema-like lung disease, chronic obstructive pulmonary disease, or COPD, compounded by a return of her breast cancer, has made it impossible for her to stay on in her apartment upstairs by herself. Wasted by her own years of depression, she is, at eighty-nine, just a husk of herself anyway, with hollowed eyes and stick-figure limbs, but she's been slowed even more by the disease that forces her to draw each breath through a tube from an oxygen tank. COPD is a progressive disease, and insidious. Each day her lungs work a little less well, fill a bit more with fluid. Eventually she will drown. She has moved at last into the room she will die in.

I think of my mother as I lie here in bed in Pamela's old room; but then, I think of my mother whenever I think of Pamela, for they have so much in common, even though they are not related. Both wellborn but timid, unsure of their powers and competence, they sought refuge in an older, previously married man who was tall, strong, commanding—and remote. Was that a setup for the depression they each succumbed to? Unable to find satisfaction through their husbands, were they bound to disintegrate? Or were they unlikely to thrive in any case, like certain fragile infants who are born without some essential enzyme? They ended up being secluded— to use an old-fashioned term to cover both Pamela's forced removals from the Old House and my mother's psychiatric hospitalizations—a total of ten times because of their mental illness, four for Pamela, six for my mother.

In writing about Pamela, in evoking her, I have drawn on my understanding of my own mother's suffering. Pamela's experience would have been unreachable otherwise, the distance in time too great, the recorded facts of her case too few. Depression, though, is eternal. Only the names of the sufferers change, and just a few relatively unimportant details, like when the agony started, how long it lasted, what it came to in the end. The interior weather—that icy, windswept quality—remains the same. And, of course, I came to know my mother's depression because I shared it.

I'd always thought she was the one to beckon me toward the great sadness. But as I lie here on the bed by the window from which Pamela thought of pitching herself, I wonder if in fact it was Pamela who was the one beckoning. As I look up at the high ceiling and feel the weight of air, it occurs to me that the Old House was not Theodore's, even though he built it. It was hers. For Pamela was the one to live here, not he—though in her living she was also dying a little more, each day, each night.

With such thoughts, sleep doesn't come easily. I see no ghosts flitting about the room, although I do look. In the Old House, I have always been more alert to spirits during the day, in a certain liveliness, a kind of shimmer, to the natural light streaming through the windows. Lying here in the darkness, I sense only the encroachment of that peculiar Berkshire coolness, an invasion of nature, not of any restless dead. But with it, I feel the vacancy, a sense of missing, of longing, of a need to fill this big space.

Conceived by his vision, and erected with the profits of his own industry, everything about the house is Theodore. Except himself. That was the secret that Pamela knew. He was never here. And in the end, she removed herself, too.

PART THREE

THE LEGACY DEFINED

———◆—◆—◆———

TWENTY

THE THIRD MRS. SEDGWICK

———◆—◆—◆———

The Judge, when he did return home to his dead wife, was more worn than grief-stricken. His spirit seemed to withdraw into him. This great orator made no heartfelt speech over the body of his long-suffering wife, nor did he weep; but signs of trouble could be seen in his rapidly aging body. He was exhausted by her long, accusing decline; by the exertions of his judicial rounds; and now by her suicide. It was not sadness he felt, not mourning, but acute disappointment that his wife should have failed him so.

Just eight months later, to his children's astonishment, he declared his intentions to marry again. His fiancée was a handsome, finely attired, well-coiffed, and—to hear the children tell it—shrewish and addle-brained thirty-nine-year-old Boston socialite named Penelope Russell. All seven children, but Catharine especially, despised her immediately. She was the eldest daughter of a pair of Tory sympathizers who'd fled to British Antigua during the Revolution. Theodore's own love for English tradition and propriety had only grown through the years, making Russell's Tory heritage, if anything, a point in her favor. After the long and bitter years of Pamela's decline, the Judge, at sixty-one, was eager for close companionship, and Penelope was a comely morsel in his eyes—small and slight, with a more than ample bosom.

The stepmother's lot is inherently perilous, and "Miss Russell," as the children coolly termed her, was probably bound to receive a rude welcome from the Sedgwicks. Twenty-two years younger than their father, with drawing-room manners that seemed as fussy in Stockbridge as they were artful in Boston, she immediately struck them as a fortune hunter, and Penelope did little to disabuse them of that impression. Shortly before the marriage, she feuded with her neighbor, Mrs. Tucker, a dear friend of the Sedgwicks who had graciously been putting up the children for years whenever they visited Boston. After Penelope fell into a vicious quarrel with her sister Catharine Russell over a Copley portrait of an ancestor that Penelope had inherited, the older sister was so incensed that the grasping Penelope had gotten it that she cut the head out of it. To Catharine Sedgwick, though, Penelope's airiness was the worse sin. "[She] had the *esprit de société*, a pleasant vivacity, and a very kind disposition by nature," Catharine later allowed, before sticking in the knife, "not worn quite threadbare by forty years of a selfish, single life of levity, and the sort of frittering dissipation incident to a single woman's social life in a fashionable town circle."

Catharine loyally tried to make her father out to be the victim of well-meaning matchmaking friends "who believed he would be the happier for [the marriage], and knew she would." But the Judge was well aware of his fiancée's shortcomings. A few months after he declared his intentions, he worried to his son Theodore that there was "much…in this solemn event, to make me thoughtful and gloomy; and it does make me gloomy." While he clung to a "high opinion of Miss R.," he acknowledged that he could imagine "the connexion a source of uneasiness and misery." That did not deter him, however.

HARRY, EMOTIONALLY THE most astute of the sons, had already sensed his father's malaise about his upcoming nuptials, and had tried to reassure him by appealing to his belief in himself as omniscient arbiter. He closed with a sly pun: "You are the sole judge with respect to the propriety of the choice." The Judge had already come to the same conclusion, and on November 7, 1808, the Judge made Miss Russell his third Mrs. Sedgwick. None of his children attended.

By then Mumbet had sized up the situation and decided to pack up her gowns, bedding, and few keepsakes into what she had termed that "old niggerwoman's trunk" of hers, and to leave the family she had served so loyally for over a quarter century.

Mumbet settled into a small dwelling—a "hut," in Catharine's telling—on twelve acres along the shore of what the deed on record at the Pittsfield Registry specifies as "Negro Pond," just south of the Housatonic near Stockbridge's chilly Ice Glen, where a few other black families also lived. Mumbet's own domestic life was a tangle; Catharine alludes to her "riotous and ruinous descendants," although she does not specify them or their precise connection to Mumbet. While the deed identifies Mumbet as a "single woman spinster," it also records that she owned the place jointly with an African-American, Jonah Humphrey, who later departed for Liberia as many freed slaves and their descendants did, leaving Mumbet to care for his daughter, Betsy. In her will, she promised half of her property to "Elizabeth," most likely her daughter, Lizzy, the one she'd protected from Hannah Ashley. (It is unclear from the record whether she came along with Mumbet to work for the Sedgwicks, or where she had lived instead.) The will also specifies two great-grandchildren with the distinctive last name of Van Schaack, suggesting that a granddaughter had married a male descendant of one or another of the Judge's great Tory friends of that name (or, more likely, freed slaves who had taken their former owner's name). Additional records suggest that Mumbet married into the Burghardt family at some point and left Burghardt descendants, the most noteworthy of them being William Edward Burghardt DuBois, better known as W. E. B. DuBois. In two autobiographies, DuBois proudly claimed direct descent from the "celebrated" Mumbet. As one of the nation's great black leaders and founder of the NAACP, he was certainly a fitting heir.

Mumbet would continue to assist at Sedgwick births, and would be an object of regular pilgrimages by all seven of Theodore and Pamela's children. But, outraged to see Pamela's place ceded to Penelope, she never would return to the Sedgwick house.

PENELOPE'S VERY OTHERNESS served an important familial function. As someone who so obviously did not uphold the Sedgwick standard, she defined what the standard was. Until this point, none of the children would have thought to codify it. But Penelope threatened their sense of themselves. It went beyond the feeling that she was leading their father away from them, although that was certainly part of it. It was also that Penelope was not like them. She was not literary, not passionate, not dreamy, not romantic, not artistic, not learned, not politically minded, not devoted to Stockbridge, not kind to servants, not appreciative of country life—of nature, of the view of the wandering Housatonic and stately Monument Mountain. And this meant that *true* Sedgwicks, with Pamela foremost among them, were all these things. In memory, then, Pamela Dwight Sedgwick would largely set the standard for the Sedgwick ideal, evolving from the weary, fragile, lonesome figure she actually was into a paragon of romantic sentiment and fine feeling.

All the same, Penelope did manage to snuggle into the Judge's affections. Charles was the only one of the children at home (or who could not arrange to be elsewhere) when the Judge returned to Stockbridge with his bride, and he was scandalized to see the newlyweds sitting close together before the fire in that best parlor, nuzzling and petting, right in front of him. "The only unpleasant circumstance...was to see our revered father caressing a woman with any other sentiment than that of paternal affection," he wrote Catharine in New York. When the couple later took off for a tour of the Judge's extensive landholdings in New York's Genesee County, Theodore Jr. moved to make sure that this interloper didn't entirely have her way with their father, and he wrote the Judge to start thinking about his will, or Penelope might end up with everything. When his father had still not acted a full year later, Harry was more forthright, insisting he divide up his estate *right now*. Theodore lacked the cash for that, but in March of 1810 he did oblige the children by writing his will, one that he hoped they would find reassuring.

For a time, the new Mrs. Sedgwick's delicate constitution looked like it might make such frantic maneuvering unnecessary. Always a nervous, high-strung person, she was prone to indigestion that left her in states of "weak health" for days, sometimes weeks, at a time. She turned to liquor—brandy, chiefly, but sometimes rum—to soothe herself, and occasionally supplemented its calming effects with the opium derivative laudanum. Neither did anything to bolster her vitality. Touchy and out of sorts, she grew sensitive to the many slights from her stepchildren. Sure that Harry was leading a family conspiracy against her, she tried to stage a preemptive strike by turning the Judge against him. But when word of that leaked out, the children locked arms in outrage. That is when Catharine finally evoked their mother, "that purity which is now sainted in Heaven," who seemed to represent the holy unity of the family, as Catharine urged her siblings to do their "filial duty" by closing ranks against this shrill usurper. "If *we* do *right*," she declared, " Mrs. S cannot invade the sacred circle of our happiness."

The familial division, along with the oncoming war with England, was hard on the Judge's own health. For the first time, he skipped a session of the state supreme court late in 1811. That freezing February of 1812, the pain from his chronic gout flared up so badly that he was unable to walk at all. Charles, now twenty, and the ever-present Ephraim Williams had to carry the Judge about the mansion in a chair. By June, when the War of 1812 actually broke out, the pain was so severe he needed opium to sleep, but he abandoned the drug when it cut into his appetite.

The Judge was determined to return to his post, and Catharine loyally agreed to go with him to Boston in December while he presided over some cases there. She secured rooms for the two of them and Penelope near Bowdoin Square, a short ride by carriage to the Suffolk County Courthouse, where the Judge would be sitting. The wintry chill inflamed his joints unbearably, and he had to concede he was unable to continue his judicial duties. He retired to his bedchamber for the rest of December, scarcely able to rouse himself even to use the chamberpot.

CATHARINE, CHARLES, AND Harry were constant fretful presences by their father's bed, and as his health crumbled, they worried about the uncertain state of his religious faith. The Judge belonged to no church. He'd never been drawn to the doctrinaire Calvinism of the Stockbridge meetinghouse. But, out of respect for Rev. Stephen West, he hadn't joined any other.

Catharine was intrigued by the new liberal teachings of the tiny, elflike William Ellery Channing, whose religious doctrine would soon fly under the banner of Unitarianism, and be a wellspring of the emerging Transcendental movement championed by Ralph Waldo Emerson and Henry David Thoreau. Warm and inspiring, Channing was the direct counterpoint of the bony, unfeeling Jonathan Edwards, whose chilly shadow had stretched over Pamela's soul. For two successive Saturdays, the tiny Channing closeted himself with the Judge, who finally relented and, to the

children's relief, "testified his faith in the religion of our Redeemer," as
Harry excitedly told his brother Theodore. For the first time in his adult
life, the Judge took communion, finally resolving the questions that his
mother-in-law, Abigail, had raised about him forty years before.

In his weakened state, the noise of Bowdoin Square—there in the
heart of the city, with carriages clattering by through the night—was a
constant irritation, and the Judge insisted on moving to quieter quarters
on Hancock Street, on the side of Beacon Hill a few blocks away. He
wished to die in peace. Alarmed by reports of their father's worsening
condition, Eliza and Thaddeus rushed into the city from Stockbridge by
carriage along the new turnpike road. Frances and Ebenezer were about
to climb into a carriage to bring them up from New York, but after much
indecision, a letter from Robert convinced them they wouldn't arrive in
time. So they gave up their seats and returned home, only for Frances,
after an anxious night tossing in her bed, to resolve to leave again in the
morning—too late.

The end finally came "almost without a struggle" at a little after seven
on Sunday evening, January 24, 1813. Frances and Theodore Jr. would be
the only children missing from the scene. The third Mrs. Sedgwick
looked on, as well, but always from a slight distance.

Catharine found it unbearable to watch death come for her father, and
when it finally did, she sobbed uncontrollably and, just as she had with
her dead mother, threw herself onto her father's corpse, clutching at him
and wailing; she could scarcely be pulled off him by her brothers.

A service was held three days later at Dr. Channing's church on Federal Street. Both houses of the Massachusetts legislature voted to attend
the funeral "as a token of their respect for the memory of an able judge
and distinguished patriot." Most of the lawyers of the city attended as
well, a band of funereal crepe on their left arms. After the service, the
mourners returned to the home of Mrs. Tucker, the family friend whom
Penelope had abused. Catharine was still light-headed with grief, and had
fainting fits. Her heart throbbing, she turned to her adored brother
Robert, just two years older and always her favorite, and clutched his arm
to hold herself up. "You must be my father now," she whispered.

"I will," Robert replied.

TWENTY-ONE

THE WILL

The children were so distrustful of Penelope that, a month before the Judge's death, Thaddeus Pomeroy delivered the only copy of the will to a Stockbridge lawyer, instructed him to place it under lock and key, and gave him strict instructions to let no one (meaning Penelope) see it without a written judicial order. So it was a matter of some suspense when, a week after the Judge's death, the lawyer—whose name is unknown—brought the will to the house and, settling the heirs in the best parlor, read it out loud to them. Penelope was there in her usual finery, sitting significantly apart from her more soberly dressed stepchildren.

They all learned that the Judge's total estate had reached a value of $60,000, the equivalent of as much as $10 million today, mostly on the strength of those land purchases in Genesee County and around Stockbridge, which amounted to almost half the town.

Before the Revolution, widows had traditionally been entitled to a third of their husband's estates. But the principle of the dower, as the policy was termed, had fallen into disfavor in a new nation that sought to discourage the creation of a titled nobility. Despite the scale of the Judge's fortune, the will limited the inheritance of "my beloved wife Penelope" to an annuity of $550, or about $90,000 in today's money, and, worse for her, considered that some English property she independently possessed be "part of the Capital" used to generate it. And the Judge gave his widow the use of just one small portion of the house, specifically, "the South West room"—the one that had once been the birthing room—"with such [use] of the cellar & kitchen as may be necessary to her...[and] of the Garden not exceeding one third as she may choose."

The "mansion house" itself would go to young Charles, in compensation for the fact that, loyally staying home to take care of his aging father, he had never gone to college, and so, unlike the others, had no lucrative profession to furnish him an income. He would also receive most of the Stockbridge land and "one horse or two cows which he may select." Catharine, the only unmarried daughter, was most amply provided for: she was to receive $8,500 outright. The married Eliza and Frances got $5,000 and $6,000 respectively. "The reason for the difference is not a difference of affection, but originates from a cause which will be readily comprehended." To wit: Eliza's husband, Thaddeus, was more prosperous than Frances's. At that point, Ebenezer had already hit up his father-in-law for a $10,000 loan he was unlikely ever to repay, and was in need of more. The other sons would divide up the remainder of the estate, with Theodore Jr. getting his father's pocket watch besides. Uncharacteristically, the elder Theodore closed his will with a prayer:

And I do most earnestly recommend to my beloved children who have hitherto lived together in harmony to continue to do so. And may God their and my Heavenly Father grant that we may all hereafter be united in the great & blessed family of peace love & holiness for the sake of our glorious Redeemer. Amen!

The children did remain allied—against the widow. Once the last of the provisions of the will had been read out, Penelope left the room without a word, retreated upstairs to that South West room that was now the full extent of her private Stockbridge domain, and closed the door behind her, while the children remained together downstairs most convivially. In the weeks that followed, Penelope stayed on in Stockbridge with Charles and Catharine, who, like her, had nowhere else to go. She joined them, however, only for dinner, taking her customary place at the foot of the table, with Charles now at the head, and Catharine uncomfortably between them. Penelope flattered herself into thinking that the servants were sympathetic to her plight as a prisoner in what she'd imagined

would be her own home. But in this she was wrong, too. All her grous-
ings about her mistreatment by the malicious Sedgwicks went straight
back to Catharine. "She…said…she had made no money by marrying,
that she lived a life of slavery, and got nothing for it," Catharine chortled
to Frances. Still, she later insisted that her and Charles's behavior toward
such a schemer had been correct. "We never varied nor abated in our respect and kindness to her, but where there is no heart, it is in vain to try
and reach it."

Penelope endured this indignity for over a month and then left in a
huff for Boston, where she engaged the law firm of Higginson & Cabot to
take the Sedgwicks on. She had no hope of receiving the house, but she
was determined to extract her English land from her late husband's estate.
It galled her that her own property was being exploited to generate the
measly $550 annuity that the Judge had bestowed upon her.

Harry, who had established a law practice in Boston, took up the challenge of defending the family. The Judge had always viewed him as the
least reliable of his children, and he was eager for a chance to prove himself. He offered to give her back her English land—if she would give up
any claim to various personal items of their fathers, like his law books,
and the Gilbert Stuart portrait that was the final result of Pamela's pleading for a "life-size" image to remember him by. As counteroffers go, this
was a gentle one, but Penelope was incensed, since it put the lie to the ridiculous claims that she'd been making all around town of how much all
the Sedgwicks loved her, and how sorry they were that she'd left Stockbridge for Boston. She suffered "a more violent set of fits than she had
ever before been known…to have," Harry wrote his brother Charles.

And she vowed revenge. She was going to sue for her "dower rights,"
meaning a third of everything would be hers. When Harry heard that, he
dashed off an intemperate letter to Higginson to go right ahead and sue.
No court would ever award her anything more than the family had already offered. Startled by Harry's aggressive tone, Higginson scurried
around to apologize. "This change in my style immediately produced an
effect," Harry exulted. And a month later, Penelope formally capitulated.
In exchange for the restoration of her English property, she would relin-
quish all claims to a dower—and let the Sedgwicks keep their precious
Stuart portrait of her late husband. With that the connection was severed,
and an uneasy peace was restored.

TWENTY-TWO

TO WORSHIP THE DEAD

The Judge's body had been buried in the Granary Burying Ground by the Park Street Church in Boston, where Paul Revere, Benjamin Franklin's parents, the erratic revolutionary James Otis, the five victims of the Boston Massacre, several signers of the Declaration of Independence, and many others of the city's most prominent citizens had already been laid to rest. Despite its location at the bustling epicenter of the city, it was a gloomy spot—a dark, drab, grim two-acre patch of death itself. "To praise the dead is to praise corruptible flesh," the Puritan minister Increase Mather had fulminated in 1707. "To praise memory is to worship the dead." For the Calvinist faithful, graveyards were little more than dumping grounds, and the Granary, one of three original graveyards within the city limits, was largely a jumble of cracked, off-kilter, hard-to-read gravestones, adorned only with ghoulish death's-heads, whose placement often bore little relation to the bodies below. Cows were permitted—for a fee—to graze on the grass enriched by the bones below.

Occasionally, the corpses were piled into a common trench under a single slate marker, and the city residents had expressed little sentimentality about clearing out the old bones occasionally to make room for the new. In 1825, when descendants wished to install in a proper crypt the remains of General Joseph Warren, the hero of the Battle of Bunker Hill, who had been buried in the Granary with some fanfare, no one could find them. He finally turned up in the tomb of Judge George Richards Minot (a close friend, as it happened, of Theodore's, who had written an official history of Shays' Rebellion, highly commending Theodore's role).

None of this sat well with the Sedgwick children. That their exalted father could die was distressing enough, but that he might be *lost* was unendurable. And in Boston, no less. Even though Harry had recently moved there, Boston was well outside the Sedgwicks' orbit. Worse, it had associations with the grasping vixen they could bring themselves only to refer to as "Mrs. S."

Two months later, Robert added a postscript to a joint letter that he and Ebenezer were sending from New York to Harry in Boston. Like most of the correspondence to Harry during this period, it was mostly concerned with how to keep Penelope's hands off Theodore's estate. But at the foot of his portion of the letter, Robert scribbled: "We all prefer Henry's plan for the Monument to the one which you selected. However we wish to have the choice made at Stockbridge." Most likely, this refers to some suggestion from their uncle Henry Dwight, whose family plot already held his sister Pamela. A few days later, Catharine nudged Robert back to the subject: "Have you attended to Harry's request in relation to the Monument? This ought not to be delayed my dear Robert, as so much delay unavoidably attends the completion of such a work."

These are the only references in the letters to the fairly drastic action that the children took a month later. As soon as the frozen earth thawed that spring, they hired gravediggers to retrieve the Judge's moldering casket and shipped it back to Stockbridge by horse-drawn cart. There, they'd acquired a large plot, a perfect square nearly two hundred feet on a side, that was situated well back from the town's burial ground off Plain Street. Here the Judge's remains would be reinterred.

The trees were still bare, the air chilly, when the children gathered by the gravesite for the brief ceremony. Following custom, Rev. Stephen West offered a few Calvinist pieties about dust to dust back at the house. Here, by the grave, he was silent. The children watched, stone-faced, as the Judge's casket made its slow descent into a large hole that had been dug in the very center of the expansive plot.

A large hole, because the Judge would not go alone. At Catharine's insistence, the children also had arranged for the casket holding the remains of Pamela to be unearthed from its original burial spot just a few hundred feet away. The casket was scarred and rotted after six years underground, but it held together well enough to survive the transfer. Having spent so much of their married life apart, husband and wife would now lie together for eternity.

Atop their father's grave, the children placed a stout, ten-foot-high granite obelisk. Its sturdiness was fitting, if not quite beautiful. Of Egyptian origins, the obelisk signified the dark, funereal light that emanates from the earth toward the sun. The dark romance of that did not sit well with the many traditionalists in town; nor would it have pleased the Judge, necessarily. But memorials are for the living, and it suited the children's idea of him perfectly, capturing what they still thought of as the inspiring upthrusts of his spirit. Pamela's funerary urn was tastefully neoclassical, and of course passive, as she was passive in the face of her many afflictions. Unlike Theodore's squared-off pedestal, Pamela's is delicately rounded, suggesting a lady's slipper beside a gentleman's buckled boot.

The tight scheme ignored their father's first wife, Eliza, who remained buried in Sheffield, and pointedly left no room for his surviving widow, who would end up in the tomb of her Tory grandmother in Boston when she finally died in 1827.

By placing their parents at the very center of the spacious new lot, the children were undertaking something bolder even than their father's original audacious building of the Federal-style house thirty years before. His, of course, was a house for the living; theirs would be a house for the dead. If his was based on the Platonic ideal of the square, with its square rooms at the four corners of the house, theirs would be based on the Romantic ideal of the circle—"the sacred circle of our happiness," as Catharine termed the family when Penelope intruded. The circle of wreaths, of garlands, of crowns, of joined hands and of wedding bands. For the two parents' monuments were not placed directly on the ground, but rather on top of a circular marble platform, about six inches high. That circle dictated another one to come: a circle of the children's graves resting at their parents' feet and facing inward, reverentially, toward them.

The two bulky monuments stood in the midst of the vast burial lot like the first lonely buildings erected on the freshly cleared land of the expanding cities along the eastern seaboard, and farther into the west, awaiting others to join them. Until that point, virtually all New England graveyards faced east, toward the place of resurrection. But God was not the focus here. It is nearly sacrilegious, but unmistakable: Theodore and Pamela were to be the presiding gods for all Sedgwicks to come.

The arrangement is so reverential, it seems strange to have to point out the discordant facts: that the hallowed father, much as he was longed for, was rarely home; the sainted mother, much as she meant to all her children, was clinically insane and died a suicide, and after her death, was quickly replaced by a difficult stepmother whom none of the children could abide. But such complex and painful truths would be buried in their graves, while the simpler and more pleasing fictions would live on in the memorials aboveground.

TWENTY-THREE

CATHARINE IN SILHOUETTE

By taking up position at the epicenter of the graveyard, Theodore would be permanently memorialized as the family founder—but only because the children agreed to see him so. And of the children, the one who did the most to confer that status upon him was his youngest daughter, Catharine.

By far the longest-lived of Theodore's children, Catharine would survive until 1867, two years after the conclusion of the Civil War, which was fought to resolve the question about the meaning of universal equality that her beloved Mumbet (among others) had raised in 1781. So Catharine buried not just her parents but all of her siblings, and Mumbet, too, reverently placing her beside the spot she had reserved for herself. Mumbet is the only one in the graveyard who is neither a Sedgwick nor related to one. As the survivor, Catharine was the one to erect the monuments of the dead, and to compose their epitaphs.

The last known portrait of Catharine, done in 1842, is possibly the most expressive; it shows her in silhouette, with a quill pen in her hand, as she sits stiffly in a chair. In profile, her nose—nearly identical to the Judge's—is by far her most prominent facial characteristic; the hair that had been seductively curled in her youth is now up in a tight bun behind. She was fifty-two. Her sister Frances had died just two months before, leaving, of her six siblings, only Charles. A solid block of black, Catharine seems consumed by mourning. "I can only feel as a child that is torn from the arms of its mother—my tears must flow till my sick soul is relieved," she wrote her minister, Rev. Orville Dewey. "I look upon my *only* brother with a fear that he will vanish from me while I look at him.... Weakened as I am, the lights seem all to be going out.... You do not know...what it is to have those whose memory to you is as fresh as if they had left you yesterday almost forgotten by the world."

With such statements, Catharine proves herself very much her mother's daughter, preoccupied by absence and loss and unable to create for herself a lasting bond. For all her luster, Catharine occasionally lets slip a revealingly mournful quality, a sad clinginess, in her fretful belief that anyone who might be with her now is ever on the verge of leaving. Was it, as with her mother, a self-fulfilling prophecy, as she put off her kin by her neediness? Through time, that push-pull has become a characteristic of the Sedgwick family, as, perhaps, it is of all families. That tightness is Catharine's grip on everyone, locking them in place.

Needy or not, Catharine did triumph, exploiting the keen, if restless, intelligence her father had early seen in her. Of all the children, Catharine was the one whose fame came closest to eclipsing his own. Theodore Jr. had his law career, wrote a three-volume political treatise called *Public and Private Economy* hailing the virtues of the common man whom his father had spurned, and, after returning to Stockbridge in 1822, turned to politics and served as a state representative. A devoted Jacksonian populist, he was a close friend of President Martin Van Buren, Jackson's successor, and he succeeded in bringing a train line to Stockbridge from Boston on the way to Albany. But he never attained higher office, failing in bids for lieutenant governor, governor, and his father's old seat as U.S. congressman. Harry had the liveliest mind of all the brothers, and achieved some renown as a polemicist, writing vituperative essays for journals in Boston and New York. He joined with Robert to form a law firm in New York whose most lasting accomplishment was to train the towering legal reformer David Dudley Field. The genial Charles became the clerk of court in Lenox, where his father had made many appearances as a supreme court justice. The remaining sisters, Eliza and Frances, confined themselves exclusively to the domestic sphere.

But in a time when American literature was still largely a wilderness, Catharine achieved considerable renown for her ten highly observant, socially conscious novels, especially *A New-England Tale, Redwood, Hope Leslie,* and *The Linwoods,* of which all but *Redwood* are still in print today. She was assiduously promoted by the poet William Cullen Bryant, whom Robert and Harry had known at Williams and who remained in their New York circle when he became editor of Hamilton's old paper, the *New-York Evening Post.* She became passionate friends with the flamboyant actress Fanny Kemble, who eventually settled in Lenox to be close to Catharine after the final breakup of her stormy marriage to the fabulously wealthy Georgia plantation owner and U.S. senator Pierce Butler. She attained sufficient eminence to converse freely with the leading literary, political, and social figures of her day, from Ralph Waldo Emerson to the early suffragette Lucy Stone, and despite her remote Berkshire locale was considered an essential destination for traveling dignitaries like the Irish novelist Anna Jameson and several members of Queen Victoria's court.

One vignette conveys her national standing. On a visit to Washington in 1831, Catharine was introduced by her brother Theodore's friend Martin Van Buren to President Andrew Jackson, whom Van Buren was then serving as secretary of state. Although her politics had turned away from the elitist Federalism of her youth, she still appreciated Jacksonian populism more in theory than in fact, and her feeling of innate superiority to these powerful men is astonishing. Of Van Buren, she noted favorably that he came for her "in his beautiful coach, [with] servants in livery, elegant horses, and two most beautiful dogs." Of the president himself, however, she was dismissive, saying only "he has a wooden face, but honest and pretty good." Such was her own allure, though, that Chief Justice John Marshall went out of his way to pay a call on her—"a compliment I pay very few ladies," he assured his wife in a note he dashed off shortly after the visit—and appraised her as an "agreeable, unaffected, not very handsome lady."

Shortly afterward, Catharine was selected for the first volume of *The National Portrait Gallery of Distinguished Americans*, a compendium of the nation's thirty-five most important figures in the half century since its founding. Many of them were presidents, starting with George Washington. Catharine is one of only two women (Martha Washington is the other) and one of just four writers, along with Washington Irving, James Fenimore Cooper—to whom she was often compared—and the now-forgotten Hartford poet Joel Barlow.

Yet when she came to write her memoirs—the ones published posthumously in 1871 as *The Life and Letters of Catharine M. Sedgwick*—she did not give the standard autobiographical account of, say, a Benjamin Franklin, recounting the inside story of her rise to greatness. Instead, the book is styled as a lengthy letter to her grandniece, Alice Minot, the granddaughter of her brother Charles. Again, the reminiscences consist exclusively of her girlhood in the family. There is almost nothing of her publishing career in the volume. Unjust in one respect, it is completely fitting in another. Never marrying, she forever defined herself within the family of her birth, as a daughter and a sister. This was both a torment and a point of pride. She was always the one true Sedgwick, the keeper of the flame, the priestess to memory.

Not that she lacked opportunities for matrimony. The most glittering of the Sedgwick daughters, and by far the most worldly and knowing, Catharine would have been a prized catch for any ambitious young man, and several contended for her hand. A society portraitist named Ezra Ames captured some of Catharine's early sparkle in a portrait from the winter of 1812, when she sought refuge with her brother Theodore in Albany from the company of her stepmother in Stockbridge. She is every inch the eligible young socialite, her hair down in fashionable ringlets and a glossy silk dress, fringed with lace, riding daringly low off her shoulders. "A pretty little thing in petticoats," Harry called her. "Pretty, amiable, and accomplished." It was a look designed to attract suitors, and it drew the fancy of twenty-five-year-old William Jarvis—with his "deep blue" eyes and "English" complexion, but "rather hitchy" gait in her later description. Catharine toyed mercilessly with him for months, allowing him to write letter after letter babbling of his love for her, while she wrote none in return. She finally bestirred herself to report a dream that had come to her one night, in which they had been set to marry in that "best parlour" of the Sedgwick mansion, and she'd told him, tearfully, No, she couldn't. She let her dreaming self speak for her: *No*. Distraught, Jarvis quickly married someone else, a Harriet Marsh, in Pittsfield. "A commonplace girl," Catharine calls her, giving in to her father's snobbishness. She didn't hear of Jarvis again until a quarter century later, when she learned he'd put a bullet through his heart.

In Albany, she attracted the advances of her brother Theodore's law partner, Harmanus Bleecker, whom Catharine likened to an "old Roman" for his adherence to the "sterner virtues." They came to an "arrangement" in 1819, but she broke it off shortly thereafter. Doubtless, there were other men who didn't even get that far. And so, unlike virtually all the young gentlewomen of her day, and absolutely all of the most desirable ones, she resigned herself to a lifetime of being, as she put it in her more melancholy moods, "first to none."

With her sisters' examples before her, however, Catharine had reasons to be leery of marriage. The most compelling fear stemmed from her father's meddlesome attempts to secure for his children that social elevation he had always desired for himself. Eliza's husband, Thaddeus Pomeroy, proved somewhat colorless, as Catharine had feared, and less successful with his Albany drug business than Theodore had hoped. Over the course of the marriage, Eliza turned into a ghost of herself after delivering twelve children, all but one of whom—Pamela's namesake—survived to adulthood.

But Frances had been punished much worse by her marriage to the man Theodore had handpicked for her, Ebenezer Watson. Although he aspired to be a publisher like his parents, Ebenezer continued to make his living largely by peddling cider and other comestibles. By 1811, ten years into the marriage, his merchant offices in Albany and New York had both failed. Trying publishing again, Ebenezer borrowed that $10,000 from the Judge—on top of a previous $9,000—to join with the more established Samuel Whiting in producing a Hebrew Bible that Ebenezer, as a confirmed Calvinist, assumed would set him up in this world and the next. But the press run of 2,000 went largely unsold, pushing Whiting and Watson to the verge of bankruptcy at the Judge's death in 1813.

This complicated Harry's role as executor immensely, for the $10,000 loaned to Watson was needed to create the $8,500 in capital that Catharine was due as her legacy. Worse, Whiting & Watson's largest creditor declared bankruptcy, and *his* largest creditor went after Whiting & Watson. Only some emergency loans from various Sedgwick brothers prevented the total collapse of Ebenezer's firm while Robert and Harry hunted for hidden assets of Whiting's to keep the partnership afloat. Frances wrote abjectly that Ebenezer had become one of the miserable souls "ground down by excessive cares—whose spirits are allways agi-

tated by defeated plans—whose days & nights are given to fearful antici-
pations of the future." Frances, however, scrupulously did not say how he
responded to these pressures.

For some time, Harry had detected an alarming tone of contempt in
Ebenezer's voice whenever he addressed Frances. Visiting the Watsons
in New York in the spring of 1815, Harry discovered something more upsetting: a bruise on Frances, or possibly he saw a blow actually being
struck—it's unclear in the record, but something alerted him to the dark
secret of their marriage. Returning to Stockbridge, Harry picked up his
pen and tore into his brother-in-law: "Remember that your marriage vow,
your oath to God to love and cherish her—is upon your soul. Its continued violation will, be assured, meet with awful retribution."

Harry worked through four drafts of the letter, trying to find the right
tone to express his outrage without worsening the situation for his sister.
In the end, he never sent any of them. He kept the dreadful realization to
himself.

In early 1816 Frances could take it no longer, and fled to Stockbridge
with the children. Ebenezer didn't dare follow her there, where his many
Massachusetts creditors could have him jailed for his debts. For weeks,
she wouldn't tell anyone why she'd left. But finally, late that summer, she
poured out her heart to Catharine, detailing the many cruelties she had
suffered at Ebenezer's hands.

"Poor Frances!" Catharine wrote Robert in the fall of 1817, "my heart
bleeds for her, when I think to what thraldom her noble spirit is subjected." It was a few years before she passed word on to her older sister,
Eliza. "Mr Watson is brutal in his conduct to [Frances] and does and has
for a long time rendered her miserable," she finally confided. "It is one
uniform system of degrading oppression—and she bears with scarcely
ever a word of reply—patiently and uncomplainingly. What is to be
done—I know not—but something must be done....He has a stubborn
oppressive temper—that temper essentially diabolical which delights to
trample on that which is in its power."

Divorce was not yet allowed in Puritan Massachusetts, but Frances
could have legally remained apart from him on the grounds of cruelty.
And she still had the protection afforded by those angry creditors. Nevertheless, she continued to allow Ebenezer to pay her occasional clandestine

visits. The abuse never did stop, though. Frances remained in her husband's thrall. When Ebenezer tossed her out of a house they set up for themselves in Albany, Frances meekly returned. Whenever she screwed up her courage to leave him, she always came slinking back. The marriage remained a complete misery for her until her death in 1842. When Ebenezer died five years later, Catharine had him buried beside his wife. But her inscription is tepid: "His memory is warmly cherished in many hearts."

EVEN IF CATHARINE had had more inspiring examples of matrimony among her sisters, she was still not likely to have followed them with a marriage of her own. For the truth was, she was already committed. As nuns are said to be brides of Christ, she was a bride of the family, as countless devoted letters to all six of her siblings attest—but especially of her next older brother Robert, which is why she first shared the grim news about Frances with him only. "My favorite brother," she unabashedly calls him in her autobiography, remembering the "most loving dependence" she felt for him from as far back as she could remember. Hunting for eggs in the barn as a toddler, she'd slipped under a mound of hay, and was convinced she might have suffocated if Robert had not heard her "faint cries" and run to her father to haul her out. While the Judge is literally her rescuer in the story, Robert was her "protector" then and ever afterward.

Just as she had turned to Eliza when Pamela was incapacitated by her insanity, she turned to Robert when her father's health started to fail. She heaped her feelings on Robert, then a slender, clear-eyed lawyer at twenty-four, as she could never do with the ill-fated Jarvis or the "old Roman" Bleecker. "I can never express by words the degree of affection & interest I feel for you," she wrote Robert in May of 1811. "You deserve it all & much more my dear Robert for your kindness has been uninterrupted & unvaried & it is written on my heart not with *indelible ink* but in colors as immutable as the virtues that have produced them." A month later, she meditated further on the meaning of their familial relationship. "*Brother* is the most comprehensive word in my vocabulary. The definition of it with the various readings, occupy at least a dozen folio pages in the volume of my heart." She started taking a rivalrous interest in the young ladies vying for his attention. "Why," she asked disingenuously in the same letter, "do you say nothing about your sweet favorite Gertrude?"

A year later, she pined openly for him. "The poor sofa is quite neglected since your departure," she wrote from Stockbridge after he had left again for New York City. "To me it looks like the ghost of my departed joys. I can *almost* see with fancy's eye the form of my dear Bob *superincumbent* upon it." She copied out lines of poetry that evoked the sunset over the Housatonic so he could share the scene vicariously with her, as if, in fantasy, they might gaze upon it together, arm in arm. She confided she read his letters "over and over and over again," clutching them as if she were holding him. Nothing separated them, she insisted. Her devotion "binds my soul to yours and directs its most ardent aspiration to an immortal union."

After the Judge died in January, Catharine went further: "I do love you, with a love surpassing at least the ordinary love of woman." No sister had ever felt more for a brother, she insisted, had "more ample experience of the purity of love, and the sweet exchanges of offices of kindness that bind hearts indissolubly together." Or, as she finally told Robert, he was "as much a part of me as the lifeblood that flows thro' my heart."

For his part, an obviously uneasy Robert tried at first to turn such sentiment back to her as simply a compliment to her own sterling character. "You are now engaged in the best of all employments; conferring happiness on those who love and whom you love," he replied in March of 1813. But as her letters continued to pour forth, he had to acknowledge that her "affection" did touch him. It had "an irresistible power to improve and to elevate—to lift above love attachments—to separate from unworthy associations—to cheer me when I am sad—to rouse me when I am inefficient." In November of 1813, he wrote the words she longed to hear: "My dear Kate, I know not how I could live without you." In reply, she started in again on her theme of the "sacredness in the love of Orphan Children" (for each other) before she had to stop for weeping. "Such feelings as at this moment make my eyes to flow with tears cannot be expressed." She left it at this: "I long to feel the kind pressure of your hand."

To Catharine's frustration, Robert still did not feel quite as much longing for hers. Robert had begun his search, already long-delayed as he neared thirty, for a wife. In the summer of 1815, on a retreat to a mineral spring at Ballston in upstate New York, Robert met Jane Minot, the younger sister of his college friend William. Robert found Jane entrancing, and he was overjoyed to bump into her again at some friends' a few weeks later in New York City.

There, Jane took note of Robert, too. In her journal, she unknowingly conjured his imperious father when she noted that Robert became irritated at the gathering that someone had broken out a flute when he was trying to have a conversation. Still, Jane wrote to a friend that he was "very sensible & good natured & so droll in conversation that you may be sure of laughing when you see him."

Orphaned, Jane lived with William and his wife, Louisa, in Boston. Although its bays and coves were starting to be filled in, the city was still a tiny spit of land, scarcely a mile across from the banks of the Charles River on the west to the often-raging Atlantic to the east; only a narrow causeway to the south kept Boston from being an island. Socially, it was still island enough, with only a few thousand homes. Most of the better ones were on the slopes of Beacon Hill, on land that had been owned by John Singleton Copley, just down from the peak dominated by Charles Bulfinch's new, copper-domed State House (the current gold came later). In November Robert made plans to stay with the Minots at the foot of Beacon Hill, on Charles Street, a fashionable piece of the new land by the rope works along the river. And he invited his feisty brother Harry, from whom he was virtually inseparable.

Harry had been sulking in Stockbridge after a Boston courtship of his own had turned disastrous. He'd set his sights on Abigail Phillips, the youngest and reputedly most fetching daughter of the richest man in Boston. Her father, William Phillips, was not only president of the Massachusetts Bank, the region's largest bank, but lieutenant governor of the commonwealth besides. Harry had gallantly proposed marriage to Abigail twice, only to be flatly rejected both times by her father.

Eager for a chance to be with Robert, Catharine came to Boston, too—only to find herself drawn into Jane Minot's circle of eligible young women. Calling themselves the "friendlies," the circle had come together largely for the purpose of ensnaring an eligible bachelor. At twenty-one, the cautious, observant Jane was the youngest of them. But the circle also included the free-spirited Margaret "Gretty" Champlin Jones, the daughter of a wealthy Boston merchant; Gretty's moon-eyed and pious cousin Caroline Danforth; and Gretty's three sisters, Elizabeth, Mary, and Martha Ellery Jones. In the fall of 1815, when Harry met them, the friendlies gathered regularly at the Joneses' handsome Federal-style house on Pearl Street, with sweeping views of the harbor, where they would entertain a few carefully selected gentlemen.

Once Harry and Robert were regulars, Catharine saw no choice but to join the friendlies, too, if only to keep a close eye on developments. She watched Robert come into the group somewhat attached to Jane—only to break free and form temporary, casual alliances with each of the friendlies in the coming months. For the brothers, the friendlies offered safety in numbers: neither Sedgwick had to worry too much about any individual woman getting the wrong idea, or the right one. But to Catharine, it meant endless agony, as Robert's attention spun like a roulette wheel, its ball jumping from one friendly to the next. Catharine never knew which was her enemy.

The rest of the family were no less transfixed by these romantic developments. While Catharine was visiting Susan and Theodore in Albany, the three of them pulled the chairs up "around our dear, ugly, comfortable close-Stove," Susan wrote Harry. And then? "We then talk over *you & your affairs;* discuss Bob...and at length retire in as sweet a delirium of hopes, fears, & impression, as if we were *ourselves in love*." Of course, what Susan might be able to chat about with some amusement, Catharine could only contemplate with nearly debilitating anxiety. She freely urged Harry on: "What are you doing? Sucking your thumbs, & building castles while all the birds of the air are building their nests?" But she urged caution on Robert, so much so that the slightest expression of enthusiasm for any friendly that Robert liked was "enough to send me thro' fire and water."

Robert started with Jane, moved on to Mary Jones, to Caroline Danforth, and finally, in June, to Gretty Jones. Although he had sneered at her as recently as March of 1816, telling Harry there was "an air of repulsive desolate majesty abt that old earth," in June he and Gretty were snuggling on the sofa, indulging "a good deal in a kind of wild play of imagination almost without knowing whither it tended," much to Catharine's consternation.

When Robert moved away from Jane, though, Harry moved toward her. Catharine had touted Jane to Harry from the beginning as having "more intrinsic worth" than any of the other friendlies, although she had to grant that the others were possibly more beautiful. Significantly, Catharine did not praise her to Robert.

Catharine's plotting came to fruition one night in June, when Robert was due to leave once more for New York. With Harry in tow as always, Robert came around to the Minots on Charles Street to say good-bye to Jane. At the door, Jane shook Robert's hand in parting, but did not take Harry's. By the exacting standards of Boston etiquette, this was only proper. A grasping of the hands between eligible unmarrieds was permitted only in cases where the two parties could reasonably not expect to see each other for a good while. But, having freely touched Robert, Jane's eyes misted in frustration at not being able to press her hand into Harry's, since he would not be parting from her.

"Occasionally I believe [Harry] has felt something like tenderness for me," Jane wrote pensively in her journal. Robert could tell his brother did, writing Harry that he, Robert, had detected "a deeper Interest [in Jane] than I can express." A clever way of putting it. Robert meant that he was incapable of finding the words to articulate his sentiments. But there was an emphasis on the word *I* as well: Harry felt more for Jane than he, Robert, did. Besides being perceptive, it reflected an unusual truth about the brothers, which was that each genuinely wanted nothing but the best for each other. While both brothers inevitably provoked rivalrous feelings among the friendlies, the reverse was never true. Every other relationship among the friendlies was threatened at one time or another—friend turned against friend, sister against sister—but the brotherhood remained intact. Or was this due to Catharine's intriguing, in her efforts to reserve the prized Robert for herself?

The unswerving loyalty was all the more striking because Harry was plainly not the easiest person to get along with. He could be fidgety, distracted, nettlesome, loud, unruly. As Robert finally revealed in a frank letter, Harry had "ten thousand little ticks" besides. Like:

> *picking yr teeth with penknives & forks—running strings thro' them &tc &tc putting yr hands on a certain part of yr pantaloons (by the bye this which was the most favorite dinnertable trick you had, you have got almost entirely rid of.) &tc &tc. Now Hal you know you can avoid*

these things, but you do not try. You seem to have a fatal indifference on
the subject. Why can you not constantly bear in mind till habit shall
supersede attention, that you must have the manners of a gentleman?

All the same, in mid-July of 1816, liberated and encouraged by Robert's reassurances, Harry proposed to Jane, and he was accepted. The news "went thro' and thro' me," Robert told Harry, bringing "bounding joy and exultation" to his heart. With Jane's father long dead, it fell to her brother William to approve, which he did without delay.

By then, though, Robert had made Harry an offer of his own, to join him as his law partner in New York City, and Harry had accepted. Thus, no sooner had he found a mate than he'd resolved to leave her. But before he departed for New York, he and Jane sat on a bench behind the houses along Charles Street, looking back toward the Boston Common. It was a chilly night, and he wrapped his coat around Jane to warm her and spoke his heart to her with, as she recorded, "all the tenderness which the dread of parting gives to tones of love." Afterward, when she watched him leave her Charles Street house for the last time, she felt an "agonizing sentiment of loneliness." To break it, she strolled alone to their favorite bridge over the Charles. "The mild freshness of the air...gave a melancholy tranquillity to my mind."

Now that she was committed to Harry, she would rarely feel such tranquillity again.

As for Catharine, her letters continued to bleed for Robert. For years after Jane, Robert found no one else worth pursuing, and he and Catharine lived for each other's company. "Whenever you marry, my dear Brother, I shall have a good deal to suffer," she wrote him, hoping to forestall just such an eventuality by the pronouncement. In 1819, she cautioned, "You may love another better—you must not love me less."

In the summer of 1821, when they toured northern New England together, Robert took her by the hand to view Niagara Falls from under Table Rock, just the two of them, the great torrent of water crashing just above their heads. And later, at Thousand Islands, along the St. Lawrence River in upstate New York, they gazed out together at sunset one evening at the many verdant islands, "some stretching for miles in length, and some so small that they seem destined for a race of fairies," as Catharine wrote. Robert drew Catharine under his cloak for warmth and whispered, "Kate, we shall remember this a great while."

And then, that very fall, Robert did the unthinkable. He became engaged to marry. His fiancée was Elizabeth Ellery, the daughter of declaration signer William Ellery. When he confessed the news to Catharine in Stockbridge the following winter, she went stone-cold and told him curtly, "We cannot walk together as we have done." The recognition dug into her soul. "No one can ever know all that I have, and must feel, because no one has ever felt the sheltering love, the tenderness, the friendship that left me nothing to desire." She could think of Elizabeth with nothing except revulsion, although she bravely insisted to Frances that she had "tasked myself to the duty of resignation with more fortitude than you would expect." Still, her obvious resistance to the match was enough to cause her brother to ponder the difference between the love of a sister and the love of a wife. "It is a very common sentiment that a sister must give up her place in a brother's heart when his wife takes possession of it," he began, giving Catharine a pang. "If this were so, I should be sorry to see you ever reconciled to my marriage. But, if I know aught of true love, instead of contracting the heart, it gives new strength to all its best affections."

But Robert knew that just to write this way, to make any sort of comparison between Catharine's and Elizabeth's love for him, was only making the pain worse for his sister. The night before his marriage, just as he was leaving New York, he dashed off more lines of reassurance to Catharine, words that she could read only as farewell: "I can not leave town without telling you that my heart never turned toward you or leaned upon you with more pure, faithful, ardent, and confiding affection than at this moment. God reward you, my beloved sister, for all you have been to me, and enable me to cherish a tender and unalterable sense of all I owe you."

Nevertheless, Catharine was the one sibling not to attend the wedding. She remained alone in Stockbridge.

TWENTY-FOUR

I HAVE LOCATED MY HEAVEN

After that calamity, Catharine settled into the pattern she would follow for the rest of her life. She spent her summers with Charles at the family mansion in Stockbridge, occupying her old room at the southeast corner, overlooking the Housatonic. She passed her winters with one or another sibling, although rarely Robert, either in Albany or in New York.

But it was Stockbridge that she was always drawn to, time and again. Staying with Theodore once in Albany, she dreamed she was actually in Stockbridge. "And when I fairly opened my eyes, . . . I could have cried because I could not see silver beams playing on our own little stream, and shining through the naked branches of dear Charles' trees," she swooned in a letter. "I sometimes think my love for that spot is, for these philosophic, enlightened times, too much like that of the savage, who thinks his heaven is to be on great hunting-ground. There I have located my heaven."

When fall finally did come to Stockbridge, and the temperatures dropped, marking the time for her departure to New York City, she would wait until the last possible day before boarding the carriage to Claverack Landing for the boat ride down the Hudson. More than once she waited too long and found the river frozen up, necessitating an arduous and time-consuming trip by carriage overland.

But soon Charles, too, had a family of his own, having married their distant cousin Elizabeth Dwight in 1820, and as their children came along, he found it increasingly difficult to meet the expense of keeping up such a large house. Without legal training, he supported himself as that clerk of court in Lenox, and in 1822, he announced his intention of moving his family to smaller and less expensive quarters there.

It was unthinkable to put the Sedgwick house up for sale on the open market. It was a shrine for all of them, not just Catharine. Theodore agreed to buy the house from him, and to move his family down from Albany. Still, Catharine was almost inconsolable. "I need not tell you, for you know already, how I feel the contemplated change," she told Charles. "Your presence here has been to me like the spirits of our parents, and it never seems home to me when you are gone." Still, she determined to resign herself to the change. "Wherever you are, I must have a home, and Lenox must be to me, when you and yours are there, the dearest spot on all this earth." No less upset, Charles cried out to Robert: "I have parted with Catharine....Oh God, is it possible that my home and hers must be forever different? I have always endeavored to avoid selfishness, but the last sacrifice has cost me pangs that no earthly being knows of, or can know. She has been my second mother."

Charles and Elizabeth initially had only rather spare lodgings at a Mr. Whiting's in Lenox, then a bleak, hillside town some six miles on from Stockbridge, but after a couple of years they were able to buy a proper house, where Catharine could visit them, much as before, but never for a full summer. For that, she turned to Theodore, who cheerfully put her up in her same southeast bedroom.

All the same, 1822 proved a decisive year for her. Now separated not just from Robert but from Charles as well, she created another world for herself and started writing fiction. Harry's encouragement was critical. Back in the war year of 1812, when he was in Boston he had gotten in- volved as part-time editor of the city's *Weekly Messenger,* and, taken by what he termed a "scrap" of hers "on the sacred subject of a pastor," he'd had her enlarge it for an article. That was her first publication; she was twenty-three. "I proffer to you the fairest portion of my dominion," he grandly told her. "Nay the royal palace—the imperial Seat itself. You shall reign the sole empress of the *Poet's Corner.*" Now, almost a decade later, Harry wished to promote her further. He'd seen a short tract of hers on the advantages of the new Unitarianism over what she usually termed "the gloomy tenets of Calvinism," and recommended that she turn it into a "tale."

Stifling her doubts, Catharine bent to the task, and the result was *A New-England Tale; or, Sketches of New-England Character and Manners*. Harry found a publisher, the New York firm of E. Bliss & E. White, and it brought out a thousand copies of the work that May. By contemporary standards, the novel is a patchy, rather conventional romance of a heroic orphan girl, blurred by subplots pointing up the hypocrisy of the exemplars of the old religion. But it did capture the American domestic landscape that had not yet found its way into the more celebrated fictions of James Fenimore Cooper and Washington Irving. Its author had an unexpectedly sharp ear for Berkshire chitchat—about "pokerish weather" and the need to "musquash along"—and a keen eye for the little putdowns of rural society, as if she were a backwoods Jane Austen.

Following the conventions of the day, the *Tale* was published anonymously, with only a quotation from Robert Burns where the author's name might go. Unlike other writers, male ones especially, Catharine would cling to putative anonymity for her entire book publishing career, placing her name only on a collection of stories that had already appeared under her byline in various magazines. It was an attractive pose for Catharine, who continued to profess a "perfect horror of appearing in print," even as the books started to pile up. At one point she declared that she'd rather have a tooth extracted than publish again—no small assertion, since by the time she died, she did not have a single tooth left in her head. While the attitude has a genteel, ladylike quality, it also recapitulates the formulation that her father settled upon for his political career, in which he worked furiously to engineer his election—without ever deigning to appear to do anything so vulgar as to seek public office. Any fame must seek one out; one must not openly chase it. That was his legacy, and her distinction.

And, far more quickly than she could have expected, fame did come. The *North American Review* hailed *A New-England Tale* as "a beautiful little picture of native scenery and manners, composed with exquisite delicacy of taste, and great strength of talent." Harry reported that Messrs. Bliss and White had informed him that, while sales "had been dull at first," the first press run "was going off very rapidly…and would soon be entirely exhausted." Harry took it upon himself to draft a new preface for the second edition. He told her he would send it on to Theodore, since "he has a good knack at such things," before showing it to her. He wound up: "I think, dear Kate, that your destiny is now fixed...so have done with these womanish fears."

Thus fortified, Catharine soon put her mind to another domestic novel, a longer and more ambitious one, filled with incident and set in the fashionable resort community of Lebanon, New York, and, for spiritual contrast, in the austere Shaker village of Hancock, Massachusetts. Appearing in 1824, *Redwood* sold briskly, and since the author was again unnamed, set many readers thinking it the work of James Fenimore Cooper. "It is to be hoped that Mr. C.'s self-complacency will not be wounded by this mortifying news," Catharine teasingly wrote Charles. Within the next few years it was translated into French, German, Swedish, and Italian, and was such a success in the United States that Catharine was subsequently identified as "The Author of *Redwood*" for the next twenty years. Harry was exultant. "Redwood sells very well," he wrote her from New York. "About 1100 are gone." Two months later, he reported, "The booksellers are all teasing me to know when another work will come from the author of 'Redwood.' They say it will go as well or better than one from Cooper or Irving."

Three years later, Catharine ventured into contentious Puritan history to retell the remarkable tale of Eunice Williams, the daughter of Catharine's illustrious great-granduncle, the Reverend John Williams, who'd been the minister of the Deerfield church in 1704 when a marauding band of Abenaki Indians had broken into the town, captured the Williamses, and dragged them back to Montreal. The Indians killed Williams's wife and two of his sons. Williams himself was freed by the French governor, and he in turn was able to "redeem" all his children but one, the darling seven-year-old Eunice. Her captors, he was told, "would sooner give up their hearts than part with her." Another attempt on a return trip was no more successful, and a look told why: she wore deerskin moccasins, woolen leggings, and wraparound blankets pinned by porcupine quills, with bird feathers for adornments. She dabbed her face with vermilion paint, greased her hair. She spoke the Indians' language, forgetting her own completely. More distressing for her Puritan father, she took her captives' Catholic religion (taught them by the French Jesuits). Reverend Williams returned again and again, trying to win his daughter back. But it was no use. At sixteen she married a Mohawk, and they had two daughters. She had turned.

On a research trip to upstate New York, Catharine had visited a reservation for the Iroquois, hoping to meet up with Eleazar, Eunice's great-grandson, "a far-away cousin of ours," as she explained in her memoir. A deacon in the Episcopal church, Eleazar was by then serving as a missionary to the Oneidas, but he was not to be found when Catharine passed through. Eunice herself reappears in *Hope Leslie* as the eponymous heroine's sister, Faith, who is cruelly taken from her Puritan colony by rampaging Indians, raised by them, and turned Catholic. Like Eunice, Faith loses her ability to speak English. If Catharine's ancestors had shuddered at the prospect of losing Eunice to savages, Catharine herself did not regard Faith's marriage to the Indian Oneco as any sort of horror. In the book's preface, she turns unusually assertive in defending this point: "The elements of virtue and intellect are not withheld from any branch of the human family," she declared. "The difference of character among the various races of the earth arises mainly from difference of condition." Her father would have been scandalized by such an assertion, but Mumbet, now just a year from her own death, would surely have been pleased.

More novels followed: *Clarence*, a snide view of contemporary New York, in 1830; a historical novel, *The Linwoods*, about a Tory family during the American Revolution, in 1835; a trilogy of didactic tales, *Home, The Poor Rich Man and the Rich Poor Man,* and *Live and Let Live,* in the two years after that; and finally, after a long hiatus, in 1857, at the age of sixty-seven, her last novel, *Married or Single?* exploring the central dilemma of her own life and resolving it for her heroine in favor of marriage. "The great law of Nature," she finally called matrimony, "by which, in every province of her infinitely various kingdom, all 'kindred drops are melted into one.'"

She continued to write short stories throughout, and one of them, quite unexpectedly, ended up having greater social repercussions than almost anything else she wrote. It initially appeared in a compendium called *The Token* in 1835, where it might have been obscured by Nathaniel Hawthorne's two remarkable contributions, "The May-Pole of Merry Mount" and "The Minister's Black Veil." Titled "New Year's Day," Catharine's story details a holiday gathering of a family very much like Robert and Elizabeth's in New York, where Catharine herself was accustomed to pass the Christmas holidays, once she finally reconciled herself to Robert's marriage. In the story, as in the Sedgwicks' fashionable town house, the family receives an endless stream of visitors, but all of them the useless, silly, too-rich gentlemen of the sort Catharine had never made time for. It is in a back room—a kind of inner sanctum well insulated from the public front parlor—that the true spirit of the holiday season reigns. For there stands a new Yuletide innovation, recently brought to America by German immigrants: a Christmas tree, laden with handmade gifts (placed on the branches, rather than underneath) that are the authentic expressions of the love the family members feel for each other. The story's young heroine, Lizzy, gives her father "a pair of slippers…beautifully wrought by her own hands," and presents her brother with toy soldiers she has painted herself. The gifts are a perfect expression of Catharine's familial devotion, albeit from a child's perspective, and the bountiful tree in which these gifts nestle like fruit is an emblem of her loving beneficence. As the first Christmas tree to appear in an American publication, let alone in a gift book designed for fashionable New York in the holiday season of 1835, it did much to create the popularity of the Christmas tree in the city and, from there, to spread it across the young country.

A SOCIETY PAINTER, the Irish refugee Charles Cromwell Ingham, had Catharine sit for him in those years, and it's startling to see the emergence of a figure of such self-possession. If the previous portrait showed a young woman eager to be taken up, this one displays a poised romantic icon enshrouded in a swirling black cape that slips seductively off her bare shoulders. Although she looks away pensively, she is fully aware that she is herself an object of scrutiny, if not veneration, and for the first time, she welcomes the viewer's gaze.

TWENTY-FIVE

THE GREAT CENTRAL FIRE

The publication of *Hope Leslie* in 1827, when she was thirty-eight, would prove the high point of Catharine's novelistic career, a fact that she recognized even at the time. To her, the reason was clear: it was the last book that Harry oversaw. "I miss excessively, more than words can tell, the light and repose of Harry's criticisms," she would tell Charles in 1830. "I felt a reliance on him that I can never feel on another; a confidence that I should not expose myself to any severe criticism." While he offered Catharine shelter, he reserved none for himself. Harry was exposed to every wind that blew.

Physically, all the children—from Eliza down to Charles—bore the distinctive Sedgwick physiognomy, tending toward long, loose limbs, wispy hair, long-bladed noses, and limpid, romantic eyes. But the true commonality was interior, in an intensity that made nearly every interest of theirs a passion—Theodore Jr. for politics; Frances for romance, no matter how dangerous; Catharine for fine feeling. It was, for all of them, an extravagance of emotion, but for all except one their behavior remained well within the bounds of normalcy.

Then there was Harry. If the other children received a light, relatively harmless version of the dark seed passed down by their eccentric great-grandfather, Ephraim Williams, Harry received the full, unadulterated black one, just as Pamela had. Like Ephraim and Pamela, Harry, too, lacked that insulation that keeps a tender soul safe from the vicissitudes of fortune. Ephraim was lacerated by guilt over having stolen from the Indians he was supposed to protect. Pamela took her husband's detachment hard. In both cases, the dominant feeling was the piercing anguish of being unworthy, of not measuring up to a certain standard. Ephraim could not live up to his Calvinist principles; Pamela imagined she'd lost the respect of her husband. For Harry, it was more about upholding the family name. His disease was not the straight, unipolar depression of his joyless, fretful mother, but what moderns would call manic depression, that variable mood state that alternately plunges a person into dark despondency and vaults him into a frenzy of unwarranted exuberance. In the early years, Harry's excesses were endearing. Whether it was a love match, a legal case, a business proposition, a theological dispute, an essay, or a social cause, Harry would throw himself into it with a winning ardor. Gradually, though, his passion became more a matter of frantic desperation and, when it ultimately fixed on his own fundamental notions of self-worth, a mortal hazard.

THE FIRST INDICATION of Harry's dangerous tendencies came in the summer of 1815, when he arrived in New York to join his brother's law office and discovered a city in near pandemonium. With the close of the War of 1812, British ships had finally lifted their stifling blockade of Manhattan ports, and the city had exploded into madcap activity. New wharves were springing up overnight; so were boardinghouses, inns, and the occasional proper hotel. Horse-drawn wagons clattered through the cobblestone streets, splattering slop as they delivered to massive warehouses the bales and barrels that had been disgorged from the high-masted ships that ringed Manhattan like an encroaching forest.

Harry found digs in one of the new hostelries on Pearl Street, not far from the East River, its wharves bristling with masts, near the southern tip of the city. The brothers' law office was around the corner at 47 Wall Street, but Robert pointedly took his rooms in a quieter portion of the city, a half hour's walk away. At first Harry professed to be appalled by the unlettered crassness of America's commercial mecca—even though it was the Almighty Dollar that had lured him there, too. "This city is nearly a generation behind Boston in general intelligence & enlargement of mind," he sniffed in a letter to Jane's sister-in-law Louisa Minot. "The citizens call it the London of America, but it is much more like Liverpool."

Still, the brash excitement of the place plainly suited the wilder swings of Harry's mood. Nothing thrilled him like the fire that broke out late one night in a cabinetmaker's shop around the corner from his boardinghouse and then, snapping and hissing as the flames leapt skyward, consumed an entire city block's worth of fresh construction. His words racing across the page, he reported the sight to Jane, devouring impression after impression like the fire itself:

> The astonishment and distress of the inmates of the houses who were forced into the street shivering & half-dressed in a cold night—the nonchalance—the gaiety of a part of the citizens—the active zeal of the others—the expertness & courage of some of the firemen—& the stupidity of the rest—the great fury of the element —its rapid extension—the vivid blaze & strongly illumined faces of thousands of gazers—the breaking furniture, crashing beams, & falling walls, the hurry, the confusion, the thwarting flames, the press of some to save furniture, of others to extinguish the flames or to satisfy their curiosity, & all in different directions…

Harry pitched in with the fire brigade, tossing buckets into the blaze, and once darted into the flames himself to rescue a lamp and a mattress. When the fire finally burnt itself out shortly before dawn, he repaired to the nearest pub with all the other impromptu fire fighters, his passion not at all spent, and, for a "merry half hour," plunged into the meat and drink of this luscious and degenerate city:

> We had no chairs or spoons nor any knives or forks except for the carvers—our handkerchiefs were around our throats, & one napkin served for all—but what of that—we had fine coffee & chocolate which seemed the better for being without milk—excellent beef, & most delicious salmon, and appetites upon which I need not expatiate.

Married at last in 1817, Harry and Jane settled into cramped lodgings on Greenwich Street. A son came, only to die a year later, but a daughter, also named Jane, survived. They moved into a row house on Warren

Street around the corner, but the accommodations, while larger, were scarcely better. There Jane would bear him three more children—two daughters, Frances and Louisa, and a son, Henry Dwight Sedgwick Jr., my great-grandfather. Along with the ever-present Robert, the magnetic Harry turned the house into a leading New York salon, with the incoming *Evening Post* editor William Cullen Bryant as one of its nodes. "The resort of the best company in New York," Bryant called the Sedgwick digs, with "cultivated men and women, literati, artists, and, occasionally, foreigners of distinction." Thomas Cole, then painting in a Greenwich Street garret, and Samuel F. B. Morse, just arrived from England, were frequent visitors. Catharine's rival, James Fenimore Cooper, attended—when he wasn't holding forth in "the Den," as the back room of his publisher Charles Wiley's bookstore was dubbed.

Most of Sedgwick & Sedgwick's legal cases were, at least at first, penny-ante matters, chasing down debts and the like. But gradually the cases built into causes. Ever his father's son, Harry took up the cudgel against relaxing the bankruptcy laws that pushed debtors into prison, arguing the point in six essays in the *New-York Evening Post* in February 1817. When the editor, William Coleman, replied with a six-part rebuttal of his own, Harry insisted on a seventh. When Coleman published that, he added a note about how he himself would not write another word on the matter. Harry took that as a challenge and scribbled yet another essay, which Coleman refused to publish. Harry then sent it to the *Evening Post*'s rival, the *Commercial Advertiser,* which did print it, and five more besides.

AND THEN THERE was the Greek frigate case, a phrase that the Sedgwicks would come to utter only with a weary sigh and a sad shake of the head. It was Harry's finest hour, but afterward there would scarcely ever be a single good one. The details of the case are too much to take in, but at heart it was a simple price dispute over the construction by a New York company of a frigate called the *Hope,* built for the Greeks in their war for independence against the Turks. In short, the New Yorkers seemed to be gouging the Greeks. Keen for causes, Harry saw this as far more, as his chance not just to champion the gallant Greeks in their fight for freedom but to play a part in an American-style revolution, thereby creating for himself a heroic role that might rival his father's. In fairness, Harry was not the only one to fall hard for the Greeks. A passion for the Greek idea of democracy swept through the Northeast and manifested itself in the templelike outlines of the popular Greek Revival architecture; and Lord Byron, among many others, gave his life for the cause.

That year, Harry was only forty-one, but his health was already failing. For some years, he'd been plagued by problems with his eyesight. If he read too much, especially in the dim candlelight, the whites of his eyes flared scarlet and the eyeballs themselves bulged painfully in their sockets. His father had suffered from the same condition, but rarely so severely, in part because, unlike Harry, he was able to moderate his habits. But now Harry was staying up to all hours in the Wall Street office poring over documents and firing off broadsides to the newspapers. In his dedication, he also suffered alarming tremors from what his wife termed "palsy" and a generalized weakness on the left side of his body.

At first, arbitrators sided with the builders' merchant houses, who'd supported the exorbitant rates. But when Harry threatened to appeal the decision to the court of chancery, as the higher court was named, the merchant houses agreed to roll back the fee after all. That should have been the end of it. At the port, however, the officious collector of customs, following the letter of a noninterventionist statute called the Neutrality Act, would not allow the Greek owners, as belligerents, to sail their newly built *Hope* off to an armed conflict in which the Americans had no part— thus creating for Harry another stage for this drama. Exacerbating his sense of urgency, Harry's eyesight suddenly dimmed further, making him frantic with worry that he would soon go completely blind, and large red blotches broke out all over his body as well. At Jane's insistence, he consulted a New York physician who found a cataract on what Harry had always thought was the better eye. He recommended an eye operation with Dr. George McClellan, the former personal physician of Thomas Jefferson (and father of the future Civil War general), who had recently established the Jefferson Medical College in Philadelphia.

But Harry would not take time out now. For the mania had kicked in—great raving fits of self-delusion, fueled by his prodigious reserves of energy, and masked by a quick-witted brilliance that seemed all the more scintillating the higher his mood soared. Harry Sedgwick became the agonist, not the Greeks, and he decided to resolve the customs issue accordingly: *he* would take ownership of the *Hope* as a private citizen. But before he could do that, he'd have to make his case in the court of chancery after all, and this time both the arbitrators who'd decided the previous agreement *and* the representatives of the merchant houses would be united against him there, so angry were they after weeks of his animadversions against them in person and in the press. The lead arbitrator, Jonas Platt, a former New York State Supreme Court Justice, was so irritated that the moment the court of chancery convened, he lit into Harry for violating his "professional duty and honor."

"Sir, *you* have grossly violated your duty," Harry fired back, an astonishing slur on such a legal personage.

Platt loudly retorted with more complaints about Harry's "prevarication and falsehood."

Harry faced Platt down, even though the justice was just a blur in front of him. "Sir, I shall only say that you had done all that was in your power to ruin one country and disgrace another."

With that, pandemonium. In the gallery, Judge Platt's son Zephaniah started screaming at Harry, but Harry blithely turned his back on him. "Your standing is not such that I can take any notice of you," he proclaimed. Infuriated, Zephaniah charged down the aisle to try to hack at Harry with his fists, but some burly patrolmen kept him away. Spit flying, he challenged Harry to a duel. Remembering Hamilton's fate, Harry would not give him the pleasure. "I will peril my life as freely as another for good cause," he declared, "and of that I can produce some evidence arising out of this very transaction—but I will fight no duels. I am a brother, a husband, and a father."

Eventually, order was finally restored, and remarkably, Harry did indeed win title. No sailor, he delegated the job of taking physical control of the *Hope* to a stout young navy lieutenant, Francis Gregory. But when Gregory showed up at the shipyard, he discovered that the man hired by the merchant house to keep charge of the ship, a pugnacious navy captain named Wolcott Chauncey, would not yield. Gregory summoned Harry, who rushed by carriage to the dock. There he found the two would-be commanders facing each other down, each one backed by several dozen rough-necked sailors. Hardly an imposing man, Harry stormed aboard, pushed Chauncey and Gregory apart, and, drawing himself up to his full height, waved the paperwork in front of Chauncey's face, every inch the self-proclaimed hero of this tiny drama, proudly declaring: "*I,* sir, am in possession of this ship."

Chauncey looked the papers over and, with a snarl, ceded the ship to him and withdrew.

When he heard the reports of this set-to back in Stockbridge, Harry's brother Charles could only snicker as he pictured his brother "patrolling the deck all covered with bristles, pistols & side arms & his spirit on fire with his new vocation."

The ship sailed the next month. Once his own part was played out, Harry's manic surge finally ebbed, and he collapsed with exhaustion. In winning, he had lost. He desperately needed that eye operation now, but he was terrified it would fail, consigning him to blindness. Unable to live with him when he was so overwrought, Jane had taken the children to Stockbridge. From there, she wrote him letter after letter in pathetically oversized handwriting, pleading with him to try surgery. Finally, at the end of October, he went under Dr. McClellan's knife. The procedure was protracted and, without anesthesia, torturous. But weeks later, when it was finally time to unroll the bandages that had covered his eyes, his sight had not improved.

THE TRUTH WAS, the damage to his eyes could not be reached from the outside, no matter how much Dr. McLelland scraped away. For Harry was most likely suffering from hyperthyroidism, and possibly from Grave's disease. The bulging, inflamed eyes are the classic symptom, but he endured other elements of the contemporary checklist, as well—tremors, and generalized weakness, and the jittery sort of nervousness that Harry also experienced as his mind wheeled about from one terror to the next. The cause is still unknown, but stress is believed to be a factor in its onset. All of this was worsened by this separate strain of manic depression. Nowadays, the preferred treatment is to remove the thyroid surgically, or disable it by radiation. Once the operation proved a failure, the boisterous passion that had fostered the radiant personality of the good years gave way to black despair, and he lay for hours on the sofa of the Warren Street front parlor, listless, barely breathing, bestirring himself only to groan that he was ruined.

He tortured himself with worry, and then, after months of anguish, conceived a salvation that testified only to his dementia. He decided that he would be saved by coal. He would create a small company to extract coal from the shallow mines near Newport, Rhode Island—"Pine coal," as he specified in the prospectus he issued shortly thereafter. "It burns with more flame, vehemence & brilliancy than any Anthracite I have yet seen." His mind burned similarly. He exploded with manic enthusiasm as he tried to draw his friends into investing with him, sure that this scheme would make them all rich beyond imagining. One of the friends, William Cullen Bryant, responded to the overture with a poem, "A Meditation on Rhode Island Coal," gently mocking the absurdity of the venture. ("Thou shalt be coals of fire to those that hate thee,/ And warm the shins of all that underrate thee./—Yea, they did wrong thee foully—they who mocked/Thy honest face, and said thou wouldst not burn.")

When, inevitably, there proved to be few takers for his scheme, Harry plunged in all the deeper, assuring himself of the very ruin that he was so desperately trying to avoid. He committed $7,000 to buying a tumble-down steamboat, the *Chancellor Livingstone,* to burn the coal as it plied the Taunton River that angled toward Boston from Rhode Island's Narragansett Bay and offered a shortcut for ocean-going passengers between Boston and New York (who would otherwise have to travel all the way around the Cape). It also added a fledgling transportation company to his would-be coal concern, immensely complicating the entire business.

Harry had closed the prospectus with an unusual concession. "I am well aware that Rhode Island Coal has been a proverb, a by-word, a reproach. It has exercised the wit of some good jokers, and a great many poor ones." Sadly, the whole gambit was delusional—undercapitalized, wildly overambitious, and completely unsuited to his talents. And it went nowhere, exposing the few friends who did invest to a share of his own ruin.

As his hopes plunged, Harry became disconsolate. Jane broke up their house in Warren Street and brought him back to Stockbridge that summer of 1827, hoping the warm breezes and happy associations of the Old House would bring him around, but they did not. Jane tried to distract him with the task of condensing a biography of Columbus that had just appeared. But he could scarcely make out the words on the page, either to read or to write. Jane hauled him to Philadelphia for another hideously painful operation at the hands of Dr. McClellan. At first, this seemed to help, sending Harry into such a giddy paroxysm of hyperactivity that he stayed in New York afterward to plunge back into the Rhode Island coal morass, only making matters worse.

Jane rescued him once more, returned him to Stockbridge, but the amorphous fatigue had given way to sleepless mania. Through the long night, Harry paced about the town, plotting ever more outlandish schemes to restore his fortune.

Jane, however, was pursuing a scheme of her own.

NOT FAR FROM where the original Robert Sedgwick had first settled in Charlestown, an energetic, Harvard-trained physician named Dr. Rufus Wyman had opened an asylum for the insane in 1818. Originally the Charlestown Asylum, it was renamed the McLean Asylum in 1826 when a Boston merchant named John McLean unaccountably left the institution a healthy benefaction in his will. As conceived by Wyman, the asylum followed the precepts of "moral treatment" that Philippe Pinel had started to establish when Harry's mother, Pamela, began her decline. McLean stood proudly atop Charlestown's Cobble Hill on an English-style estate of nearly twenty acres that, when acquired from the estate of the bankrupt previous owner, came complete with hundreds of ornamental plants and European fruit trees; dovecotes, greenhouses, stables, and a handsome manor house designed by Charles Bulfinch. A few modifications had been necessary to accommodate the new inhabitants. Bulfinch himself had expanded the manor house into a residence for as many as a hundred inmates off two wings that would be divided by sex, each side with a "strong room" for the isolation of "raging" patients.

Wyman encouraged "agreeable occupation"—walks, boating, horseback riding, sewing, gardening, taking tea, chess-playing, and the like that were the hallmarks of moral treatment. Wyman himself had a playful side, and was not above throwing shadow rabbits on the walls for the amusement of his female patients. His essential idea was to divert the patients from their emotional distress by shifting their concentration from their troubles to cheering pastimes. Wyman boasted a high cure rate. (A close inspection of the outcomes, however, shows that many of the unsalvageable patients took advantage of the openness of the place by "eloping," which is to say, running off—and thereby removing themselves from the books.) Others were immune to the kindness they initially experienced and, despite Wyman's philosophy, had to be physically restrained by straps, chains, straitjackets, and "muffs"—the heavy mittens that provided the extra benefit of restricting masturbation, which, in these first years of Victorian sexual anxiety, Wyman believed a "probable cause" of madness.

Still, when Harry arrived on a bitterly windy day in January of 1829, he sensed his life was about to be improved. Wyman divided his patients in part by social standing, and Harry was given a rather grand room in the original manor house, albeit one without the adjoining servant's quarters, since Harry lacked the funds for that. The care was fastidious. Before breakfast every morning, Harry was bathed while his room was washed down, floors and walls, with lime-water, and his soiled clothes were whisked off to the laundry. Outdoor exercise was limited in the winter, but he did have a chance to tramp about in the snow. For the most part, his own activity consisted almost exclusively of feverish writing. While the winter winds swirled about outside, he put pen to paper and poured forth page upon page of a venomous screed, part rant, part legal brief, part prayer, that seems to have been intended as a defense of his behavior but instead ends up providing a haunting portrait of lunacy in black and white. Sometimes written in tight, exacting, lawyerly prose and sometimes in a heavy, sprawling ramble, the screed runs every which way over the page, is heavily blotched with spilled ink and angry cross-outs, and adds up to nothing. There are tantalizing hints about the source of his anguish, however.

> *Who then is to judge of the soundness of the human mind—the multiply sided mind—and emanation from the Great Central Fire——*
> *JEHOVAH THE ETERNAL—Who shall dare to assume this Jurisdiction against the express command of Jehovah—announced from on High, to all the world & to all ages—this is my beloved son—Hear ye him!*
>
> > *25 min before 7*

It irked him that, in this madhouse, Wyman had set himself up as the god who decided who was sane and who was not. As he wrote:

> *14 min before 7 Who shall dare assume the Jurisdiction Shall it be Rufus Wyman acting on his individual . . . or his professional character—certainly not—Does he act as Superintendent of McLane asylum—if so an appeal lies to his master the trustees—from them to*

their masters the People of the Commonwealth—from them to
enlightened public opinion—to future ages—to conscience—& to
God—now hear the Holy one—Judge not that Ye be not judged
                              H D Sedgwick
                              10 min past 7

Although moral treatment was intended to provide a healing sympathy, Harry clearly could not escape the feeling that he had been, as he says repeatedly, subjected to the judgment of Superintendent Wyman, and been found wanting. Nothing could sway him from this belief. The whole idea of judgment was bound to be a sore point for a lawyer like Harry, who always saw himself on the right side of every issue. And it carried an extra kick to be himself the son of a supreme court justice—*the Judge*— who had always singled out Harry, of all his sons, for particular criticism.

For all its comfort, McLean Asylum did nothing to soothe Harry. Instead, it freed him only to chase after his obsessions all the more fanatically. After three months, it was reported to Jane that her husband had become a "most violent maniac." He raged at his attendants, wept profusely, threw his food in their faces, and, in her later summary, proved so "unreasonable" and exhibited such "fury" that the attendants believed they had no choice—moral treatment or no—but to enclose his legs in manacles and stuff his hands into those humiliating muffs. Now that he was unable to take solace in his screeds, his physical symptoms worsened as well. The blotches that had plagued him before had burst forth into frightening sores on his wrist and his ear, and he also had "dropsy"—the same painful swelling of the lower legs that his father had suffered from before he died. Wyman himself wrote Jane to prepare herself that the end might be near. But, still committed to a belief in the value of total separation, he would not permit her to see her husband. A visit, Wyman was sure, would only agitate Harry further.

Harry's brother Theodore was twice allowed in over the next few months. He steeled himself for each encounter, and even so he was "grieved" to see his dear brother in chains, and incoherent besides. Each time Theodore found it nearly impossible to conduct his business in the city afterward.

Finally, after nearly a full year of such reports, Jane could take it no longer, and against Wyman's wishes, she drove by carriage to Charlestown shortly before Christmas in December 1829 to see her husband. They met in his rooms, with an attendant present for her safety. She scarcely recognized him. Harry had aged at least a decade that one year. His eyes blazed red; he had to squint to see anything at all. His face was lined with worry and fatigue. His hair had thinned, turned gray; he had angry sores; he was stooped, feeble; his clothes hung off him; he shuffled, and his hands and arms trembled with what Jane took to be his palsy. When Harry saw her, he was speechless, unbelieving. Then he burst into tears.

"Though I perceived a great deal of insanity," she reported later, "I also saw that much of it was produced by the exasperation which such a situation could not but occasion."

She removed him from the asylum that very day and took him in her carriage to Connecticut to be treated by Dr. Eli Todd at a rival institution, the Hartford Retreat for the Insane. In appearance, the retreat was startlingly similar to McLean. It consisted of an imposing three-story Georgian building, with wings for forty patients each. As at McLean, the inmates were segregated by gender and class, with the most well-to-do given fairly grand apartments, with room for servants; and there were nearly twenty acres of gardens and ornamental trees for all the patients to stroll about in good weather.

Big and broad-shouldered at sixty-two, Todd was of a refined temperament, with a delicately probing gaze. His interest in the insane was not academic. His father had been worrisomely erratic, and his sister Eunice suffered from severe mood swings that, despite her brother's tender ministrations, led her to kill herself by slitting her throat just months before Harry arrived. Like Wyman, Todd was a stout believer in compassionate care; unlike him, Todd actually demonstrated it to Harry. He took the time to play chess with Harry and, more helpfully, did not just allow but actually arranged for visits from the family, starting with his oldest daughter, Jane, then just nine.

Harry had always been a doting father, and the sight of his daughter— tender, hesitant—in the front hall at the high-ceilinged entranceway to the Hartford Retreat cheered him as nothing had for months. His wife had brought her, but the elder Jane merely watched as Harry's eyes met his daughter's. Father and daughter moved awkwardly toward each other, and she wrapped herself about him. The elder Jane left the two together for the afternoon. When the visit was over, Harry was glowing. "The effect of the natural interchange of the affections between the father & child was more salutary than all the restraints & all the medicines had ever proved," Jane recalled.

After that, Charles and Catharine both came down from Stockbridge to see their brother, "softening & tranquilizing his feelings," as Jane put it. And then, finally, Jane came down by herself. She had been dreading it. She'd been laboring under the delusion—encouraged by Wyman—that she would only worsen her husband's condition. Indeed, when she arrived, Harry raged at her for having left him so long at McLean, and not visiting him sooner there in Hartford. But instead of arguing, she produced a volume of Shakespeare that she had brought along and offered to read to him from some of his favorite plays.

"Would you like that?" she asked.

"Oh yes dear," he replied, transformed by the question. He then listened attentively as she read, occasionally offering a comment about the passage she selected. When his attention finally flagged, she played chess with him, although his play proved clumsy. Then they went outside for a long walk about the barren fruit trees on the grounds that were still white with snow. He moved slowly, but his gait was stronger, steadier, than it had been at Charlestown, and she felt more vigor when she took his arm.

The visit went so well that, with Dr. Todd's permission, she stayed on at the retreat herself, taking a room of her own in the women's section. From seven in the morning until ten o'clock at night she was at her husband's side, reading, conversing, playing chess, walking—"trying," as she wrote, "to restore his mind to its former track & prevent his gusts of passion by having constantly something agreeable at hand." For week after week, she stayed on, until he began to long for his old life at Stockbridge again. His eyesight was still poor, but his appetite had returned, and with it a good deal of his vitality. There was a liveliness to his personality that Jane scarcely remembered. And for the first time in a long while, his thoughts were almost entirely orderly and rational. After five weeks, Dr. Todd decided that Harry was fit to leave.

Back in Stockbridge, Harry immersed himself in the pleasures of country life. He swam in the Housatonic and walked everywhere, covering a dozen miles a day. He was still not entirely right, though. For a diversion, Theodore's wife, Susan, took Harry with her family on a jaunt to

Quebec, but when he came down with a sudden toothache on the boat, he went "wild," Susan told her husband, and could barely be kept from plunging overboard in his eagerness to get to shore for treatment. In general, Susan found she needed to think of him as a small child, someone whom she had to keep watch over every minute, what with his "uncertain footing" and "imperfect sight," not to mention the strange looks he got from strangers unused to his many "peculiarities," like his still-inflamed eyes and shuffling, uneven gait.

Back in Stockbridge, his palsy worsened over the next few months, and he began to complain again of a deadness in his limbs. He rallied to produce some fiercely reasoned polemics against the John Quincy Adams administration, and in February 1831 he delivered a fiery speech in favor of abolition. But he was not stable by any means, and was prone to ridiculous frenzies over nothing. Unable to get over his rage at Rufus Wyman, he was determined to sue the McLean Asylum for what he regarded as gross mistreatment, and he labored over briefs to that effect, ignoring his family's pleas to put them aside. Despite his infirmity, he insisted on taking part in a national political convention in Philadelphia that September. Worn by the ordeal, when he finally returned, he needed to be helped out of the carriage by his brother Theodore and nephew Theodore III, who held him up on either side.

With that, his mind went. He was unable to make himself understood, his limbs were prone to sudden spasms, and his eyes were grotesquely inflamed. Small children, seeing him stagger along the street in Stockbridge, pointed and giggled. At first he tried to protest, but, when that only provoked more stares, he retreated into silence.

Catharine, with him in Stockbridge, could hardly bear it. "To see a mind once so powerful, so effective, so luminous, darkened, disordered, a broken instrument—to see him stared at by the vulgar, the laugh of children—oh," she confided to her journal, "it is too much." Back in the house, Harry kept to himself, sitting in a chair before the fire, scarcely moving, for hours at a time. Still, he kept his faith. "It was the most affecting thing," Susan wrote her son Theodore, "to hear him, a poor paralytic, an object of disgust to himself, and of pity or curiosity to others, declare '*how good, how very good* God is.'" Eventually he was too weak to rouse himself from his bed and come downstairs. As winter descended, his breath grew labored and then rapid, and, on a Friday afternoon, December 23, it ceased altogether—"without a sigh," as Susan noted. Jane was sitting by his bed when the end came. She cried out, "Dear darling Henry!" and threw herself on him, and Catharine, who was also there, wept. "The tender remembrances of Harry during her whole life," Susan wrote of Catharine, "seemed to descend upon her, and she could not give him up."

To UNDERSTAND WHAT had gone wrong with Harry, Jane asked a Stockbridge physician, Dr. Flint, to perform an autopsy on her husband's brain, and she copied his results into her own report, carefully noting how her husband's head had been sliced open from the "frontal sinus" to the "occipital ridge" and the top of the skull lifted off; and how portions of the delicately fibrous "dura-mater" clung to the cerebrum, making close examination difficult. Dr. Flint suspected he may have discovered the source of the trouble when he located a "mass of disease"—mostly "*fleshy*," but oozing "*pus*"—in the midst of the corpus callosum, that band of nerve-rich fibers that curves over the hypothalamus and spreads down into the walnut-sized cerebellum. Flint could tell this was ruinous, although he could not have known why. The corpus callosum is heavily responsible for intelligence, and the cerebellum controls voluntary movements. If anyone needed to know why Harry was virtually comatose in his last days, they needed to look no further. As Flint summarized it for Jane: "When the brains are cut the man would die."

But the degeneration that was so evident in his brain did not entirely explain Harry's increasingly erratic behavior—the mood swings, the raving—until those final months when he finally quieted. At least, it did not reassure Theodore, who had himself suffered from a period of what he termed "a most violent and long continued fit of Hypochondria" in his late teens, and had received countless letters from his father to be sure to get his exercise and attend to his health. He had sensed for years that whatever Harry had was not his alone. "It is confirmed by medical men that this disease...is in a great degree hereditary," Theodore wrote his son Theodore Jr.

> *You are aware how great a sufferer my Mother was— Though it be hereditary, it does not follow, that it may not be counteracted— Gout is hereditary, consumption, Drunkenness &tc.—So are the passions, good*

or bad—The question therefore arises, what shall be that treatment of
ourselves, the tendency of which is, to escape the danger— The essence
of reason seems to be a calm tranquil judicious mind— A conduct
regulated by wisdom, & far removed from all extravagant hopes, fears,
phantasies.—There must be a sound mind, in a sound body— Bodily
disease brings on insanity—Health therefore is often indispensable,
where there is a previous tendency to it.

Then, having offered as sensible a prophylactic course of treatment as
modern science would endorse, Theodore succumbed to a whimper of
despair:

If only Harry had stayed away from the Greek frigate case, and from the
Rhode Island Coal Company!

HARRY WOULD NOT be the only cause for sorrow in the family in those
years. Charles's son, Charles Jr., or Charley, was a Harvard senior a
decade after Harry died. At six feet, he was taller than his father and,
blond, blue-eyed, and quite robust, in Catharine's telling, possibly more
handsome. His father had noticed a certain "hot temper" in the lad, but
Charles Sr. assured himself that his son would "fly his kite with a strong
string." Catharine worried that, like Hamlet, the boy lacked will. He'd
done fine at Harvard for his first two years, but in his third he seemed
overworked, and suffered from what Catharine termed a "nervous fever,"
or a kind of breakdown. He went to live with his aunt Jane's family on
Charles Street. His father traveled from Stockbridge to see him and was
relieved to find him sleeping well, and as quiet "as the Sedgwicks ever
are." Once the family was assured that the boy had suffered merely a
passing attack of "dyspepsia," Charley was allowed to book passage to
Liverpool by steamer with a promise that the captain would look out for
him. Catharine and Bessie, Charley's sister, saw him off from the pier in
Boston, and watched till their eyes ached as the great steamship dipped
under the horizon. It was six weeks before the family got further word of
him. Ominously, the letter did not come from Charley himself, but from
the ship's captain, Delano: he'd found their son dead in his hotel room, an
empty vial of laudanum by his bed. He offered few details, but enough:

Charley had been agitated, then depressed, almost immovably so. It had been all the captain could do, once they berthed in Liverpool, to extract the boy from the ship and settle him in the hotel, he was so distraught. Captain Delano needed to say no more. The Sedgwicks knew perfectly well: young Charles Sedgwick Jr. had died by his own hand, a victim of the family disease, the third in three generations.

TWENTY-SIX

IN THE COUNTRY BURIAL PLACE, WOULD I LIE

Harry was buried in the graveyard, occupying the spot due east of his parents, alongside his sister Eliza, who had died of a paralytic stroke four years earlier. The surviving family was gathered about his grave to watch his coffin descend. For Harry, his siblings selected a low obelisk, shorter than their father's, and Catharine provided the two inscriptions: "Blessed is the man in whose mind there is no guile," and "The end of the upright man is peace."

ALONG WITH ELIZA, Catharine's other surrogate mother was in the graveyard as well. Mumbet had died at the end of December 1829, two years before Harry. She'd been ailing for several months from what Catharine termed "a singular local disease" that she feared was of a "cancerous" nature. Her delicacy suggests that it was in her breast. Mumbet remained "perfectly self-possessed," although occasionally "delirious," as she waited for the end. Catharine adds that when Rev. Field came around to visit her, he asked if she was at all afraid to meet her God, presumably because she had failed to attend church regularly. "I am not afeard," Mumbet replied, indignant. "I have tried to do my duty, and I am *not* afeard."

The funeral was at the Sedgwick house, and Charles was moved to look upon Mumbet that last time and see "her great & noble face, tranquil & beautiful as it seemed to me." It evoked for him

> the memory of those great qualities, & beneficent deeds, which place her in our hearts by the side of our parents, our brothers & sisters & children, the living & the dead.

Afterward Mumbet's coffin was taken by a cart to the burial ground, where it was placed beside the spot that Catharine had reserved for herself. Catharine furnished the epitaph:

> Elizabeth Freeman
> Known by the name of Mumbet
> Died Dec 28th 1829
> Her supposed age was 85 years
> She was born a slave & remained a slave for nearly thirty years. She could neither read nor write yet in her own sphere she had no superior or equal. She never violated a trust nor failed to perform a duty. In every situation of domestic trial she was the most efficient helper & the tenderest friend. Good Mother farewell!

It was a moving gesture to bring Mumbet permanently into the family in this fashion, but it was not an unmixed one. For by then Mumbet had those descendants of her own, even some great-grandchildren, who would not be buried with her, devoted as she was to them. She had bestowed careful consideration on them in her will, granting half the proceeds of her real estate holdings to her daughter, Elizabeth, and the other half to benefit two great-grandchildren, and she distributed with care the bits of finery she had accumulated: a pair of "birds eye petticoats," two linen handkerchiefs, a "lace cap," two "tucked ruffles," a chintz gown, a "blue broad cloth cloak," a pair of "small gold ear rings," a "black velvet hat," "gold beads," "morocco shoes," a set of "burnt china," and a "satin cloak." The affection is palpable, and yet, going alone into the Sedgwick graveyard, she would be separated from her heirs forever. In fairness, though, she remembered one Sedgwick in her will, too. To Catharine she gave her favorite necklace, the gold one that appears in the Susan Ridley

Sedgwick portrait of her; Catharine turned it into a bracelet and wore it daily for the rest of her life.

And when she died, years later, Catharine lay down beside Mumbet forever.

AFTER THE HEARTBREAK of losing Robert to marriage, Catharine had turned for refuge to her younger brother Charles back in the early 1820s. Aggrieved as she was that he'd had to give up the Stockbridge house for Lenox, she was determined to overcome it. "Wherever you are I *must* have a home," she'd vowed to him. And so she did, claiming for herself a bedroom in the house he built for himself on a hill there, and she thought of Charles, as she had once thought of Robert, as "more than [a] brother." In 1829, she told him, "I know nothing of love—of memory—of hope— of which you are not an essential part." Unlike Robert, Charles responded to all of these overtures with unfailing hospitality. He promised to "make my house, myself—my all as conducive to your happiness as it is possible it should be." She became extraordinarily attached to Charles's daughter Kate, whom Charles and Elizabeth had named Katharine Maria after her. Catharine sent her endless numbers of chatty letters signed "Aunt Kitty." Like a doting mama, she fussed over her "darling Kitty's" winter apparel, wondering if she'd be warm enough in a waist-length "muslin spencer" jacket of the sort she was seeing on the New York streets. In 1826, after Eliza and Harry both took sick, she dreamed of walking together with little Kate in paradise, with fruit trees all around them, the limbs laden with sensuous fruit.

Still, in her bleaker moods, Catharine was never quite sure where she stood with the survivors in her family. In an anxious entry in 1837, she called the roll, ruthlessly examining her relationship with each of her family members, from Charles on down. Each connection, she could see, had been worn by the "rust and moths" of time. When she got to Robert, she devoted several sentences to him, but then carefully inked them all out. About Robert there was too much to say. A decade later, when she re-visited this portion of her journal, she couldn't resist saying it in the margin, about how devoted she had always been to him, and how, when he finally spoke of marrying, she "felt an aching void" at the prospect, and told him so "as she should not."

In the spring of 1838, shortly before he turned fifty-one, Robert had a paralyzing stroke. Catharine rushed to him, moved into the room his family kept for her, and, practically shoving Elizabeth out of the way, tended him around the clock. "Every moment of my time is occupied," she joyfully told her journal. She read to Robert, mended his socks, took him on short walks, washed his face, his hands, his feet, and as much of his soft white body as decency would permit. "A thousand nameless services," she called them, and all of them filling her with an "inexpressable happiness." It went on this way for a full year, Robert getting neither better nor worse.

Unlike Harry, Robert had a talent for business, and he'd made a fortune in real estate in Chicago (where North Sedgwick Street is named after him). So it was nothing for him, now that he had somewhat recovered from his stroke, to take off with Elizabeth for a fifteen-month tour of Europe. Catharine went along; she was not to be separated from him ever again. And she brought along their niece Kate, too, making up a kind of ersatz family for herself, with Robert as her surrogate husband. In London, she dined with the "great literary people," was "twice at the Marquis of Landsdown's," and she once spotted Queen Victoria in her royal opera box. "A plain little body," Catharine sniffed. "Ordinary is the word for her."

While they were away in Europe, however, her brother Theodore died suddenly of an apoplectic seizure, which we would call a stroke—the same event that had taken his grandfather—while delivering an impassioned speech to the Jacksonian Democratic faithful in Pittsfield. When Catharine received the news some weeks later, she felt like a "child torn from the arms of its mother."

Robert rallied while they were on the tour, but shortly after he returned to New York, he took again to his bed. Catharine remained with him, and she was there at the end. Once again, she could scarcely be pulled away from the corpse. Robert was fifty-three. And, when he was buried in Stockbridge, she stared and stared at the hole into which he had been placed. "I feel as if half my life were buried in his grave," she wrote a friend.

AND SO IT was. Her literary output was nearly halted. She had published all but the last of her novels; mostly what remained were short pieces, didactic tales, and an occasional travelogue like her account of the European jaunt, which she published as *Letters from Abroad,* many of them actual letters she wrote to Charles. For Catharine, though, the shocks kept coming. The following year, 1842, her favorite niece Kate announced her intentions to marry. In Catharine's mind, she was being abandoned, just as Eliza had abandoned her long ago, and so many others had since. After Catharine received the letter, she could do nothing but climb into bed, draw the covers up over her head, and sob.

Still, after Kate moved with her husband to the Boston neighborhood of West Roxbury, Catharine eventually followed her there, and, as she had at her siblings' houses, soon had a room of her own. And when Kate had her first daughter, Alice, Catharine took her up much as she had Alice's mother, having tea parties with the little girl and playing dolls as if she weren't great-aunt at all but sister. This was the Alice to whom Catharine's autobiography is addressed.

Gradually, life narrowed for her. When Charles died in 1856, Catharine came to live year-round with Kate and her husband. As her productivity declined, so did her income. In her straitened circumstances, she'd had to give up even small pleasures, like her subscription to the *New-York Evening Post*, the paper edited by her old friend William Cullen Bryant, although, once Bryant heard she had done so, he restored it as his gift. Toward the end she carefully wrote out farewell notes to all her friends. She died at Kate's house on July 31, 1867. She was seventy-eight.

In her later years, Catharine never returned to Stockbridge without a stroll through the ever-expanding family graveyard. In 1842, when Frances died alone in Stockbridge during one of her many separations from Ebenezer, the circle of that first generation was almost filled in. Robert, Frances, and Eliza were buried to the south, Harry and Theodore to the northeast. Catharine had long planned where she would go: to the northwest, between Charles and Mumbet. In her gloom, she had meditated on the meaning of death. Was it the final separation, or a permanent union?

In an essay from this period, she wrote that she envied the Romans their practice of keeping their dead about them by preserving their ashes in monumental urns they set about their houses. It would be pleasant, she decided, to have her brothers and sisters forever by her as she read, wrote, or sewed. But she recognized that this would never work for most Americans, who, in their transiency, wouldn't bother to tote along the remains of the departed each time they moved. No, better to return the remains to the earth, she decided. Let the burial grounds be a permanent mansion.

> There, in the country burial-place, would I lie, amid my friends of all conditions, where the sod over me was freshened by the same summer showers that pattered on the roof I had loved in life— where the morning sun, as he comes over in my native hills, shining into the windows of the homes I love, shines also upon my grave, and the twilight that calls the merry boys to our village-greens sends its dewy sweetness over my resting-place.

And with that thought she completed a cycle, a revolution, returning the family to its source. Catharine had finished the work her father had begun, but in her own distinctive fashion. As father and daughter, they necessarily operated in separate spheres, and the differences between them were pronounced. But so were the similarities. The superficial selflessness of his electioneering matched her seeming diffidence in her literary stardom. He may have sacrificed his marriage for his career, but she went him one better and never married at all. But, in the end, it was a particular set of differences that united them. While he had built *up,* with that mansion that reached to the skies, she had built *down* into the underworld with that graveyard. Further, while his house remained static, bounded by its four walls and roof, her graveyard would not. By the time she entered it, the graves were already reaching, death by death, into the third circle. More even than his house, her family graveyard would draw future generations into itself. This was not the Judge's idea, but it certainly would have pleased him. Thanks to her, that obelisk of his, and the graveyard it marked, would become a kind of magnetic pole, a true north that would serve for generation after generation of his descendants— none of them Catharine's—as a permanent place of reckoning.

PART FOUR

THE PRICE OF LEGACY

TWENTY-SEVEN

THE GREAT WHEEL TURNS

There was nearly a foot of snow on the ground when I returned to Stockbridge a few weeks before Christmas in 2003. I was feeling sorrowful, having just laid Catharine to rest in the portion of the narrative you have just read. As the writer, the teller of the tales, she'd become a kind of proxy for me in that generation of Sedgwicks. To see her come to her end, especially such a bleak, solitary one, made me feel I had come to an end of my own.

As I had done many times since I started researching this book, I drove from my house in Newton out the Massachusetts Turnpike portion of I-90, rising up from the Connecticut River to ascend a 1,700-foot spur of the Berkshires that comes just a few feet shy of marking the highest point of the entire transcontinental highway. As usual, I exited at Lee, passed the familiar Hubby's Cabins, the bowling alley, the fragrant paper mill that puts the winding Housatonic to industrial use, and into Stockbridge on the thoroughfare that is now called Main Street.

On its broad plain, with its generous streets, Stockbridge looks oddly midwestern for a Massachusetts town. On my left, I passed the wide-flanked, 1950s-style Stockbridge library, whose trove of Sedgwick papers I had combed through many times; a cluster of old-timey shops, with front doors that jingle on opening and old-fashioned angled parking out

front; the touristed alleyway called the Mews; the corner that once housed
Alice's Restaurant, made famous by Arlo Guthrie's song of that name;
and the broad white veranda of the Red Lion Inn, which the widow
Bingham established over two centuries ago. I've been in all these places,
and their predecessors, countless times. It's all part of a past that never
quite retreats into the past at all—or is it that its present is never quite
now? Norman Rockwell, Stockbridge's new favorite son, painted the village streetscape so many times for *Saturday Evening Post* covers, it has
become an enduring image of heartland America for many others as well.
It's a town that seems to exist under glass.

As I drove through the main intersection in Stockbridge, marked by a
slim monument to some war dead, I crossed the route that the first Theodore had taken when he rode into town from Sheffield in 1783. The Edwards house, where he purchased the land for what would become the
Sedgwick house, is long gone. On the site now stands the Austen Riggs
Center, founded in 1919, a sprawling psychiatric hospital best known for
having treated James Taylor, who immortalized his stay in "Sweet Baby
James" and "Fire and Rain." By now, it's taken in any number of troubled
Sedgwicks as well.

Passing through town, I experienced again the queer sensation I'd had
many times as I investigated my ancestors. A bit like snorkeling, as I
thought of it. Reading so many of my ancestors' letters, and reviewing
other accounts, I fancied I'd somehow ducked under the surface of what
is, to discover a vast and teeming other world below: a world that had
always been there, but invisibly, without my knowing, existing just
beyond what I'd always taken to be the limit of shiny reality.

And so now, as I drove along in my Honda, Main Street was simultaneously the placid Main Street of the present day—with slender sidewalks set back from the street, handsome houses well removed from the
sidewalks, the whole scene whitened with downy snow—and also the
Plain Street of long ago, a dirt thoroughfare for horse-drawn carriages
and wagons, lined with a few unpainted and unprepossessing houses,
crawling with stray pigs, and, most importantly, crisscrossed here and
there by the members of that first Sedgwick family who had become so
completely vivid to me. Poor Harry, stumbling along through the dark in
one of his maniacal midnight rambles; Theodore astride his mare, taking

on the Shaysite rabble; the frightened Charles darting across to the Bidwells to get help for his crazed mama.

I might have shouted to them, and they would have turned to stare.

I WAS HEADED for the family graveyard behind the town cemetery. Imbued with the stories of that first circle of Sedgwicks, I'd become rather sentimental; going to the graveyard was like opening up a family album and getting misty-eyed over yellowing snapshots.

I'd visited many times, of course, most recently to bury my mother. She'd died at the end of the summer and taken her place beside my long-dead father, on the other side from his first wife, Helen. The arrangement didn't exactly please my mother, having to compete for her husband in this way; the only instructions she'd left behind regarding her burial was that her headstone *not be like Helen's*.

But now, as I came up on the ancient rings of stones, my mind slipped past my mother altogether and filled entirely with thoughts of that first generation of Sedgwicks. Harry, Robert, Catharine... It was as if *they* had become my immediate family, metasiblings of some sort. Yet they'd come alive only to be dead, and I nearly wept to see the tall, proud stones of Harry and Robert, each one an obelisk like their father's, looking inward toward each other from the east and west; the cross-topped column of their older brother Theodore, toward the north, with gentle Charles around to his right; the vine-enswirled cross of the hysterically devoted Catharine beside him; and the slab of proud Mumbet beside her. On the other side of the circle from Theodore Jr., the grave of poor put-upon Frances lay just to the east of due south, and exhausted Eliza to the west. And in the middle, of course, stood the Judge's thick obelisk and Pamela's dainty urn.

Walking around in my street shoes, the snow-melt soaking through the leather and trickling down inside my socks, I felt like Catharine, communing once again with her dead. Here in the graveyard, the dead were not dead, just as she said. They seemed to be standing all around me, proud, indomitable, unified in their love for each other. If it hadn't been so cold and miserable, I might have settled in, just to hang with them a while.

* * *

MY MOTHER DIED in late August, when the leaves outside her window were just beginning to turn pale. She'd been in the hospital ward of her retirement center for almost two months, in a double room: her roommate had recovered from a mild stroke and returned to her regular quarters upstairs, leaving my mother to die alone, as she preferred. Every day, it had been a little harder to draw enough oxygen out of her oxygen tank to sustain her. Eventually, just as the doctors said she would, she drowned in the fluid shed by the disease. She was down to about seventy pounds by the end, just a husk of herself.

I'd never seen a dead body up close until I saw hers. She was turned on her side, her mouth parted, as if she was just rolling over to reach for a drink from the bedside table when it was her time to go. By the time I arrived, she looked completely dead—finished, used up, empty. Had all the Sedgwick corpses looked like this before they were stuffed into their coffins and dropped down into the earth? It wasn't just that her skin had turned to leather, but her whole body seemed somehow crinkly, like crumpled-up wrapping paper, or a crushed beer can. And it—she—remained utterly still as my wife and I went quietly about collecting her things—her wristwatch, her purse, her glasses, the book she'd been reading. She was gone.

We had a small memorial service for her at the retirement center. If she had been her own person in life, a Lincoln on her father's side and an Ames on her mother's, she became a Sedgwick in death, buried in Stockbridge and gathered permanently into the clan.

It's the custom for Sedgwicks to be buried simply, with as little intervention from undertakers as possible. For generations, the Sedgwicks wore their pajamas for their long sleep in the Pie, but now most Sedgwicks are cremated, as my mother was, and they pay no attention to their clothes. Still, we do not use any fancy coffins, the ones with mahogany, gilded handles, and satin plush, but a plain pine box. The same one, in fact, is used over and over, solely as a means of transport, with the ashes of the departed placed inside it, in a small wooden urn of their own. My father had the current coffin made by a carpenter in Great Barrington back in the 1960s. When it was done, Dad climbed in to make sure it was right for a big man like himself. Even though it would contain only his urn, it still had to look right. Then he brought it back here to the barn in

Stockbridge to await its next occupant. He didn't know that that would be himself.

Now it was my mother. Many of my nieces and nephews, the ones I'd played with in the Pie as children, had flown in for the service, some from as far away as California, which was particularly kind, considering that my mother had turned so crotchety in her old age that she had not much opened her heart to them. Nevertheless, there they were, and they helped my brother and sister and me load the coffin onto the back of a horse-drawn wagon that, sans horse, resides most of the time in the barn by the house. The town mailman, Tom Carey, used to double as the horseman for Sedgwick funerals, but he had long since died himself. After getting a little advice from cousins who were old hands at Sedgwick funerals, I hired a rather crabby older couple who came around to the house with an aged pony. After placing the coffin on the wagon, we covered it with the distinctive Sedgwick pall, a red cross on a white background, and the children adorned the cart with the customary hemlock boughs and other greenery. Finally, the couple gave the pony a click, and the wagon wheeled out the driveway and down Main Street, with all of us mourners walking behind, to the stone Episcopal church at the center of town.

I did not cry at my mother's service. I'd had a long time to brace myself for it, and she herself had been resigned to dying, if not positively eager for it. Her years of depression had worn down the deep bond of love that had been there in my youth, leaving a raw spot behind at the nub. Or was I becoming inured to death? And the rituals do soften the impact, making death seem like it has been accounted for. My two daughters played the violin for the service, which my mother would have liked, and several of the other grandchildren contributed a poem or a reading. My siblings and I all spoke; I did my best to evoke the goodness in my mother. Then we loaded the coffin back onto the cart for the long walk down Main Street and through the town cemetery to the Sedgwick burial ground on the far side, with all of us mourners walking behind, trying not to feel too self-conscious as passersby stopped to gawk. For the previous Sedgwick funeral, of my distant cousin Netta Lockwood, the bereaved had neglected to have the hole dug beforehand, a mild catastrophe, and I had received any number of reminders this time. Fortunately, the hole for my mother was there, right where it was supposed to be, beside my father in the segment of the Pie that radiates outward from Henry Dwight Sedgwick I, II, and III.

The very first Sedgwick burial I can remember was of Gabriella, that step-grandmother of mine who had first led me down to the graveyard with all my nephews and nieces to play hide-and-seek. I was still in my teens, and I started to cry when the time came to drop a bit of earth on her coffin. The sound of the earth thumping against the wood had more finality than I could bear. It seemed so cruel that she should be dropped below ground, when I wanted so much for her to remain above.

This time, for my mother, I felt the tears loosen to see my brother Rob, with his long arms, reach down into the hole to place her urn at the bottom. The two of them had never been close, and yet he handled her urn with such tenderness. When he stood up again, there was a smudge of black earth on the knee of his suit pants from where he'd kneeled down. A memento.

ONLY TWO HOUSES still stand from the Stockbridge of 1785—the Mission House that Pamela's mother, Abigail, built for her first husband, and ours. The Mission House has been moved down from its original hilltop location, and placed a few doors down from Austen Riggs and diagonally across from the Sedgwick house, where it is open to tourists during warm weather. Ours remains where it has always been, looking out to the wide street and, in the back, over the Housatonic to Monument Mountain.

It is called the Old House now, and is owned by a trust called the Sedgwick Family Society, which consists of all the living descendants of that original Theodore Sedgwick, now routinely referred to as the Judge. The society is run by seven trustees, who gather two or three times a year in that "best Parlour," now a book-lined study, many of the volumes written by family members. Rob and my half brother Harry serve as the trustees representing my branch.

The house is available to any descendant of the Judge who can commit to living for at least a few years in what is still a fairly remote part of the world, and can afford the modest rent. The shortfall is covered by the society's endowment. The trustees determine who will stay there, and for how long. Now my cousin Arthur Schwartz, an architect descended from the line of Theodore Jr., is living there with his wife, Ginger, and her elderly mother, Virginia, who occupies that other downstairs parlor, which has long since been converted to a bedroom. The hallway behind—where Catharine sometimes wrote at an unusual standing desk—has been turned into a bathroom, with exposed pipes that are always in danger of freezing in winter. Arthur's architectural skills have come in very handy in making a drafty, antique house liveable through the wide seasonal mood swings of the Berkshires year.

Coming back from the Pie this winter trip, I pulled my car in through a line of hemlocks protecting the house from the street and parked on the semicircular drive. On the outside, the house looked very much as it did when the Judge left it for Boston in the fall of 1812. It was still a handsome, rather stately clapboard-sided house, whose Federalism was just as confident and striking as it had been in 1785, even though it has now come to be the prevailing architecture in town. The Judge's original white walls had been brightened to a more cheerful salmon pink, and a wing was added for servants during a brief period of sunny prosperity for the family in the late nineteenth century. One corner of the roof collapsed sometime around 1800, and had to be jury-rigged back up into position. The grand portico still stood out front, although the columns supporting the roof had been several times replaced. For a touch of elegance, the original pine floorboards had been replaced with marble tiles, set on a diagonal, but winter snowmelt had seeped between them over the years, forming ice that gradually pushed the tiles apart and splayed the columns themselves on either end. Some peonies that Catharine's bosom friend Fanny Kemble once planted alongside the house were, of course, long gone, as were some roses that Pamela had put in. Memorial trees had risen up in their place, including a graceful honey laurel that arched up by the drive, in honor of my father.

Inside, the house was alive with memories of the departed. Just inside the front door were the framed, tattered remains of a letter that the Judge wrote to Pamela from Quebec, when he was involved in that failed assault on Canada. On the wall directly across hung an etching of Mrs. Washington's ball that Pamela attended so gaily in 1794. A copy of Sara Ridley Sedgwick's tiny portrait of Mumbet was also there, along with a Charles Cromwell Ingham photograph of a fading Catharine, taken a few years after he did that swirling oil portrait of her. The gorgeous Stewart portrait of the young Catharine and her mother hung in what used to be the kitchen, now the living room, which looks out over the Housatonic

to Monument Mountain beyond. More portraits of that first generation—
the gentlemen in black cravats, the ladies in gowns—lined the stairs going
up to the second floor. A copy of the Gilbert Stuart portrait of the fatten-
ing Judge hung over the mantelpiece in the dining room.

THE OLD HOUSE is not the house it was. The wing where I passed the
night was a later addition, and it has been renovated several times. Instead of the Judge's fully reasoned, four-square Platonism, it has the air of
an afterthought, a scrawled postscript written in a different hand, in a different spirit. Then again, the original house would not have remained the
original house even if it had been left untouched, even if those eight-inch
yellow pine boards were still the ones we trod on in that "best Parlour,"
and the ten-inch white pine ones in the other parlor across the hall. No,
houses change because the world changes, and the people inside them
change, and see them differently.

Houses are seen differently by others, too. Years ago, when I was still
quite impressionable, I took a savvy Boston friend to the Old House, and
he looked around at the decor, the decrepit, unplayable clavier in the
dining room and the antique lamp with a tattered lampshade (the tattered
part turned discreetly to the wall) in the study, and in the tone of summary judgment, loudly declared: "Oh, *shabby gentility!*"

It was a shock to feel the family put in its place. But a worse shock was
the truth in it. At that point, the house *was* a bit run-down. We preferred
to see the wear as a glorious patina of age, something almost Venetian in
its decay, but obviously there were other points of view.

In the early going of the family's years in Stockbridge, there were no
other points of view, none that couldn't be dismissed as envy or resentment. No one would have dared to size up the house the way my friend
did while the Judge was alive. It would have been astonishingly rude, and
also risky, for the Judge was not one to cross lightly. And probably incor-
rect. But whatever wealth and power he held has long since been dissipated, and the cultural attainments of the family—our writers and movie
stars—in more recent times have not been enough to hold the scoffers in
check.

The great wheel of history has turned. We who were high are now
lower. More exactly, we have shifted from active to passive, from being
full participants on the national stage to being mainly distant observers of

it, noting with dry, sometimes bitter irony the developments we are pow-
erless to affect.

And that is our burden, rather like the weight of the roof on that front
portico that Arthur has been tangling with, the weight combined with the
ice-widened floor that splays the columns that support it. It is not the
weight of years exactly, but of expectations. That portico should look
the way it always has. There is a standard to uphold. Yet with time it has
become ever harder to know what that standard is, why it might be so
important, and how to maintain it.

I feel this acutely, but I am not the first Sedgwick to experience the
anxieties that stem from an awareness of the familial past. It grew through
the nineteenth century to become inescapable by the dawning of the
twentieth. The precise moment when it first became manifest can, in fact,
be dated. It was late in the year 1881, and like so much else in the life of
the family, it found its expression in the house itself, more particularly in
the wing where I stayed that night after coming to commune with the
first circle of the dead.

To SET THE stage for the family's crisis of confidence in 1881, we need to
hurry through the intervening years. We last left the Old House with
Theodore Jr. When he died in 1839, he passed it to his son Theodore III,
the one he'd lectured as a boy about the value of regular exercise in ward-
ing off the madness that had come for his uncle Harry. Tousle-haired and
spindly, that Theodore became a lawyer like his father; contributed aboli-
tionist diatribes to the *New-York Evening Post;* defended the flamboyant
Bowery actor Edwin Forrest in a splashy divorce case; produced a voluminous *Treatise on the Measure of Damages* that became a standard text for
a full century of law students; served on the legal team, headed by John
Quincy Adams, that argued the Amistad case; and led a group of New
York eminences to erect an exact copy of London's Crystal Palace along
Sixth Avenue above Fortieth Street for a world's fair. Frantically busy in
New York, though, Theodore III used the Old House only as an occasional summer refuge. Through the winter he left it closed up, the windows boarded, with dustcloths tented over the furniture. It was the first
time the house had ever fallen into disuse for such long periods of time.

On his death in 1859, Theodore bequeathed the house to his only surviving son, Arthur George Sedgwick. Also a Harvard graduate, Arthur

was commissioned a first lieutenant in the Twentieth Massachusetts Regiment of the Union Army, only to be captured two months later by the Confederates in Deep Bottom, Virginia, and clapped into Richmond's Libby Prison, a hellhole that was second only to the gruesome Anderson Prison in Georgia as the deadliest in the Confederacy. (He kept a diary of his wasting-away on the back of receipts from his Boston tailor, the only paper he possessed.) Nearly skeletal when he was released after just three months, Arthur never shook the frailty, partly of body and partly of nerve, he acquired there. He, too, trained as a lawyer but took to journalism, first editing the *American Law Review* with Oliver Wendell Holmes, a law school classmate who'd been a fellow soldier in the Massachusetts Twentieth, and then writing liberal dispatches for the *Nation,* which had been founded by his brother-in-law Charles Eliot Norton, before briefly becoming managing editor of William Cullen Bryant's *Evening Post.* Arthur married the socialite Lucy Tuckerman in a high-society wedding in 1882, with Theodore Roosevelt among the grandees in attendance. Gradually, though, his ill health turned him, and, amplified by that dark seed of the family disease that heightened the anguish of any reverse, he ended up giving tortured speeches at Harvard on the perils of democracy. Completely embittered in his old age, he shot himself in a Pittsfield hotel room in 1915.

Back in 1875, though, when Arthur was still a somewhat kindly thirty-one-year-old bachelor, his first cousin Henry Dwight Sedgwick II—the son of the dear, deranged Harry who was my great-great-grandfather—was fifty-one, and raising a family of five in a small brick house down around the corner from the Sedgwick manse, not far from the station for the trains that Theodore II had brought to town. A cheerful soul known first as Hal and then, after his father's death, as Harry, or, for clarity, Harry II, he had married his first cousin, Robert's cautious daughter Henrietta, as one last manifestation of the extraordinary partnership of those inseparable brothers. They became co-fathers-in-law—and in the process doubled my own genetic links to that intense first family. Realizing the Sedgwick house suited Harry II far better than it did a bachelor like himself, Arthur sold it to his cousin for a dollar.

Harry II joined Harvard's class of 1843, becoming the first of my direct line to attend the college, and a New York City lawyer as well, occupying his father's and uncle's old law offices at 49 Wall Street. But, of a dreamy turn of mind, he fled the city on weekends to come by rail to Stockbridge, never failing to take off his hat the moment he arrived, the better to breathe in the sweet country air. He carried a bit of the bad seed as well— not outright lunacy, but more a kind of mild, fun-loving dottiness that was likely to exasperate the more practical-minded, who feared it might become hazardous. In the warm weather, he'd float on his back in the Housatonic for a half hour at a time (to the consternation of his wife, who feared he'd fallen asleep) and tramp over the countryside in daylong hikes that would often leave his children searching for him with lanterns at nightfall. In 1880, he happily quit the law for good to live out his days in Stockbridge with his five children, although the oldest of the sons—my grandfather Henry Dwight Sedgwick III—had gone off by now to Harvard, taking up his father's old rooms in Holworthy 14 in elm-shaded Harvard Yard.

And this is when the Sedgwicks began to succumb to other points of view, for other personages—far more illustrious, worldly, accomplished, *rich* ones—started to cast a shadow over a family that had always had to itself whatever there was of the Berkshire sun. At first, Harry II and his family were blithely oblivious, caught up in the many joys of their old ways. The change snuck up rather as the Shays' Rebellion did. This time, though, the Sedgwicks were being attacked not from below, but from above. For this was the Gilded Age, when unbridled tycoons fleeing the sweltering New York City heat made the Berkshires a summertime playground to rival Newport and Saratoga, complete with vast mansions, turreted castles, and darling châteaus, all on a scale so vast they made the once-commanding Old House look puny.

Ironically, the Sedgwicks themselves had drawn these rich to their midst. Back in the 1830s, it was to see Catharine that the British writer Harriet Martineau came to the Berkshires, putting the region on the map for Englishmen, and therefore for Americans, too. The two strolled along beside the "sweet Housatonic," while Martineau, an avid botanizer, hunted up marsh flowers. Catharine's literary stature also made the area enticing to other writers like Nathaniel Hawthorne, Herman Melville, and Henry Wadsworth Longfellow, and, as writers and artists can do, their cultural gloss transformed what had always been a starkly forbidding outback into a tony vacation spot. Indeed, the Berkshires began to be thought of as the American Lake District.* Catharine had also assisted her brother Charles's wife, Elizabeth, with the fashionable Young Ladies School she established in Lenox, which attracted, among other charges, the future actress Charlotte Cushman and Winston Churchill's mother, Jenny Jerome. And finally, Catharine also enticed her dear friend Fanny Kemble to build a house near Charles's in Lenox.

Kemble was a force to reckon with. Over a half century later, one of Catharine's grandnephews, Harry II's son Theodore, still quavered at the memory of the arrival of a certain highly elegant chaise called a victoria, with a liveried horseman and a pair of well-harnessed bays, in the drive of the Old House.

A lady stepped out, not tall but with an air of great importance. Her long black veil was thrown back from her face over her bonnet. She closed her sunshade parasol, which bent sideways to shield from the sun, and coming towards me said, "who are you little boy?" Frightened by the dominant personality, I told my name. She said, "You are an ugly boy. You look like your father. Go, tell him Fanny Kemble is here."

One of the first members of the moneyed crowd to come was the Boston financier Samuel Gray Ward, dubbed "good Sam Ward" to differentiate him from "bad Sam Ward," the notorious New York playboy. The good Sam appeared in Lenox as a gentleman farmer in 1844, and that summer produced a hundred bushels of potatoes while working on a translation of Goethe's autobiography. He left a few years later, but after he had become the American representative of Baring's, the British investment house, he returned in 1870 to build one of the first of the vast Berkshire estates, Oakswood, with a colossal mansion house designed by the society architect Charles Follen McKim. In 1893 the New York investor Anson Phelps Stokes bought the property and, employing four

---

*Even though his house would give the name to one of the later cultural attractions, Tanglewood, Hawthorne himself considered the region's notorious weather nothing but dreary. "This is a horrible, horrible, most hor-ri-ble climate; one knows not, for ten minutes together, whether he is too cool or too warm; but he is always one or the other, and the constant result is a miserable disturbance of the system. I detest it!"

hundred workmen, erected upon it the largest house in America, a hundred-room turreted castle that rose up over an entire acre. He called it Shadow Brook, for the rivulet on the property where Hawthorne had once gathered his children to tell them stories at the close of day. Andrew Carnegie eventually purchased the estate.

A ninety-four-room extravaganza called Elm Court followed in 1886, and then many more: Blantyre, Eastover, Clipston Grange, Wyndhurst, Orleton...like so many stops on an English train line. The names convey the Tory longing for a country estate that would not have been unfamiliar to the Van Schaacks, or, for that matter, to the Judge himself. But these were show houses, ones intended to assert their superiority in every particular, from the firmament of crystal chandeliers over their visitors' heads down to the swarm of liveried footmen at their elbows. They were statements not so much of politics (beyond a hankering for medieval feudalism) as of personal psychology, fantasies done in stone. Mrs. George Westinghouse, wife of the corporate titan, had grown up in a grubby mining town, and insisted that everything at her Erskine Park in Lenox be an incandescent white. And so it was: the rooms were dressed, walls and ceiling, in tufted white satin; the bridges over the artificial lakes were of white marble; the driveways of *crushed* white marble; Mrs. Westinghouse herself was invariably attired in bridal white; and, as the pièce de résistance, there was on permanent display under a glass globe on the living room table a stuffed fox terrier—white.

BY 1880, WHEN Harry II decided to live year-round in Stockbridge, Lenox's hills were studded with thirty-five of these vast cottages; by the turn of the century there would be seventy-five. And, of course, with the influx of American royalty, social life was transformed. A southern visitor, Mrs. Burton Harrison, used to stop at Lenox for a few months in the fall on her way back to New York from Bar Harbor. In 1911 she recalled the old Lenox ways, when a gay life consisted of tea parties followed by a rhyming game called "dumb crambo" and, on Saturday evenings, a dance. A courting couple would go for a romantic stroll by the ledge at Woolsey Woods, the gentleman invariably equipped with a volume of verse. On Sunday afternoons, the height of excitement was a walk to see "Mr. Goodman's cows." But then, as she writes, the countryside was bought up "at fabulous prices," and turned first into reasonable tasteful "modern villas" and then into ghastly "little palaces, some repeating the facades and gardens of royal dwellings abroad." She went on:

> Instead of the trim maidservants appearing in caps and aprons to open doors, one was confronted by lackeys in livery lounging in the halls. Caviar and *mouse aux truffes* supplanted muffins and waffles. Worth and Callot gowns, cut low and worn with abundant jewels, took the place of dainty muslins made by a little day dressmaker. Stables were filled with costly horses, farmyards with stock bearing pedigrees sometimes longer than the owners', the dinner hour moved on to eight o'clock, and lastly came house parties, 'weekends', the eternal honk and reek of the motor car.

But again, this was Lenox, and Lenox was not Stockbridge. Not at first. As Veblenian ostentation became stylish across the border, Stockbridge clung all the more resolutely to its country habits. To close the summer season in Lenox, there was a glittering processional of every manner of conveyance, from two-wheeled pony carts up through flower-bedecked chariots pulled by a half-dozen garlanded stallions. The Sedgwicks' Stockbridge, by contrast, offered only a chilly moonlit scramble through the Ice Glen, with all the villagers in homemade costumes and bearing torches. Stockbridge did have its Casino, designed by Stanford White and erected across the street from the Old House, but it was a place for nothing racier than tea parties and an occasional Saturday-evening dance.

But inevitably, these Lenox potentates crept over the border into Stockbridge, too. Joseph Hodges Choate was one of the first. A partner in the eminent New York firm of Choate, Butler and Southmayd, which handled the legal work for Standard Oil, Choate first summered in Lenox in 1874, and then moved just across the line into Stockbridge, where he built an Italianate villa on the ridge overlooking the Sedgwick graveyard. He was drawn there by the Sedgwicks, too, albeit indirectly. His law partner, Charles E. Butler, had won the hand of another of Robert's daughters, Susan Ridley Sedgwick, as his second wife. To please her, he'd built a grand summer house for the two of them in Susan's ancestral village along the Housatonic, and named it Linwood after the novel by Susan's aunt Catharine (not realizing that her portrait of the Tory Linwoods was not entirely affectionate). Butler, however, could be hateful toward his

Sedgwick relations, a cigar-chewing ogre. "Gruff, rich, self-satisfied," in one Sedgwick remembrance, he used to greet his timorous Sedgwick nephews and nieces when they arrived for dinner with their favorite aunt with a blast of ill humor. "Well, didja bring your dinner with you?" he'd roar at these Sedgwick mice. "You can eat it in the coal hole!" Nevertheless, Choate had followed him there.

MUCH OF THE day, Harry II closeted himself in that better parlor, now a book-lined study, behind a door marked "H. D. Sedgwick's Lair," to wile away the hours with the Sir Walter Scott novels he'd loved since childhood, and try to insulate himself from the potentates who were closing in around him. He'd tell the romantic tales on hikes or picnics, between pulls on a cigar.

The Sedgwicks tried to make sure their life in Stockbridge remained as it had. When she wasn't riding herd on her husband, Henrietta sat on the porch beside the house, singing out hellos to passersby along the street. Dinner was an endless feast, with claret and port for adults and children alike, and highlighted by squash pie from a recipe that Pamela had given to Mrs. Washington almost a century before. Among the many guests who rounded out the table was Booker T. Washington, the founder of Tuskegee Institute, who once came to dinner with Stockbridge's General Samuel Chapman Armstrong, the founder of the Institute for Negroes that Washington had attended. It was the first time a black man had ever sat for dinner, and the event confounded the Irish maid, Nellie. "Waiting on colored people!" she muttered, with a shake of the head. But the Sedgwicks themselves found the occasion unexceptional.

After dinner, the whole family lit kerosene lamps to pass the evening acting out the charades that were deemed outmoded across the border, and a diversion of their own called the Animal Game, which involved generating as many species as possible for each letter of the alphabet. Ichthyosaurus; ibex; iguanodon; ignumen fly; ibis; inchworm...Until finally, at midnight, Harry II would shovel over the embers to stifle the fire, leaving the only heat to come through isinglass windows from a back hall stove, and, bearing kerosene lamps to light their way, everyone would trundle off to bed.

Steeped in the family's reverence for the dead in the graveyard, the children extended it to their pet dogs. When Harry's oldest daughter

Jane's snappy white poodle, Zozo, died, she placed him in a box, covered it with myrtle and flowers, and solemnly bore the tiny coffin outdoors, with all the other children following, in order of seniority, leading up to Harry and Netta—as Henrietta was sometimes called—at the end, to a small grave that had been dug at the foot of the lawn, on the lip of the high bank overlooking the Housatonic. Later, a small white marble marker was raised over it bearing the inscription, "Zozo, 1875." So began Dog's Acre. The little cemetery attracted notice in Baedecker's Berkshires guide, which brought some of the first tourists to the house.

But this eccentricity came at a rising price. While they had once ruled the Berkshires as powerfully as the Williamses and Dwights did before them, the Sedgwicks soon found themselves on the receiving end of a subtle form of showing-up that was new to them, and quite unnerving. Ever since the Judge had arrived in town, he had set a certain imperial standard that, while frequently challenged by his political opponents, was never topped. Catharine, too, had always held her head high. But the prominent display of staggering wealth in their little corner of the world made the latest Sedgwicks feel smaller than they were used to, and disturbingly beside-the-point.

The first indication that the Sedgwicks might become the butt of a joke came with the arrival of the fast-talking fourteen-year-old Frank Crowninshield, who hailed from various European locales, and who started to tease the younger generation mercilessly about their country customs, which he found hilariously provincial. Crowninshield would go on to become the editor of *Vanity Fair* and, as a visiting Sedgwick cousin named Nathalie Sedgwick Colby later nervously recalled, "brought out the bumpkin in all of us." But it was the lordly Joseph Hodges Choate who had the most fun at the Sedgwicks' expense. An inexhaustible source of japes and jests of all kinds, he took in the Sedgwicks' peculiar circular graveyard from his ridge-top aerie and, with a hearty laugh, called it the "Sedgwick Pie," and then added the fillip that the plan had to be that on Judgment Day the Sedgwicks would rise up and see no one but fellow Sedgwicks. The resident Sedgwicks, of course, were not pleased to see their hallowed traditions become the object of such merriment, but the painful truth was that the notion was droll enough, and apt enough, that it was repeated endlessly by townsmen, visitors—and, before too long, by the Sedgwicks themselves.

As those enormous cottages crept ever nearer, Harry II betrayed a new anxiety about his own social position. He started to investigate his ancestry, writing up learned accounts of that first Sedgwick immigrant, Robert, and he took an interest in the efforts of a distant New York cousin, also named Robert, who had hired a genealogical firm to trace the Sedgwick lineage back several generations more on the English side, in search of connections to royalty. (The closest they could come was to find some Sedgwick royal brewers.) He joined with other Sedgwicks deriving an ancestral coat of arms, and he contributed one of the competing versions that had begun to circulate before the family settled on a lion astride a book with the motto "Confido in Domino," "With faith in the Lord," which was duly inscribed on Sedgwick cutlery.

Most revealing of all, Harry started to think the once-grand Sedgwick house was looking a little dowdy. He was the one to add the wing to house a larger household staff, and while he was at it, he replaced the Judge's original six-over-six windows with more stylish casement windows (which have leaked ever since) for the front parlors; laid down slim walnut floorboards in elegant patterns over the original broad pine boards; replaced the plainer ornaments around the doorways and along the ceilings with far more refined millwork; and expanded the house by pushing out the two back rooms a good ten feet into the meadow behind.

The stress became considerably more acute after the final bills for enlarging the house came due, for Harry discovered that he had no money to pay them. He'd taken the family to Europe while the work was being done, leaving his fortune in the care of a business agent in New York, who in the late spring of 1881 absconded with it, leaving Harry destitute. He was only fifty-seven, but, retired, he had lost whatever moneymaking instinct he possessed, making the Sedgwick fortunes ebb all the lower in comparison to the plutocrats in their baronets on every side. Harry turned frantic with worry, with shame. Like his father before him, he plotted harebrained schemes to restore his fortune, investing what little money he had left in a writing paper company that quickly went bankrupt, leaving his family with a vast supply of black-bordered stationery, waiting— townsmen chuckled—for a corpse. In his distress, Harry went on and on about his straitened circumstances to his youngest son, Ellery, then just entering his teens. "Bills, mortgages, notes of ninety days crept into my dreams," Ellery recalled later. "Of course I did not understand them, but

I felt the aura of horror that surrounds such things, and my sympathies could always be relied on." Money was so tight that Ellery's older brother, Harry III, had to withdraw from Harvard and press himself into service as a clerk at the small Boston law firm of Bangs and Wells.

Harry II put off his creditors for as long as he could. Terrified he would lose the Old House, he decided that the only solution was to borrow from a wealthy neighbor, Charles Butler's other partner, the stout, proper Charles F. Southmayd. On Sundays Southmayd could be seen strolling to the new Episcopal church, a gray beaver hat on his head, his stiff collar up in points, and a silver-topped cane in his hand. For years, Southmayd had pined for the young Rosalie Butler, his partner's daughter by his first marriage. It was for her that he had come to Stockbridge, and he had run a path between his house and her father's, hoping that one day she might come gaily tripping down along it to him. To lure her, he even made her a present of a brooch in the form of a cross set with pearls. Rosalie gladly wore the brooch, but she never did oblige him with a visit.

Still, it was to the lovelorn Mr. Southmayd that Harry II came. To bolster his courage, Harry brought young Ellery along in the rickety phaeton, drawn by the family's broken-down old pony, Conrad. Fifty years later, Ellery could still recall the scene.

> Papa's spirits were at an all-time low, so I held the reins while he bit his nails. The crunch of the gravel before the door is still in my ears, and I see my father with the desperate look of the debtor doomed, visible across his face. In a moment he had gone within and I was left to loop the reins about the whip and sit back shivering in the warm sun. Visions of the Old House slipping inexorably into the river were before my eyes. We were homeless; I had little doubt of it. Then suddenly the door was flung open and out came my father, his ruddy face glowing with happiness. There was no need for words, but as we trotted home, my father kept exclaiming: "What a kind man! What a good man!"

Southmayd had given Harry a check for $7,000 in exchange for some meadows the Sedgwicks owned on the far banks of the Housatonic that could not have been worth half that much. A godsend, and more: on Southmayd's death in 1912, he willed the land back to the family.

* * *

HARRY II's YOUNGEST son Alexander, known as Alick, was the owner then. Actually, Harry II had willed the house to all five of his children, but when that arrangement proved unmanageable, Alexander bought the other four out. Of the five, he was the only one interested in living in Stockbridge year-round. He'd been badly scalded as a child, and remained so frail he had scarcely been able to attend school as a boy, and then spent little more than a year at Bishop's College in remote Lennoxville, in southern Quebec. Still, he was devoted to Stockbridge, and represented the town for two years in the state legislature as a Democrat. In 1916 he geared up to serve in the army, but as he departed in his car for the train station, he turned back to wave good-bye to his family and smashed into the stone pillar marking the entrance to the drive, delaying his departure for several months. His wife, Lydia, stayed on in the house after his death two weeks before the great crash in October of 1929, and the Sedgwicks at last entered the modern world.

TWENTY-EIGHT

BABBO & ELLERY

While the rest of the family largely recovered from Harry II's financial plunge, his oldest son, Harry III—my grandfather, nicknamed Babbo—never would.

A shade under six feet, with a handsome, chiseled face, alert eyes, and lustrous hair, Babbo was an immensely attractive man, with a lively mind, a fluency in many languages (modern and ancient), and an abundance of charm that survived to the end of his days. My sister Lee, whom Babbo always called by her proper name of Emily, recalls seeing him in Stockbridge, where he customarily sat in his favorite chair, a dog at his feet and his two canes at his side, when she was a toddler. Whenever Lee trundled into the room, Babbo would break into a big smile, and sing out, "Happily! Merrily! Joyfully! Emily!" But there was always a subtle inward frailty that could be traced to that early sense of dispossession, one that was accentuated by the fact that he had not inherited the Old House, to which, as the eldest son, he might legitimately have felt exclusively entitled, since that had been the custom since Theodore II took possession of the house from Charles two generations back, and would be the pattern going forward. For most of his life, in fact, Babbo never owned a house at all, boarding in rooming houses in the hard years after college and, in his long widowerhood later, taking shelter with one surviving son or the other, following what he called his "vagabond life" as a writer of books.

There is a curious neutrality, neither entirely active nor entirely passive, to Babbo's life that makes it difficult to evoke, pleasurable as it seems to have been to live. Unlike the Judge, Babbo almost never imposed himself on the world. He was a being strangely without force. All the vigorous activity that makes the Judge's life a gift to biographers, as he won elections, freed a slave, wrote laws, defeated his opponents (until the end, of course, when his opponents defeated him)—none of it lived in Babbo, who for the most part merely observed life, and enjoyed it.

While he himself was not a force, he was not unconscious of the forces acting upon him—economic ones in particular. Born in 1861, Babbo had grown up with wealth, which made it all the more noticeable to lack it. He was just four when Lincoln was shot, and never forgot the shouting about the terrible news in the street. Yet his earliest memory concerned social rank. He was sitting bare-legged on the floor of the sewing room of the family's three-story house on Twenty-fourth Street in New York, peering up at the family seamstress, Susan Points, while she worked. Although he was literally looking *up,* he somehow sensed he should be looking *down*. For he had a feeling that she lived "in a little room somewhere, in a dingy boarding house, with nobody very near to love." What's more, he had become aware of the Sedgwicks' rather lordly family crest over the door to the pantry from the front hall. Sure there was "nothing of the like over Susan Points' door," he'd concluded that he belonged to a "class superior" to hers. However innocently arrived at, it was a chilling perception. It was, he said, his "first lesson in snobbery." It would not be the last.

By disposition, Babbo was normally too ethereal to pay much attention to such social distinctions. His mind usually went elsewhere. In a burst of prosperity when Babbo was eight or nine, his father moved the family to grander quarters on Forty-eighth Street near Fifth Avenue. Walking home from school one day shortly thereafter, distracted as always, Babbo didn't notice that a city crew had freshly tarred the street, and the soles of his boots were coated with thick black goo. Sticky black footprints followed him up the café-au-lait-colored steps of the newly acquired Sedgwick house, into the grandly tesselated white marble front hall, to the parlor—a "hideous room," he recalled, "formal, awkward, never occupied except by callers, and suitable to a period when decorators had lost their reason"—and on to the "pride of the room," a vast, extravagant, plush rose-colored rug, where, several steps in, he encountered his mother, who had by now turned a pair of saucer eyes toward him, utterly aghast.

It is typical of Babbo that, in his memoir, he distanced himself from the horror of the moment by turning to a bit of poetry to express her mortification, and thus his. He cited a scrap of Longfellow that he could no longer recall without wincing:

> And departing leave behind us,
> Footprints on the sands of time.

Babbo claims that this misadventure imparted upon him only one "fundamental social lesson," namely "the elementary principle of brushing off the tar upon your boots on the door mat." But, of course, he is being disingenuous. The real lesson lay in the look of mortification his mother directed toward him, that he could have been so oblivious to his surroundings, especially elegant, expensive ones like the interior of the Sedgwick town house. It was a tension that he felt for much of his life, as he could never be quite sure where he truly belonged.

When Babbo had to withdraw from Harvard after three years because of his father's business reverse, he tried to salvage his degree by petitioning directly to Harvard's President Eliot to take his senior examinations even though he had not yet done the necessary course work. "[Eliot] replied," Babbo recalled, "that I was asking a favour and that I was not entitled to any favours." It was, he says, a "hard blow," and a painful lesson in the cruelty of the world toward those of insufficient means. So, lacking any other options for gainful employment, Babbo passed three years of drudgery in that Bangs and Wells law office. By then, he'd read enough law to pass the bar, but to get anywhere as a lawyer he needed a proper law degree. Somehow he scraped together the tuition for a year of Harvard Law School, but he had to give up after a term when a case of measles nearly blinded him. This time, Stockbridge's sharp-tongued Joseph H. Choate rescued him, much as his partner Mr. Southmayd had rescued Babbo's father, by giving him a job at his illustrious New York firm—now, in the early 1880s, constituted as Evarts, Choate and Beaman—at 52 Wall Street.

Despite his truncated education, Babbo was a witty, erudite young man, always ready with a quote from Shakespeare or Horace, although he himself always claimed to have wasted his Harvard years playing football, leafing through back issues of *Punch,* and quaffing rum-and-ginger at his club. But his personal charm carried little weight in the hurly-burly of a New York law firm. A former secretary of state, William Evarts was a U.S. senator from New York whom Babbo instinctively disliked for luxuriating in his own power. He noted a "look of conscious superiority" that he later recognized in "Rigaud's portrait of Louis XIV hanging in the Louvre." Choate he rather liked for his peppery wit, and Babbo enjoyed seeing him take on the rogue ex-senator Roscoe Conkling in one court case, Choate's witticisms clouding Conkling "like a swarm of mosquitoes." But Babbo himself was relegated to the lowest orders of the city courts, where in his first contested case his star witness was a candy seller.

Under such circumstances, Babbo quickly grew to despise the law. "I was mentally and morally uncomfortable," he wrote later, "as if I were swimming in glue. I did not understand the law. It seemed to me to create most of the difficulties it professed to settle." More particularly, he objected to the fact that the law did so little to temper the misery he saw everywhere about him in a city as socially stratified as New York. Strolling in either direction from the social heights of Fifth Avenue, he could not help noticing that the neighborhoods descended steeply in quality of life. To the west, one soon encountered "cans, barrels, refuse in the street, enormous numbers of progeny," and to the east, the horror was existential. After Third Avenue, he wrote, "it was hard to understand why people preferred life—life, composite of bitter cold, of sweltering heat, of all kinds of dirt and all sorts of unpleasantness in every sense—why people preferred this to nothingness."

On only one occasion, however, did Babbo take it upon himself to use his legal talents to try to intervene. That came when, as a junior man in the firm, he was appointed a poll watcher in "one of the stronger Tammany districts" in lower Manhattan. It was the pivotal mayoral election of 1886, when the grubby Tammany Hall forces that had controlled the city since the days of Boss Tweed at midcentury turned to a wealthy Democrat, Abram Hewitt, as its candidate. Hewitt, however, was challenged on both sides: on the left by the Labor Party's Henry George, the brilliant, rampaging author of *Progress and Poverty,* and on the right by the hard-charging aristocrat Theodore Roosevelt, just back from a hunting trip to the Dakota Badlands. With George leading a battle cry against the high rents charged by "the heirs of some dead Dutchman" (a not-so-veiled reference to Roosevelt's ancestors) and other "leeches," the usual ethnic tensions in the immigrant-swollen city were giving rise to an out-and-out class war of the sort that the Judge would have found very familiar.

The issue came to a head on election day. Tammany had, as always, stocked the polling place with its own burly enforcers. Nevertheless, once the voting was over, Babbo was shocked to discover that the number of ballots stuffed into the ballot box exceeded the number of eligible voters. When Babbo pointed out this discrepancy, there were loud guffaws from the Tammany henchmen, who ceremoniously removed all the ballots from the box and then stacked them up with the George votes on top, the Roosevelt ones in the middle, and the Hewitt votes on the bottom. They then peeled off the "extra" votes from the top down, thereby only increasing the Hewitt ratio, to great hilarity from the Tammanys. Swelling with indignation, Babbo hauled out his law books and in his reedy Harvard accent started to read aloud from the relevant statutes. "It was freely suggested that I should go to hell," he recalled. The Tammany captains mocked him as "Mary Jane," among other less gentle epithets.

Hewitt was elected, needless to say. But Babbo pressed charges, and remarkably, he secured convictions. One Tammany boss was sentenced to a year's imprisonment, another was fined, and a third fled to Canada. It should have been a solid victory, a time to light up a fat cigar of his own, build a career on it. But unlike the earlier Sedgwick men, he realized that he was no crusader. He was too private, too self-contained, and too uncomfortable with power. He preferred to be an observer only, a recipient of impressions, preferably cheering ones. He'd become, as he put it, an "epicurean," one interested primarily in his own pleasure. "The social whole and I are disparate," he concluded. "It is *my* nerves that suffer or rejoice; it is I, I, I, that find life joyful or damnable, and I realized that… the welfare of the rest of the world meant nothing to me in comparison to my own happiness." Put so baldly, the statement is hardly commendable, but it is a mark of his sense of removal from the world that he found so little point in any broader engagement with it.

BABBO'S YOUNGER BROTHER Ellery, though, went another way. Thickset, blustery, he was the last exemplar of the Judge's confident muscularity. Despite the fact that, at heart, they were both serious literary men, Ellery and Babbo could hardly have been more different in personality. Eleven years younger, Ellery emerged unscarred by his father's near-bankruptcy, since the family fortunes had revived by the time he crossed into adulthood. He was one of the first graduates of Groton School, a Massachusetts boarding school just emerging as a bastion of masculine virtue, and had a full four years of Harvard besides. He bounded through life, unburdened by any of the self-awareness that so hampered his older brother. As it happened, both brothers published their autobiographies within a few years of each other in the 1940s. The titles reveal their divergent styles. Babbo called his *Memoirs of an Epicurean;* Ellery, focusing on his publishing career, *The Happy Profession.* Although their lives touched at many points, both men scarcely mention each other. One detail they skipped was that, after they were widowed, both brothers fell for the same Englishwoman, Marjorie Russell, whom Ellery had hired as a nanny for his children. Babbo had taken her on a long and delightful motor tour of the English countryside, which was the sort of thing Babbo did quite a lot of in those years. (He was quite the gallant, often squiring women—some of them married—about Europe for months at a time.) But Ellery won her. Marjorie later explained that, while Babbo was "the most charming man who ever lived," Ellery had "more solid virtues." Still, when Ellery announced their engagement at a Sedgwick gathering, one mystified cousin was heard to exclaim, "What? *Babbo's* Marjorie?"

While both brothers brimmed with literary dreams, Babbo had to suffer through those two woeful decades in the law first. Ellery charged at his own right away, securing the post of editor of a struggling New York magazine, *Frank Leslie's Popular Monthly,* a few years out of college. When Ellery swept in to the magazine's upper Broadway office to take over, he was startled to discover that the previous editor, Henry Tyrrell, had not been told that he'd been sacked. He remained steadfastly in his editor's chair, refusing to cede it to Ellery, and continued right on with his work.

What to do? Ellery simply drew up a chair across the desk from Tyrrell and started going through some of the editorial papers, too. Outraged, Tyrrell leapt to his feet. "If you want to be editor, try it!" he yelped, and then swept up all the papers on his desk and shoved them into the office safe. Then he added all the other business papers in the office—manuscripts, letters, address books, publishing schedules, everything. He slammed the safe door shut, locked it, pocketed the key, and marched out in triumph, tossing over his shoulder as he departed: "*Now* you can try your hand at getting out the magazine."

Now what? It's hard to picture Babbo puzzling that one out, but Ellery simply summoned the office boy and told him to go out and fetch a locksmith. Within an hour, he'd "opened the safe competently as any Jimmy Valentine," referring to the safecracker made famous by O. Henry. Ellery got out the issue, and many more, and by saving over half his eighty-dollar salary he eventually acquired sufficient capital to buy controlling stock in the publication, now renamed *Leslie's Monthly,* putting him in position to offer himself a raise.

Shortly after the turn of the century, the *Atlantic Monthly*—the vaunted literary magazine that had been started in 1857 by Emerson, Longfellow, and other eminent Bostonians—briefly considered hiring Babbo, who had contributed several high-toned articles on subjects like the place of the Catholic Church in society, to succeed Horace Scudder as editor. Babbo knew he was not suited to such a worldly enterprise. Instead, he became the temporary headmaster of the New York girls' school Brearley, a job for which he was possibly even less well suited, and where he remained only a year. (He later said that he charmed all the girls right away, the faculty eventually, and the board of trustees not at all.)

By 1908, Ellery had married his first wife, Mabel Cabot, of the Boston Cabot family. With her money and connections, Ellery was able to snap up the *Atlantic,* installing himself as editor for the next thirty years. And he was good at it. He gave the magazine remarkable stability, raising its circulation tenfold to 135,000 during a period that included World War I and ran through the Depression. Ellery acquired considerable respectability himself, as he bought an imposing brick house on Mount Vernon Street, near the top of Beacon Hill, and later a fine country place called Long Hill in the horse country of Beverly Farms on Massachusetts' North Shore. He established himself as a Boston autocrat and something of a national power, a quality that had eluded the family since the death of Theodore III over half a century before. In the pages of the magazine, Ellery championed Grover Cleveland, whom he'd first met as a Harvard undergraduate, and then Woodrow Wilson. In a charming bit of derring-do, he passed himself off as a member of the president's Secret Service retinue in order to sneak onto the floor of the House of Representatives to hear

Wilson address the nation on the eve of World War I. "Never was his voice sterner, stronger," he wrote of his hero's oration that evening. "The sentences marched like fate."

By instinct, though, Ellery remained a popular journalist, a glutton for scoops of all sorts. At *Leslie's Monthly,* he'd gleefully run an entirely spurious story about a race of pygmy horses in Kentucky. While he tried for a more elevated tone for the *Atlantic,* he could not resist the extraordinary tale of the imaginative Opal Whiteley, who had come to him at twenty-two as a writer of nature stories for children. He wasn't interested in the stories, but he found her enchanting—"very young and eager and fluttering, like a bird in a thicket." When he asked if, by any chance, she had kept a diary, she said, Yes, she had, starting at age six, but it had been torn up by a jealous sister. She'd saved the pieces in an enormous hatbox, though. When Ellery expressed interest, she sent for them by telegraph. Back came thousands upon thousands of scraps of paper, some as small as the tip of her finger. Ellery installed her at his Cabot mother-in-law's house in Brookline, and for the next eight months, Whiteley painstakingly pieced them all together. When fully assembled and transcribed, the finished manuscript ran 250,000 words and gave a fantastical account of her childhood in a series of Oregon logging camps, where she'd ended up, she claimed, as the kidnapped daughter of a French prince who'd died in India.

Ellery printed long excerpts of the reconstructed journal in the *Atlantic,* where they created an international sensation—and widespread doubt about their veracity. In his memoir, Ellery insisted that the young Opal had most likely been nothing more than the possessor of an unusually hyperactive imagination, one that too easily absorbed the details and impressions of other lives into her own, and then lost the distinction between the truth and her fabrications. In his credulity, Ellery demonstrates something similar, an innocence about the source and nature of duplicity. Such innocence is probably a hallmark of a generation that had no use for Freud, but it also reflects a certain willful ignorance about his own family history. Like Babbo, Ellery refused to see anything but charm in his father's aberrant behavior, and pointedly has nothing to say of the dark legacy of the grandfather for whom both their father and Babbo were named. Of that first Harry, Ellery reports only that he, like his brother Robert, was one of the most "upright men" ever to tread the "devious paths of the law." Such an assertion demonstrates the limitations of the seemingly carefree charm on which both Ellery and Babbo thrived, and the denial it represents provided an escape from a truth that would otherwise have threatened them.

A cultural conservative who was put off by modernity, Ellery's literary tastes were still shaped by the Sir Walter Scott romances his father read and recited on those Stockbridge tramps. He was especially disdainful—often defensively so—of the modernist writing coming into vogue. He put off Gertrude Stein for years before finally agreeing to run a portion of *The Autobiography of Alice B. Toklas* in 1933, near the end of his tenure. In 1919, he'd rejected a batch of her poems as too much of a "puzzle picture" for his readers. "All who have not the key must find them baffling," he lectured her, "and—alack; that key is known to very, very few." She retaliated in her letters to him by occasionally addressing him as "Ellen."

Likewise, the poet Amy Lowell's sparkling "Amygist" free verse gave Ellery fits. He recognized that, in the war-torn new century, he needed to appropriate a little of the newness of Harriet Monroe's *Poetry* magazine for the *Atlantic,* and as a Boston Lowell, Amy Lowell was his sort. (She had been introduced to him by his Boston Cabot wife, who'd been a friend of Lowell's since childhood.) But why couldn't Lowell write poems that rhymed and scanned? Starting in 1910, he accepted a few poems but rejected many more, often with exasperation. "This is vivid," he wrote her, sending back two more of her verses in October of 1914, "but is it...poetry?" Lowell fired back: "O Ellery, You are a dear, kind, non-understanding thing." She tried to educate him, and his readers, with an essay on Imagism, but he threw that back at her, too. Furious, she assailed him at home by telephone; he snapped that he found her remarks "censorious and discourteous in a high degree." In retaliation, Lowell shot off a poem, "Fireworks," to express the rage that had exploded between them.

> You hate me and I hate you
> And we are so polite, we two!
>
> But whenever I see you I burst apart
> And scatter the sky with my blazing heart,
> It spits and sparkles in stars and balls
> Buds into roses, and flares and falls...

And when you meet me, you rend asunder
And go up in a flaming wonder
Of saffron cubes, and crimson moons,
And wheels all amaranths and maroons....

Personally affronted as he was by their set-to, Ellery could not, professionally, help but be attracted to the poem's colorful pyrotechnics, which were not only gorgeous but for once rhymed and scanned, and had the added advantage of transforming a difficult interpersonal exchange into art. The editor in him won out. He replied to "Miss Lowell" in just one crisp sentence: "I am glad to accept 'Fireworks' which seems to me original and effective." He printed the poem a few issues later. With that, an uneasy standoff was reached.

TWENTY-NINE

MARRYING UP

Just as Ellery depended on his Cabot wife to buy the *Atlantic,* Babbo depended on his wife to fund the transition to a writing career. Until she came along, he lived in a boardinghouse on Twenty-second Street, eating at a dinner table for six, two of them faded French ladies, another pair a retired Presbyterian minister and his daughter. "As a place to live in, a boarding-house is odious, sad, disconsolate, demoralizing," he admitted. But family connections led to invitations to better, more fashionable places, such as the home of Mrs. Francis George Shaw, who lived in a wooden town house on Thirtieth Street that shared a wall with another one owned by her widowed daughter, Josephine Shaw Lowell, and Josephine's daughter. Josephine's husband had been the heroic "Beau Sabreur" Charles Russell Lowell, who had been killed at Cedar Creek during the Civil War. Babbo used to visit Mrs. Shaw of a Sunday evening in the summer months, and as they sat together on the little veranda overlooking the garden she shared with her daughter, she would pour out tales of her only son, Colonel Robert Gould Shaw, who had also died in the Civil War. Shaw had famously led a black regiment, the Fifty-fourth Massachusetts, against the South in a fatal charge on the heavily fortified redoubts of Fort Wagner in South Carolina. The family lived on Staten Island then, and Mrs. Shaw used to get the news of the war from her husband when he came home. "One evening," Babbo reported, "he walked a little less steadily up the graveled path, without a word, holding up his newspaper, and she knew that her son was dead."

Mrs. Shaw had another daughter, Susanna, who'd married the New York shipping merchant Robert Bowne Minturn. One of the most elevated members of New York society, he'd been an early organizer of Central Park, a founder of the Union League, and, with a partner, the owner of a sleek clipper, the *Flying Cloud,* which, on its maiden voyage west in the Gold Rush year of 1851, sailed around Cape Horn to San Francisco in eighty-nine days, breaking the record by a full week and establishing a mark that would stand for over a century.

Those Minturns, in turn, had four enchanting daughters, of whom the oldest was Sarah May Minturn, called May, a pale beauty who might have been carved from Italian marble, and although he breathes not a word of this in his memoir, Babbo was desperately in love with her. Robert Bowne Minturn had died by then, though, and the family had suffered a serious financial reverse. By a quirk, a business partner had done to May's father almost exactly what the unscrupulous business agent had done to Babbo's. Taking charge of some Minturn sugar plantations in Cuba, the partner had made off with three-quarters of a million dollars, a prodigious sum. This occurred just as May was coming of age; instead of coming out in satin and pearls, she came out in cotton. Young and idealistic, she claimed to be unbothered, but she did become a tyrant on the subject of the family's finances. Happily, by the time Babbo came along, the fortunes were still sufficient to support a kind of family compound for the Minturns of four adjoining brownstones on Twenty-third Street, near Gramercy Park, and a glorious summer house in Murray Bay, along Quebec's Saint Lawrence River, with a broad porch out front, and lawns leading down to the frigid water.

Respectable, secure, beautiful, elegant, May represented everything Babbo craved. But now that her father was dead, she needed a far stronger man in her life than a modestly successful lawyer, no matter how charming, and she wouldn't have him. Babbo proposed to her on bended knee, with heaps of poetry. But she *still* would not have him. The rejection drove Babbo to distraction, summoning up that feeling of dispossession that, in him, was always present, but buried. He went to a gun shop, bought a pistol and some bullets, and then returned to his miserable boardinghouse, where he wrote out farewell letters. Bending to the grim task of ending his existence, he tried to push the bullets into the chambers of the gun. But they would not fit. He never knew if the clerk in the gun shop had guessed his intentions, or if he had simply blundered on his own. Either way, he came to his senses and gave up his plan to murder himself. Instead, he redoubled his efforts at courtship. And one morning his treasured May was sitting at her mirror, brushing her hair in her bedroom of the Twenty-third Street brownstone when, through the open window, she heard Babbo come up the front steps. He was whistling, and that made all the difference. "I can't disappoint him," she murmured to herself. "I just can't." She placed the hairbrush on her bureau, went downstairs to open the door for him, and let him into her heart.

THEY MARRIED IN 1895, and set sail for Europe three years later. Babbo had risen by now to assistant district attorney for the Southern District of New York, but he cheerfully abandoned the post for four months of holiday. Europe was his liberation. The beauty of it, the tradition, the light, the history—everything about it touched him, enchanted him. Just to be aboard the steamer was a joy, and then, when the ship finally docked in Italy, he was in complete raptures. "From what I have read, and heard say, other people have fallen in love with Italy, but never any, I believe, with so full a soul as I." He quotes Browning: "Did my heart love till now?"

The statement reveals a startling truth about his marriage—that there was a secret inner love within the outer, apparent one. He did not love May only for May, but for the chance she brought him to escape from a dull existence into a beautiful one. It took the form of Europe—Italy especially, but also France and Spain. But it really was just Away. Still, it was to Italy that Babbo would return, over and over. As he wrote:

I was dazzled, like an Aladdin in the magician's cave. The street cries— *Aranci! Aranci!* (Oranges! Oranges!)—were purest music, Verdi Bellini, Cimarosa, Pergolesi, and I know not what. The dirty little ragazzi, the tatterdemalions in all attitudes of recumbency, the kerchiefed women, the black eyes of the girls, the donkey carts, the jaunty little carriages with their light-footed little nags, the poor old women, the poor old horses, all changed one glorious chorus, "Italia, O Italia, thou that hast the fatal gift of beauty." *"Ecco Signori!"* We had reached our hotel. I trod on air, I was in a delirium.

This was Naples. From there, he and May went on to Rome, Pompeii, and Verona. They traveled on the cheap. Babbo was too genteel to write directly of money; he left financial affairs largely to May, who was exacting about them. When she discovered their bank had misquoted the exchange rate on the small sum of money they'd withdrawn on a Brown Brothers letter of credit at a bank in Verona, she insisted on taking up the matter with the president of the bank personally. Told the president was engaged, May waited several hours for him and then, prowling about the halls, discovered him idly smoking a cigar as he read the paper in a back office. There she assailed him with all the Italian she knew. "Cicero, Burke, Wendell Phillips could not be more eloquent," Babbo reported admiringly. The president coughed up the difference, which amounted to about $1.50.

Then Florence, and finally Venice. It was there, toward the end of this trip later that spring, that in a hotel room overlooking the Grand Canal, May conceived my father. It was also there that she gave her husband the name Babbo, Italian for "father," reflecting his new parental status, and his devotion to Italy.

BESIDES PROVIDING FOR European travel, May's money also allowed Babbo to become a full-time writer shortly after their return. He wrote largely for his own pleasure, but he wrote prodigiously, thirty-eight volumes in all, nearly all of them plump, airy editions printed on yellowing paper that now feels slightly spongy to the touch. Published mostly by Bobbs-Merrill, they sold a few thousand copies apiece. For the most part, Babbo favored biographies of lively, garrulous, odd-lot Europeans— Alfred de Musset, the Black Prince, Lafayette, Madame de Récamier— whom he selected as members of a kind of private club. But there are also extended essays, many of which appeared first in the *Atlantic,* that reveal the underlying idiosyncratic attitudes. *In Defense of Old Maids,* "Pro Monastica Sua," "The Art of Happiness," each a tight little volume, like a prayer book, that feels wonderfully snug in the hand.

I often take them down from the long shelf where I have many of his books, just to hold them and leaf through the pages. As with so much of the Sedgwick literary output, I have read few of the books through, but I have always enjoyed the fact of their existence. Located, as he was, at the exact balance point between the heroically tradition-minded and the hopelessly fuddy-duddy, Babbo's essays in particular have a museum quality, an exhibit that is perhaps best placed under glass. Still, "In Praise of Gentlemen" remains, for me, a curio favorite, with its many darling, evocative sentences like this one:

> The old order was aware how large a part casual relations play in social life, how many a little makes a mickle, how even momentary meetings with friends and acquaintances, a lifting of the hat, a smile, a wave of the hand, a few steps out of one's way, a batch of conventional good wishes, an outward solicitude for one's health or prosperity, drop sufficient pleasantness into the fluctuating scales of ordinary existence to make the day a happy one.

I like this because I like *him,* and this captures him, perhaps more even than his autobiography. As a writer, Babbo became a pathfinder for me, just as Catharine, whom he knew, had been for him. I would not otherwise have imagined a writing career as a legitimate occupation, and in truth I have always identified with him, far more even than with my own father. It wasn't just becoming a writer; it was thriving on the gregarious solitude that is the writer's schizoid life. Babbo was always the one to call for champagne in company, yet he could be quietly contemplative in the long stretches of time spent alone. While the whole notion of epicureanism seems to me somewhat extreme, he did make the most of his existence, by engaging in whatever conviviality came his way in his abundant social life, but also by pondering the deeper aspects of life as one can only do in solitude. He once closed a letter to my half brother Harry by summing up his philosophy, "Squeeze hard the flask of life, and drink it to the dregs."

In marrying a Minturn, just as his brother had married a Cabot, Babbo was also setting a pattern that would be repeated by his sons, one that explains one of the essential conundrums of the modern Sedgwick family. Given the detachment, the ineffectuality, the otherworldliness of the Sedgwicks as they shifted from the active stance of that first generation to a more passive one, how were they able to maintain their social position?

The answer is, by marrying up. My father did this, twice. And so did his brother Francis.

The pattern had actually been set long before, in that marriage of Theodore and Pamela. But Theodore had provided a certain dynamism in exchange for Pamela's social prominence. In marrying Babbo, May gained no economic advantage, and this would prove decisive. Strip away their cultural gloss, and the Sedgwicks were a comedown for her. When other tragedies struck, as they shortly would, she took her fall hard.

For their part, the Sedgwicks quickly became very proud of their Minturn legacy, not that Babbo himself ever celebrated it, and most Sedgwick houses have a picture of the *Flying Cloud* on one wall or another. My father placed a reproduction in the living room at our summer house on Cape Cod. We also named our tubby Beetlecat sailboat after her.

To Babbo, though, May's very material gifts allowed him to retreat into a dreamscape. A photograph shows the two of them in a canoe on the still waters of the Saint Lawrence, Babbo in the stern under his ever-present fedora, May under a bonnet up in the bow, as they disappear into a fine mist. They might be in a Winslow Homer watercolor, and not in the world at all.

Judge Theodore Sedgwick, as painted by Gilbert Stuart in 1808, the year after Pamela's death.

Pamela Dwight, in an engraving done when she was in her late teens, before her marriage to Theodore.

Pamela Dwight Sedgwick and Catharine Maria Sedgwick, done by an itinerant portrait-painter in 1795 when Catharine was six; the elegant Mission House of Pamela's parents is barely visible in the upper right.

Catharine Maria Sedgwick, by Charles Cromwell Ingham, at about forty-five, when she had fully emerged as a celebrated American writer.

Catharine's "favorite brother," Robert Sedgwick, in his early thirties.

Catharine in silhouette in 1842 shortly after the death of her sister Frances.

Mumbet, as painted by Theodore Sedgwick II's wife Susan in 1811, four years after Mumbet left the service of the Sedgwicks because she could not abide Theodore's new wife, Penelope Russell.

Mumbet's gravestone, with its inscription by Catharine Maria Sedgwick, in the Sedgwick Pie.

ELIZABETH FREEMAN
Known by the name of
MUMBET
died Dec. 28 1829.
Her supposed age
was 85 Years.
She was born a slave and
remained a slave for nearly
thirty years. She could neither read nor write yet in
her own sphere she had no
superior nor equal. She neither wasted time, nor property
She never violated a trust, nor

Halla, Babbo, May, with the young Francis, and Minturn at Murray Bay in 1913, the summer before Halla died. Minturn was fourteen.

Reverend Theodore IV, Babbo, and Ellery before the fireplace in the Old House.

Minturn in a charcoal sketch by the Boston artist Polly Thayer Starr.

A newspaper photo of Emily Lincoln shortly before she met Minturn Sedgwick.

John Sedgwick at seventeen when he was at Groton School.

Edie Sedgwick with Andy Warhol in a publicity shot.

THIRTY

GOOD NIGHT, SWEET PRINCE

When Babbo and May went to Italy, and my father was conceived there in Venice, they had one infant at home already—yet another Henry Dwight Sedgwick, numeral IV, called Halla, a dear boy whose voice was, in Babbo's adoring account, a "sweet, little, woodland treble." They'd left him in the care of May's mother, Robert Gould Shaw's sister Susanna, whom they called Madre. By this time, poor Madre had been much shrunken by grief—three of her seven children had died in childhood, leaving her with the four daughters. Even more than the Sedgwicks, the Shaws were afflicted by their own line of manic depression, one that had been amplified by various intermarriages with another leading Boston family, the Parkmans, who were bearers of the disease. An ancestor, Dr. George Parkman, had been murdered by a Harvard professor of anatomy named Dr. George Webster in a sensational society crime in 1849, after Parkman, whose grim obsessiveness may have revealed his own bipolar characteristics, had hounded Webster mercilessly about a debt. The dead Parkman's nephew, the historian Francis Parkman, suffered so acutely from his manic depression—which he termed "The Enemy"—that often-times he could write only with his eyes shut, so explosive were any visual impressions on his brain, requiring him to rely on a frame of tautly stretched wires to guide his pen. Most likely Susanna was a bearer of the illness, too, for she took her grief terribly, terribly hard. A photograph from this time shows her on the porch of the house in Murray Bay, alone in a rocking chair in her widow's weeds, her white hair like a ghostly flame swirling around her skull.

This precarious soul was Halla's sole caretaker while his parents were away.

Halla was light and ethereal, like his father. His brother Minturn, who arrived at the end of January 1899, was far more earthbound. Supposedly, the physician who delivered him took one look at him and declared he didn't need a nurse, but a trainer. A daughter followed later the following year, in 1901. Named Edith, she lived only a day. In Stockbridge, her two brothers picked flowers for her grave in the rain. In his memoirs, Babbo mentions this tragedy only indirectly. A few years later he was on a train to New York, when he struck up a conversation with a clerk who told him that he and his wife had recently succeeded in adopting a baby. The clerk

> was leaning over towards me like a burning bush; I was leaning towards him, also all ablaze. I was thinking of a little girl—
>
> > Et rose
> > Elle a vecu ce que vivent les roses
> > L'espace d'un jour—
> > > (A rose,
> > She lived the time that roses live,
> > The period of a day—)
>
> and how for a time my wife and I had longed to adopt a little girl in her place, and the fire of human sympathy burned bright in me. Had not the table, plates, dishes, tumblers and carafe intervened, our arms would have been round each other's necks.

After Edith died, May's mood darkened terribly, just as her mother's had after her losses, and possibly for the same reason. May took to her bed, but was still fiercely vigilant regarding her surviving children, rousing herself to remarkable severity if her boys misbehaved. She washed

Minturn's mouth out with soap after he called Halla "a smelly skunk" one time, even though he pleaded with his nurse, Mary, to persuade his mother to spare him. "I'll die, Mary, if I have my mouth washed out again," Minturn cried out piteously. "I shall die!" Still, May did not relent. Another time, she smacked Halla on the side of the face with the bristle side of her hairbrush so hard, it left an angry mark that was still visible when the family gathered for Sunday lunch at her mother's a few days later.

May was especially hard on her youngest son, Francis, when he finally came along. He was named for her own brother, who'd died of diphtheria at age six, and Francis, too, was a sickly child. He was born with an umbilical hernia, essentially a hole in the abdominal muscles, which left him screaming with pain, and he was still weak at age two. When Halla also fell sick with pneumonia that year, May insisted on taking the whole family west to Santa Barbara, California, in hopes that the sunshine and dry air would do everybody good. Babbo felt marooned out there in a land that lacked all the vestiges of civilization—museums, cafés, opera, street life—he'd grown used to from his European travels, but Francis was restored, physically at least. He became deeply attached to the place, and when his health was threatened again years later, it was to Santa Barbara that he would return.

When Halla turned eleven, he and Minturn were dispatched to an innovative London boarding school called Bedale's, while Babbo took May and the young Francis with him to Rome, where he was researching a history of Italy. Halla had turned into a bookish, dreamy boy by then, very much his father's son, and it was a hard parting on both sides. When Babbo first told Halla his plans, the boy tugged furiously at his hair, but said nothing. After his parents visited that fall, he started to bawl again when it was time for them to leave. Two years later, he was shifted to Groton, where his uncle Ellery had gone.

Groton School was a forge of manliness in that post-Victorian era, hardening the sons of the rising industrialist class. Founded by Endicott Peabody in 1884 when he was just twenty-seven, Groton had been built on the rigorous English public school principles of the high-minded Thomas Arnold of Rugby, which Peabody had experienced secondhand at the Cheltenham School, northwest of London.

Peabody, known as the Rector, was the chief exemplar of the muscular

Christianity he espoused. His previous ministry had been in Tombstone, Arizona, where he arrived shortly after the Earp brothers' famous shoot-out at the OK Corral; bank tellers routinely kept a gun trained on their customers as they doled out any cash. He stayed long enough to build the first Protestant church in the Arizona territories, and to organize a base-ball team, the two accomplishments summing up his notions of much of what civilization should entail. At Groton, Peabody used to stand on the Circle, the central lawn of the campus, and lead the entire school in calis-thenics every morning, vigorously hoisting dumbbells in each hand, as the supramanly paterfamilias. He had any boys suspected of indulging in masturbation—which he primly termed "sentimentality"—marched out of study hall and then nearly drowned under an open spigot. John Singer Sargent caught the Old Testament power in a portrait later. In it Peabody rises up, arms crossed amid a swirl of billowing, black robes, to the top of the immense frame, where an imperious head stares down, as if from the clouds, with the commanding, disdainful eyes that pierced schoolboys and presidents alike.

Despite its strain of intellectualism, Groton did not take kindly to aes-thetes, and it was hard going for Halla, who was always happiest reading a book under a tree. Still, he devoured volume after volume: the *Rubaiyat* of Omar Khayyam (which he memorized), *The Eumenides, Paradise Lost*...Unathletic and retiring, Halla was not popular, but he could be talkative when drawn out. He'd bend his head forward, speaking quickly, on his face a "little social, deprecatory smile" his father always noticed, as he twisted his fingers about his jacket hem. Halla lacerated himself for being so sensitive to the other boys' persistent teasing about his airy ways, and vowed to take their "rather cutting jokes in the right spirit." Never-theless, he was several times "fired out" for talking back to a master he found imbecilic, which landed him hours of detention. And, most dar-ingly, he had a "tremendous row" with the Rector himself in midwinter over Peabody's insistence on throwing open all the windows to let in the frigid air he considered clarifying.

It did not make things any easier for Halla when Minturn arrived two years later. Tall, strapping, athletic, uncomplicated, he was greeted by the Rector like a son. The two brothers could not have been more different. When they saw a production of *Hamlet* in New York, Halla wept through the tragic final scene, while Minturn watched stone-faced, bored. Because of their English accents, both boys were called Duke. To Halla, the word was teasing—"Dook." To Minturn, though, it was a quasi-royal title he was proud to carry for the rest of his life. Although Minturn was three years younger, Halla began to think of him as the older brother, feeling an "increasing tendency," Halla confided to his father, "to lean on [Minturn's] strength, and noble courage." A photograph of the family taken at Murray Bay the summer of 1914 shows the eighteen-year-old Halla looking nervously at the camera in ill-fitting clothes. Little Francis, just nine, is proud in his sailor suit, their mother, May, in white, clinging to him from behind. Babbo is dapper as a dancing master. At fourteen, Minturn, already a lofty six-four, gazes almost protectively at the whole brood.

Even as a third-former, or freshman, Minturn became a force on the varsity football team, and that winter he took on the school's champion boxer, a powerful senior named Francis Lothrop, while Halla and the rest of the school watched, awestruck. Minturn dispatched Lothrop with a vicious uppercut that struck him so hard, the first part to touch down on the canvas as he tumbled backward was the back of his head. Halla had never seen anything so impressive. "I'd barter my chances of everlasting salvation for the ability to box and play football and row [like Minturn]," he told his father.

Late that spring, just a few weeks before graduation, Halla and Minturn rowed together one morning with some of the other boys on the river that cuts through the campus. Afterward they all raced back up the mile-long path for lunch. As always, Minturn won. Halla finished third despite the "considerable incline and blooming hot sun," as he told his parents. But the next morning, he felt slightly ill. At first, he thought it was just indigestion. Ravenous from all the exercise, he'd bolted his food the previous day. But the infirmary nurse discovered a mild temperature. When it rose through the day, Halla slept in the infirmary that night as a precaution. A doctor looked in on him in the morning, discovering a mild case of pneumonia, but nothing worrisome. The school sent a reassuring telegram to his parents in New York. Still, May took the first train from New York.

Halla was delirious when she arrived. She summoned Babbo, and the two stayed by their son's bed while his fever raged and raged. His face glistening with sweat, his hair mussed, Halla tossed on the soaked sheets and murmured words his parents strained to interpret.

"Are you comfortable?" the nurse asked Halla at one point.

"Yes, thank you. But so dull."

The fever continued to burn. Minturn did his best to reassure his mother. He took her in his arms, kissed her, and declared, "We'll *win*." He stroked his brother's hand and kissed it, too. Increasingly fearful, Babbo and May summoned Ellery from Boston, and the Rector kept a vigil by Halla's bedside, too. Days passed, and fevered, anxious nights. Babbo relates the end with a cool, Pre-Raphaelite beauty:

> That night we gathered about his bed, Minturn on one side kneeling and holding his hand, their heads near together, May and I on the other side, Ellery and Mr. Peabody at the foot. Mr. Peabody read from the prayer book and we repeated the daily prayers we had always said with the boys. It was a bitter cold night, and the windows were wide open to give Harry air. Toward morning a bird sang on the little tree close beside the window, and then, as the day was dawning, his spirit left us.
>
> He looked very handsome as he lay there in his white linen, with sprigs of many coloured snapdragon about him.... The coffin was covered with a deep red pall, and lay in the chapel. A dim light was burning as Minturn and I went in to bid good night. Ellery was there. The chapel looked solemn and beautiful, full of traditional feeling, and of Harry's sentiment for it. Horatio's words burned themselves into me:

> Good night, Sweet Prince;
> And flights of Angels sing thee to the rest.

Wracked by grief, May cried convulsively, and Babbo was so distressed, blood oozed out of the pores of his nose.

May wrote to her mother:

> *Darling, darling, Mamma—All that last night, I thought of you so often, and after the end, all the first day, I kept saying, "Poor, poor Mamma,"—I don't mean because of your loss of our darling boy, but because of a new understanding of your loss of Frankie. Every year, I felt more and more I understood but it is not until our own hearts are*

pierced that we can begin to know the suffering.....All Saturday, I
thought of you and I often said to Harry, "Poor, poor Mamma, I don't
wonder she nearly went crazy."

Before the year was out, May, at fifty-one, was nearly blinded by the
strain, and she suffered a series of strokes that crippled her. In 1916, President Wilson brought America into the war. Minturn, the oldest now,
went on to Harvard, and spent the summer after his sophomore year in
Gainesville, Florida, drilling to be a soldier. He returned north to see his
ailing mother at a rented house in Greenwich. He scarcely recognized
her. She had aged terribly, and her face had reddened from the blood
thinners she had taken for her strokes. Babbo sat beside her, inconsolable.
"How I wish you had known your mother when she was young and
beautiful," he told Minturn. May could not stop crying.

Minturn tore himself away to return to Florida, but shortly after he arrived, he received a cable from Babbo that May was dying. Minturn hurried back home, and his mother expired almost the moment he came
inside the door. In death, the blood in her face drained away, leaving her
skin a beautiful ghostly white.

For the burial service in Stockbridge, Babbo carried May's ashes to the
burial ground in a student's green book bag.

In *Memoirs of an Epicurean,* Babbo describes Halla's death only briefly
before quoting Robert Louis Stevenson—"O stricken hearts, remember,
O remember/ How of human days he lived the better part." He recites the
bleak chronology that followed. "Then came the Great War. And in January, 1919, my wife died."

AFTERWARD, BABBO LIVED with Minturn in an apartment in Harvard
Square. If he wasn't traveling in Europe, he continued to live with one
son or the other for the next forty years, passing most of his winters in the
depths of Harvard's Widener Library. He always wore knickers, which
he insisted on calling "knickerbockers," and, along with his tweed jacket,
a sprightly necktie, which he termed a "cravat."

He never spoke of his wife again; it was almost as if she had never existed. In his memoirs, though, he writes of a wine-soaked trip he took to
Europe in 1924, five years after his wife's death, with his great friend the
writer Owen Wister, who'd been his classmate at Harvard. Together they

toured Spain—the Toledo of El Greco, the Segovia of the "King Maker" Alvaro de Luna—for a book on the country Babbo was writing. But after a few weeks, Wister needed to return to America, leaving Babbo to continue on to France alone. He climbed high into the French Pyrenees to find a little town called Luchon by the headwaters of the Garonne River that reminded him of Murray Bay, where the Minturns had that summer house that May had purposely not left Babbo in her will, not trusting him to manage the property. She had given it to Minturn instead. It must have reminded him of being passed over for the Old House, too. Still, Babbo packed a lunch, along with a notepad and pencil, in his knapsack, and a volume of the French poet Pierre de Ronsard, and followed a slender, trickling brook called the Lys up into the mountains. In his memoir, he imagined the brook a woman. Was it May? Or was it the freedom, the liberation, she had represented? He did not expect her to "take me...to her white breast," but the water seduced him all the same. "She was lovely, sometimes dancing and clapping her hands down precipitous descents, or gliding slow through a poppied field, and her waters were as crystal clear as the eyes of the Blessed Damosel." When he'd climbed high enough, he lay down on the grass beside the brook, had his lunch, made a note or two, read a few poems, and then lay back and closed his eyes. He half expected the finger of a "beautiful princess" pressing gently on his lips, and a voice telling him not to stir, not to stir forever. But much as he longed for it, he felt nothing, and heard no one at all.

THIRTY-ONE

THE FAMILY PATRIARCH

As he had been at Groton, Minturn proved to be a tremendous athlete at Harvard, and Babbo gloried in his football triumphs, crowding into the locker room to watch as his son regaled the press with his accounts of the game. Minturn played left tackle on both defense and offense, and he was a mainstay of the storied, undefeated Harvard team of 1919 that went on to play Oregon State in the Rose Bowl in Pasadena. In those years before professional football, the Rose Bowl attracted some of the national fascination that now goes to the Super Bowl, and frenzied sportswriters billed the game as East vs. West. The geographical loyalties were so fierce, even the Yale team could not be persuaded by Oregon State's coaches to divulge the details of the Harvard attack. On the six-day train ride west, the Harvard squad disembarked periodically to do calisthenics by the tracks, and when the train slowed, occasionally ran alongside. Douglas Fairbanks, who starred as Zorro that year, gave the Harvard team a tour of his Hollywood compound and in return received a seat on the Harvard bench—right beside Charlie Chaplin.

Before a crowd of 32,000, under a broiling sun that would drain Minturn of more than ten pounds of sweat, Oregon took an early 3–0 lead, then Harvard came back to score a touchdown, and an extra point— drop-kicked in those days. Twice, Oregon tried a diabolical play in which its quarterback, pretending to be injured, lay prone on the field until a teammate said, "Are you all right?" That was the signal to hike the ball to a running back who flew down the field while Minturn gave chase. Oregon came back with another field goal to close the score to 7–6, and then three times tried to boot the ball through the uprights, but missed each time. On the last attempt, in the waning minutes of the game, Minturn, exhausted, dropped down on his knees to pray to God to send it wide, which He did.

Minturn was the Harvard boxing champion, too; when the Harvard football team arrived in Los Angeles, an eager press agent arranged for him to pose with his fists up against Jack Dempsey, then the world heavyweight champion, for a publicity shot. He often boasted that Jack Dempsey had kept track of his progress afterward, and told a mutual friend that "Duke" Sedgwick would someday contend for his heavyweight crown. True or not, he clung to the image of himself as a practitioner of—as he often put it—"the manly art of self-defense," a phrase spoken with a growl, as he raised his fists in the classic pugilistic pose.

MINTURN, OF COURSE, was my father. He told me the story of the Rose Bowl game over and over, with great relish. Although he never mentioned any particular play that he had made—no game-saving tackle, no key block—the game was easily the high point of his life, simply because it had been such a great victory. By the time I came along, his business career was on a downward course. To me, his life, consequently, has a disappointed, backward quality, as all the best things came early, leaving him nothing to progress toward later. There are tales of his getting down into his three-point stance at cocktail parties to demonstrate his football prowess well into his thirties. He was so eager to re-create his glory days that he didn't see he was being made sport of. His athletic triumphs evoked for him the greater grandeur of being the man of the family, looming over his slender, unworldly father, supporting everyone else in their grief, and so never having to surrender to his own.

After Harvard Business School, Minturn tried a year of teaching at Groton, which bored him silly, and then went to work for the Boston investment firm of Scudder, Stevens & Clark, where he milked his social connections to assemble an impressive roster of Old Boston investment clients. These connections came largely from his wife, the former Helen Peabody. She was the Groton Rector's oldest daughter, whom he'd first met at twelve. She was then an alluring golden-haired beauty of twenty-one, the object of erotic fantasy for any number of Groton boys. She was also the daughter of the man he loved like a father—and, as a Peabody, a step up the Boston social ladder for the Sedgwicks. She had a sunny charm that would win the hearts of bank presidents and porters alike, chatting up both with equal abandon. Such allure added to her magnetism for Minturn; he was thrilled that Paul Cabot, the star of Boston finance, was fiercely jealous of him for winning her. Nine years older, she was thirty-four to his twenty-five when they married in 1924. They quickly had three towheaded children and, in the climactic year of 1929, settled into a big yellow house they built on the banks of the Charles River in the woodsy Boston suburb of Dedham.

Minturn designed it—with the aid of Ham Robb, an architect friend from the Somerset Club—to look exactly like the Old House, which represented the rootedness he was desperate for. Built in the same Federal style, the Dedham house was also lordly, confident, clapboard-sided, with two windows on either side of the formidable doorway entrance, five windows across the second floor; the same "hipped roof" that Theodore had specified; and identical chimneys, besides. Like the Old House, it also had a view of a meandering river, the Charles, out the back. The only variation is the proud touch Minturn placed over the front door: the Sedgwick coat of arms. As a widower, Babbo often stayed in the Yellow House, enjoying Helen's lively, wide-ranging conversations at the dinner table. While Babbo and Helen carried on with their birdsong at their end of the table, Minturn was an increasingly silent, brooding presence at the other end. He certainly cut an impressive figure, though—part Roman senator, part prizefighter. A society portraitist in Boston, Polly Thayer Starr, was so taken by his head that she drew it in charcoal. Large and imposing, it is more sculpture than sketch; the head seems to have been chiseled from Berkshire granite. Thick, close-cropped hair, full lips, iron brow, unbreakable jaw.

Inwardly, though, Minturn wasn't quite so solid. While his social connections left him safely ensconced at Scudder's, he was not prospering there. Several of the bright young men who had joined the firm well after him had already leapfrogged over him, making partner while Minturn stayed essentially at the level at which he'd started almost fifteen years before. Then in 1939, his pride was dealt a heavy blow when the much younger Ronald T. Lyman was hired to head up the "following depart-ment" that dealt directly with clients. It was the position that Minturn, as a senior man, had set his cap for.

There were internal contradictions, too, that were puzzling. While the Judge had started out a political Whig but social Tory, he combined the two in the Federalism he ultimately championed. Minturn remained riven all his life. Politically, he was a fierce Democrat, utterly devoted to FDR, but socially he was a Republican who adored his clubs, the more ex-clusive the better. Besides the Somerset, the granite social bastion on Beacon Street, Minturn also belonged to the "B and T," or Badminton and Tennis Club, where he played furious squash matches all winter, and later The Country Club in Brookline, too. But the Porcellian Club ranked highest, since it represented so much of what he prized—lineage, Har-vard, a late-Victorian manliness, an exclusivity that amounted to a kind of brotherhood, and the easy conviviality that comes with tribal ritual, not to mention prodigious quantities of drink. In homage to the club's emblem, he delighted in pigs of every substance and description—ceramic, carved wood, embroidered—except the genuine article. They overran his bureau, side tables, mantelpieces. It was childish, I suppose, but to him completely irresistible. Babbo's friend Owen Wister, who composed the club song, once said the bonds of loyalty to the Porc "could be felt but not analyzed." That was the beauty of all such clubs for Minturn.

As a fellow Grotonian, FDR was probably bound to win his allegiance, even though the Rector voted for Hoover. But Minturn admired the pres-ident's sly combination of patrician flair and ward-healer shrewdness, and worked hard on each of his presidential campaigns, ultimately serving as a member of the Massachusetts electoral college pledged to Roosevelt. In gratitude, FDR dropped the tantalizing possibility of naming him assis-tant secretary of the Treasury in his first administration. But Minturn, a bit short on confidence as his career languished, never took him up on it.

Still, he was thrilled to be intimate with the president. In his memoir, he details his few interactions with FDR and other heroes (Churchill was another) in a few sections he titles "Glimpses of the Great," a phrase that reveals possibly too much of the limited nature of his encounters with the powerful. In 1933, he escorted the crippled president down a hall at Groton, taking the "opportunity to feel the quality of the muscles particularly the biceps." Later, when he and Helen were invited for a night in the White House, staying in Lincoln's bedroom, Minturn swiped all the room's stationery so he could write jubilant little notes to all his friends.

On that visit, he saw Roosevelt only once, when he was wheeled out of the elevator in his "chair." In his telling, the president greeted him enthusiastically.

"Hello, Duke!"

"Why, hello, Mr. President," my father returned.

So ended that encounter. Still, it was enough to make him swoon like a schoolgirl. "It was as if he had been eagerly awaiting me," he wrote. "His charm in his own house was fantastic. You felt the birds would come flying in the windows if he whistled."

IN THE FALL of 1939, in a classic symptom of marital distress, Minturn at forty developed an overfondness for the family's au pair, the marvelous, twentyish Jean Phippen. Beyond, I'm guessing, a particular kiss of greeting that lasted slightly longer—and felt more meaningful—than either expected, the relationship was not sexual; but it was a violation all the same, as they both knew. A recent Radcliffe graduate, Jean was bright-eyed, energetic, and leggy—his son Harry V, just eleven, could hardly stop staring. Besides helping out with the kids in the Dedham house, Jean had come along for several weeks to the summer house at Murray Bay the previous summer, and won everyone's heart—except that of Helen, who naturally regarded her with suspicion. Minturn was a man of boundless energy on these trips. Frustrated at the office, he lived for his vacations, and turned them into great knockabout extravaganzas of hiking, canoeing, picnics, tennis, and cocktails. Nearing fifty, Helen had turned retiring and often pleaded ill health that sent her to bed and left the vivacious Jean as her stand-in at these family outings, taking the bow paddle on the family canoe and even filling in occasionally as Minturn's mixed-doubles partner. The more Jean shone, the more Helen faded. Her once yellow hair had dulled to a straw color, and she needed more than a hint of rouge to restore the original brightness in her cheeks—or did it just seem that way to Minturn?—and she turned listless, too, preferring to read magazines on the wide porch rather than to engage in all this tiring activity. There was never an angry word between them, but the tension was manifest in other ways. That summer, young Harry cursed after missing a shot on the tennis court when he was playing mixed doubles with his mother. To register her disapproval, Helen simply walked off the court without a word. That was how she sent her messages. By silence.

So it had gone, that last summer before the war came. As Hitler's armies swept across Poland in the fall, Minturn took the family that October to Stockbridge for a respite from the bitter news from abroad. He always found the Old House restorative. Its large dimensions, expansive views, and long history enlarged, or perhaps just renewed, his sense of his own possibility, revitalizing him. By then, ownership of the Old House had passed to his cousin, a Melville scholar named William Ellery Sedgwick, who mostly treated it as a country house while he taught English and American literature across the Vermont border at Bennington. William Ellery, his wife, Sally Cabot Sedgwick, and their children joined Minturn and his family that weekend. And if it hadn't been gay, there had been enough cocktails consumed to generate a semblance of festivity.

Until Sunday morning, that is, when Minturn returned from church. He found Jean, who had not come with the family to the service, in the pantry of the servants' wing. She'd been putting away the dishes from the night before, but she was perched on a stool, her shoulders slumped, her face clouded with unhappiness. Even Minturn, who could be a fool about such things, could see that something was terribly wrong. He asked what the matter was.

"It's that I know I can't have you," she told him, and then burst into tears. It was the first time they had spoken of what was in both their hearts. He took her into his arms and held her, stroked her hair, told her what a fine young woman she was—but could not say the words Jean longed to hear. Then he pulled away.

He spoke to no one about this, of course, but Helen sensed trouble—a distraction, a reserve, a coolness—which only served to divide them further. There was too much to say, and therefore there was nothing to say. To speak of their alienation from each other would only deepen it. Back in Dedham the following weeks, Minturn turned quieter than ever, left even more of the dinnertime conversation to Babbo. Helen likewise retreated into silence, at least where Minturn was concerned.

Minturn wasn't in love with Jean, not in the sense that she was with him, but he was in love with the *idea* of her, of the freshness of possibility that she represented, of youth, of starting over, of loving a vibrant, enthusiastic woman who did not, on some level, represent his dead mother. And this drove home to him a cold truth: he no longer loved his wife.

Babbo was not the sort of father a son would turn to for any serious difficulties. Babbo had never breathed a word to anyone about the difficulties in his own marriage, apparent as they became after Halla's death. Babbo's response to any unpleasantness was to ignore it. Minturn had friends, loads of them. He was a very clubbable man. But in Boston one simply did not discuss love troubles openly.

As winter came on in December of 1939, Minturn took the bulbous-nosed Ford out of the garage one morning, and instead of driving to the subway station in Forest Hills, as was his workday custom, he followed Route 9 to Worcester and from there snaked west to Stockbridge.

He was not going to the Old House, though. He was going to the psychiatric hospital diagonally across the street, the Austen Riggs Center.

THE CENTER HAD been established by Dr. Austen Fox Riggs, Harvard class of '98, in 1919 for the "study and treatment of psychoneurosis." Riggs himself had never trained as a psychiatrist, only in internal medicine. Practicing in New York, he'd come down with tuberculosis, a disease that had killed his father a few years before, and he'd sought respite at his father-in-law's Stockbridge farm. There he'd begun to ponder those nebulous medical cases that did not respond to conventional treatment, and once he had recovered, he'd decided to devote himself to the gentler kinds of psychiatric disability that Freud called neuroses.

Initially, Stockbridge villagers had been aghast at the prospect. Shortly after the first Riggs building went up, someone planted a sign on the lawn: "Look squirrels! The nuts are here!" But the Sedgwicks had regarded their new neighbor benignly, and the frail Alick Sedgwick, Babbo and Ellery's younger brother, had been one of the center's original trustees. In 1926 he placed his daughter Christina there after she'd sunk into a hopeless lethargy that had many of the earmarks of the family disease. That summer, on a trip to Europe with her husband, the writer John P. Marquand, she'd suffered mysterious fits of anxiety intermixed with crushing fatigue. She spent a month in bed at the American Hospital at

Neuilly-sur-Seine, where her husband was expressly forbidden to visit her. She emerged in a wheelchair, a flimsy, unhappy figure with a fashionable cloche pulled down over her ears.

Back in America, she was whisked to the Riggs Center, where Dr. Riggs himself took charge of her treatment. As Christina reported to her husband, Dr. Riggs located "the emotional conflicts of my life" in the class tensions between her aristocratic-minded parents and the workaday Marquand. She'd hit on the essential question: Were the Sedgwicks members of an elite—or were they no better than anyone else? It had to be one or the other, and yet somehow neither seemed entirely correct. The ambiguity of their standing came out all the more ferociously in regard to the literary pursuits that had always been a hallmark of the family. To Christina's parents' way of thinking, it was vulgar to write for money. And the disdain they felt for Marquand—a whirlwind of productivity, wonderfully proficient in every genre (from entertainments like the Mr. Moto detective series to a dozen literary novels)—was unmistakable. Once, while Marquand was laboring over a story for the *Saturday Evening Post* in the dining room of the Old House—the only space he could find to work— his mother-in-law charged in to insist he take her son Shan's lapdog Choux-fleur for a walk, explaining, "Shanny's busy upstairs with his novel." Never mind that Shan's novel would never see print, while the Marquands would live off the proceeds from John's story.

Such issues could not be resolved by any psychiatrist, although Riggs at least helped Christina identify the source of her strain. The marriage did not last, unsurprisingly, and Marquand later took his revenge by lampooning the quaint pretensions of the Sedgwicks—and the Boston gentry like them—in *The Late George Apley*. Much to the Sedgwicks' irritation, the novel was not just a best-seller but won the Pulitzer Prize as well.

Minturn's much-younger brother Francis was another Sedgwick inmate of the Austen Riggs Center. A gorgeous man of great promise and sensitivity, he was also spoiled, deeply insecure, and plagued by homosexual inclinations that he was desperate to fight off. His own vulnerability to the family disease had been intensified by his father's marriage into the Shaw-Parkman line. Although Francis had been a fine arts major at Harvard, he rather jauntily decided he'd like to become a tycoon after graduation. A fine-arts scholarship he'd won had been sponsored by the international financier Douglas Dillon, and Babbo thought to enlist his benefactor in Francis's quest. After a meeting in New York, Dillon took Francis on as his private emissary to various European banks. The junket in international high finance proved somewhat daunting to someone with so little training, and Francis soon retreated to a desk job at Lazard Frères in London, which was too much for him as well. One morning, Francis collapsed to the floor, scarcely able to draw breath.

He took refuge in an English manor house rented by the wealthy parents of a Groton friend, Charles de Forest, and he stayed there long enough to fall in love with Charles's shy, artistic sister Alice. Once again, the prerequisites of the de Forests, immensely wealthy from their railroad connections, were possibly more enticing than Alice herself, whose beauty was often overcast with the gloom of self-doubt. The sojourn didn't cure his distress, in any case. He took a Cunard liner back to America. When Minturn met him at the dock in New York, he thought his brother looked all right, but he wasn't. So Minturn arranged for him to go to Riggs, and Francis spent three months there in a large private room, where he stretched out before the fireplace on an immense chaise longue. A maid turned down his bed every evening, and left him a pitcher of fresh milk on the bedside table.

He was treated by a psychiatrist, Dr. John Millet, who diagnosed Francis as suffering from "manic-depressive psychosis," a term that had only recently come into the psychiatric lexicon, after the pioneering German psychiatrist Emil Kraepelin identified the disorder in his 1921 classic *Manic-Depressive Insanity*. He'd noticed that, in the psychiatric institutes he had run in Munich and elsewhere, there was a certain class of patient who experienced what he termed "periodic and circular insanity," "simple mania," and "melancholia" as phases of a single disease, which he believed to be hereditary. (He placed the "pass rate" at 80 percent, meaning that eight children out of ten were likely to receive some form of the disease from an affected parent. Modern researchers drop that to about 10 percent, which is still several times the risk for the general population. Still, if straight depression is included, the likelihood that a manic-depressive will pass on some form of his mood disorder to a child rises to nearly 30 percent, making manic depression one of the most inheritable of the common psychiatric disorders.) Kraepelin was a close student of mania, noting its grandiosity, "flight of ideas," "excitability," hypersexuality (as demonstrated by some male patients' tendency to "kiss strange ladies in the street"), theatricality, and "exalted mood," and he was keen to perceive the frequently edgy quality of depression, too, in that fidgety nervousness that can often underlie the more pervasive, and dominant, gloom. For Francis these characteristics coalesced into an exaggerated sense of his own glamour that was never quite neutralized by his heart-fluttering anxieties. For him, though, the disease came and went. That is to say, the oscillating mood itself oscillated, making it all the more difficult to grasp.

When Francis told Dr. Millet he was thinking of marrying Alice de Forest, Millet did his best to dissuade him, but added that if he had to marry, he absolutely should not have children. Even setting aside the possible genetic consequences, with his mood swings Francis could not be counted on to handle parental responsibilities. To Minturn that meant, *Do not get married under any circumstances*. But Francis would not be deterred. Shortly after his release from Riggs, he married Alice in a big society wedding at New York's Grace Church, with the Rector presiding. Francis was as awed by Peabody as his brother was; it was like receiving God's blessing. He took Alice west for much of the year to a vast ranch in Santa Barbara, where he had gone as a child when he and Halla were so sick. Having ignored Minturn's advice, he likewise ignored Dr. Millet's. By the end of 1939, Francis and Alice had five children, and there would eventually be eight: Saucie, Bobby, Pamela, Minty, Jonathan, Kate, Edie, and Suky. All would suffer acutely, either from their own versions of the disease, or from Francis's, or both.

AFRAID THAT HE, too, might be suffering from some mysterious psychiatric malady, Minturn felt some trepidation as he turned in to the drive of the main Riggs building, an imposing white-brick mansion that might have been built for a Lenox plutocrat, and parked the Ford behind, where it would not be seen from the street. Hatless, his overcoat open to the Berkshire chill, he strode around to the wide steps in front and was ushered into Riggs's private office off the entrance hall. There he clasped hands in a firm greeting with the "great man," as Minturn termed such luminaries, each taking the other's measure.

For a psychiatrist, Riggs was an unusually masculine figure—even taller than Minturn, with a lot of the supple leanness of the rower and fencer he had been in college. Unknown to Minturn, Riggs was afflicted with the prostate cancer that would kill him just a few months later. But he was intolerant of pain, meaning he simply put it out of his mind as he did all other needless impediments. He directed Minturn onto a leather chair while Riggs himself settled in behind his desk, and reached for his pipe as he asked Minturn, pleasantly, what he'd come to see him about.

Here, history doubles back on itself. By coming here, Minturn had returned to the exact site of Jonathan Edwards's house, the one that the Judge had ridden to on horseback at the onset of winter in 1783 when he was just a few years younger than Minturn was now. Theodore had come to stake his claim to a portion of the new country that had just been claimed from the British, and then build upon it a house that could be stocked with children to extend the Judge's legacy down through the generations—to Minturn and beyond.

Now Minturn was seeking a title of his own. Most immediately, he wanted Dr. Riggs's reassurance that he had missed out on the least desired portion of the Sedgwick legacy, namely the madness that seemed to be claiming his brother. But more importantly, that stamp of sanity would also bequeath to him a legitimacy as the Judge's true heir, the good son who would return the family to the position that the Judge had himself always intended for it. As the eldest surviving son, he would save his family from the dissolution that he had sensed was coming ever since Babbo, himself the eldest son, had failed to receive the Old House. Minturn would rectify all transgressions, reclaim the Old House as its rightful owner, and in so doing restore the family legacy.

This was the deeper subject, but nothing he could discuss with Dr. Riggs—at least, not explicitly. Those aspirations ran too deep for words. No, the topic that afternoon was his marriage. Should he stay, or should he go? But it was on the larger question that all his dreams for the future depended.

Between puffs on his pipe, Riggs probed into none of Minturn's distress. Unlike the Freudian analysts, Riggs was practical-minded and respectful, never expecting his patients to assume any servile or indolent posture on a couch, and he was not going to bother himself unduly about childhood memories about Father. He preferred to hand out to his patients sensible little booklets that outlined his no-nonsense philosophy. About worry, for example, his advice was: Don't.

Worry is a complete circle of inefficient thought whirling about a pivot of fear. To avoid it, consider whether the problem in hand is your business. If it is not, turn to something that is. If it is your business, decide if it is your business now. If so, decide what is best done about it. If you know, get busy. If you don't know, find out promptly. Do these things; then rest your case on the determination that, no matter how hard things may turn out to be, you will make the best of them—and more than that no man can do.

Buck up, in other words. It was the message that Minturn listened to—or was it listened *for?*—all his life. And he seized on it this time, too. But Riggs also offered Minturn a deeply pertinent sociological observation, one that can be stated so briefly that its import can easily be missed: When parents divorce, he said, the children suffer.

That was all that Minturn had to hear. Riggs knew his man. He'd identified the essential truth Minturn had believed all along, without quite knowing he believed it, or how much it meant. No less than for Theodore, the future was the important thing for Minturn, not the present. It was his children, not his wife, who mattered most to him. He was father more than husband. He would continue to be protective toward Helen, even though she would have much preferred his love. For his part, the fatherly role came possibly too naturally. To him, it *was* love. For years he had been father to his father, and before that to his mother, too, wrapping the whole family in his big strong arms. He would now be a father to his children, and to his wife as well. A father to all, thereby taking on the duty he had always wanted: the family patriarch.

THIRTY-TWO

EVASION AND ESCAPE

So Minturn returned to his marriage, and did his best to commit to it—until December 7, 1941, when he heard on his scratchy car radio that the Japanese had bombed Pearl Harbor and brought America into the war. He would be saved by the call of duty no less than the Judge had been. It offered the perfect excuse to withdraw from a marriage that had become almost pointless, despite his best intentions. He'd already been politicking furiously to unite the United States behind the Allied cause, taking a month's leave from Scudder's to barnstorm around New England campaigning for the Boston chapter of Defend America, an organization determined to end U.S. isolationism. He'd joined another Boston committee to help raise sympathy for Finland after it was overrun by the Soviets, and dined at a smart Back Bay town house with the Finnish ambassador and his British wife, who gave him the " 'honor' of pressing her knee in an affectionate manner" against his. He did not "reciprocate," but the incident, so carefully recorded in his memoirs a quarter century later, may have represented the true liberation he was fighting for.

Because of his age—he was now forty-two—Minturn didn't find a way to join up until the following spring, when he discovered a program to take five thousand businessmen as old as fifty and commission them as nonflying officers in the air force. Drilling with his fellow officers at Fort Devens, not far from Groton, Minturn threw out his back stooping to enter a supply tent and was laid up for almost three months. Even so, he pulled every string he could—a banking executive friend, a cousin of Helen's who was an air force general—to get the best post he could overseas, and finally secured a job as a supply officer in the Eighth Air Force outside London.

Before he could leave, though, he received a jolt that nearly threw all his plans into disorder. Harry V, now a skinny, cowlicked second-former at the Westminster School in Simsbury, Connecticut (at the last moment, he'd been judged too poor a reader—too dyslexic in today's parlance—to hack Groton), had been stricken with osteomyelitis, a rampant bone infection that could still be fatal in 1942, the last year before penicillin became widely available. He'd been seized by a stabbing pain in his right shin, and the school doctor sent him by ambulance to Hartford for immediate surgery. He was fourteen.

The surgery involved drilling a series of holes into the shin to drain the infection, and it went well, but for long days afterward Harry remained in intensive care, delirious with fever, and shouting so loudly from the pain in his leg that his mother, who had rushed down with Minturn to be with him, tried to shush him for the sake of the other patients on the ward. ("Oh, let him holler," a dying woman in the next room told her. "It sounds like life.") After nearly a week, the fever broke and Minturn returned to work in Boston, but Helen stayed on. With the memory of his own brother, Halla, dying at a similar age, Minturn simply could not bear to remain.

It was six weeks more before Harry V could get out of bed. When he did, he stripped off his pajamas for a bath and scarcely recognized the ghostly stick figure in the bathroom mirror. Doctors told him he'd never be able to run again. Football, hockey, tennis—all his favorite sports were out. Helen bravely dismissed all talk of that. She took him around to doctors in Boston, all of whom repeated the same gloomy assessment—all except for Dr. Edward Hamlin, the orthopedic surgeon for the Harvard football team whom Minturn had recommended she see. Hamlin told Harry he'd be fine, and right then hacked off the cast the Hartford surgeons had fitted him with. The leg that emerged was a scrawny, bloodsmeared stick, but Hamlin was right. Harry played football that fall.

Confident his son would recover, Minturn packed his father off to stay with Francis in Santa Barbara and rented out the Dedham house, settling

Helen and the children in her parents' house at Groton, where the Rector had finally retired at eighty-three. Before shipping out, though, Minturn spent one of his last afternoons on the beach at Ipswich with a glamorous socialite named Isadore Smith, recording the event in his memoir. Again, I doubt the encounter was sexual, or he wouldn't have mentioned it. But Minturn obviously felt guilty all the same. In the memoir, he calls this move "very thoughtless of Helen." *Toward* Helen, he means.

FOR MINTURN, IT was a lovely war—the perfect respite from a tightening domestic situation. After a stint with the supply office, securing the officers' liquor in London proper, he was transferred to the Eighth Air Force's command headquarters at a former girls' school in High Wycombe. There, Minturn was attached to the 328th Bomber Squadron, a fleet of B-17 "Flying Fortresses," P-47 Thunderbolts, and other bombers that were being scrambled into the air to rain destruction down on Germany. His job was in "Evasion and Escape," instructing the pilots in the use of the parachute and in avoiding capture if they were shot down behind enemy lines. He was proud later to have come up with the innovation of offering IOUs to any enemy families who assisted them.*

From Groton, Helen wrote Babbo in Santa Barbara that she was "a spot scared" that the air force might send her "M." back home on leave and "forget" to ever return him to England. This confession was all backward. Like Pamela, Helen wanted her husband home, but even more, she wanted *him* to want to be home. Her real fear was that, if Minturn was left out of the war, he would feel marooned with her. Still, she went so far as to ask a visiting British officer if there was any chance that the air force might simply forget about her husband, and dutifully reported the answer: "Oh no sez he—the Duke's job is too important," she crowed to Babbo. "So—that's really the first time I have had any real 'feel' of his job. Pretty exciting."

Pretty exasperating, too. On top of her other anxieties, Helen had ample reason to fear her husband wasn't doing all that much for the war effort. Her inverted statements to Babbo, fearing what she didn't dare

---

* Years later, he opened up a trunk down from the attic and showed me the spoils of his war—cloth maps, a few medals, some pins, and, a particular treasure, brass buttons that opened up to reveal tiny compasses concealed inside.

hope for, were perfectly matched by Minturn's contradictory messages home. He addressed her as his "Dearest Beloved," but sounded overjoyed to be apart from her. Indeed, he behaved overseas as if he was not married at all. In the evening, if he wasn't dining at Claridges with a woman named Frida Scharinon, he was "theatering" with the stately Virginia Ellis, and he told his wife all about it. In a single page of one of his many letters to Helen, he mentioned five different women he was squiring about. Plain as his flirtations might appear to others, they were a mystery to him. In one letter, he told Helen of an odd dream in which he was surrounded by girls "May's age"—referring to his older daughter, then eighteen—all of them eight feet tall. He claimed only to find it "funny craning up at these gorky young things." But the dream hints at a far more complicated state of mind, in which he is entranced by these forbidden beauties who both stimulate and unman him. They were dazzlingly young (in comparison to Helen), and yet they made him feel small, presumably because he didn't quite dare view them sexually. Indeed, the fact that these encounters were technically innocent probably freed him to describe them to his wife.

More threatening to Helen, though, Minturn had been completely taken in by another family—the Jameses, a fine old English clan who lived in the town of High Wycombe just a few miles from headquarters. "Fingest Grove" they called the house, a lovely, wisteria-draped place that looks, in the watercolor he brought back, like something out of E. M. Forster. Sixtyish Angela James ran it, her husband having died before the war. She was soon "Aunt Angela" to Minturn. With all the men off at the war, the household quickly made a place for this tall, handsome American officer. A friend of Angela's had spied him at air force headquarters, and Angela had immediately invited Minturn over to play tennis on the overgrown lawn tennis court behind the house. Handsome and powerful, Minturn must have cut quite a figure, swatting away on the ratty old court in his short sleeves, for he was invited again and again, and he always came, laden with illicit goodies from the larder at headquarters—chocolates, ham, luscious pieces of fruit.

A dozen children who'd been evacuated from London had taken over the top floor of Fingest Grove, looked after by a couple of matrons. But there were several James girls about, and Minturn's eye was drawn to one of them, Angela's lively, frolicsome, thirtyish daughter, Diana, who still rolled her *r*'s exotically, French-style, after being raised by a French nanny. She'd married the whimsical Jock Murray, scion of the John Murray publishing family, which had published English writers for eight generations, going back to Charles Darwin and Lord Byron. But Minturn cared little for the literary associations. With Jock off at war—first in the ranks, and then as an Intelligence officer—Minturn concentrated unduly on Diana, who was plainly intrigued by this large Yank, with all his charming naïveté. When he bragged to her he was a "combination of Apollo and Hercules," she burst into giggles that anyone could say anything so silly and conceited. As his penance, she took his photograph draped in a Greek tunic of green parachutes while standing him atop a stepladder, to make him look twenty feet tall. Minturn sent the picture on to Helen, who found it "'cute' but not booful, darling." He included a snapshot of Diana, too, which Helen pronounced "unflattering." Probably he was simply being unthinking—it was not in his character to be deliberately cruel—but he was plainly detached from Helen and her feelings. The emotional distance had come to match the geographical one.

Compounding the trouble, Minturn told Helen he read Diana to sleep from the *Arabian Nights* or, occasionally, Beatrix Potter. Again, it could not have reassured Helen to think of her husband with Diana at bedtime. "I can't tell you what a nice family this is and how much being taken in this way means to me," Minturn wrote her. He closed the letter on a more reassuring note: "Good night, my darling. God Bless You."

As if to punish him for the many pleasures he was taking in the war, Helen's letters back to him were consumed with her mounting anxieties about their children. Fan, the youngest, was starting to act up. She'd been caught shoplifting, and was increasingly prone to an alarming impulsivness and white-hot anger that would only worsen over time. Harry's leg was still a worry, even though he was able to play football again, and hockey, too, now that he was back at Westminster, just as Dr. Hamlin had predicted. Only May, off at her Canadian boarding school, seemed problem-free. Helen herself frequently complained of the nagging physical complaints that had bothered him at Murray Bay, and that frequently left her bedridden. "Wish I felt better. Boo," she wrote him on Christmas Day, 1943, underlining the word "wish" four times. "This after-math of grippe is a very weak affair."

At times, Minturn's abiding affection for his wife, compounded by sexual frustration, led him to send Helen some startlingly passionate love letters. With one, he arranged to include flowers to his "beloved," along with his request that she kiss them, touch them to her bare breasts, her "lovely ears, & face," and then send the petals back to her "lover." Overjoyed, Helen wrote right back to thank him for the "perfectly lovely" flowers and the "darling note with it. Ooo." His sexual eagerness for her mounted as the end of the war finally drew near. When she fretted she couldn't find the right dress to wear for his return, he told her to forget about any dresses. He longed for her naked— "as Eve."

DEMOBILIZED IN DECEMBER of 1945, Minturn had worked himself into a fever of desire and distress about the return. Once he was at sea, he decked an enlisted man who'd started acting up on the transport ship home. He resettled the family in Dedham, but his head was still brimming with his war adventures. His English accent reactivated, he styled himself an English country gent, and even renamed the Yellow House "Fingest Grove" on his personal stationery. After a burst of excitement to be reunited with Helen, the old disappointments returned. It had been four years, but she seemed to have aged at least a decade in his absence. She was easily wearied, frequently bedridden, and inclined toward pique. The sunniness that he had so loved in their youth was all gone.

In his restlessness, he turned his eyes more longingly than ever to the Old House. William Ellery Sedgwick—who, as the eldest son, had inherited the property from his father, Alick—had died suddenly in Philadelphia in the winter of 1942, shortly before Minturn had left for the war. William Ellery hadn't drawn up a will, which meant everything went to his headstrong wife, Sally, and their two young children. Sally felt overwhelmed by a house so big and drafty. She closed off a large part of it to conserve heat in the winter, and although she'd busied herself writing a bicentennial history of Stockbridge, she found the town irritatingly provincial. And she wasn't so sure about the Sedgwicks, either. "At least *my* family's graveyard isn't the laughingstock of the *entire Eastern seaboard*!" she once griped to her sister-in-law, Christina Sedgwick Marquand, now divorced from her husband, the writer. But it appeared that no other family members were in a position to take over the property. Babbo was not to be trusted with it; his brother Ellery was already oversubscribed with houses; and their other surviving brother Theodore IV was committed to a life as a churchman in New York City. Most of the members of the younger generation were too settled where they were to contemplate relocating to the Berkshires. Only Minturn's brother Francis, flush with his wife's fortune, expressed any interest in the purchase. But his willful impetuosity frightened everyone, and all the East Coast Sedgwicks viewed his overture with alarm.

Sally was an indomitable woman, a tyrant really, but she was brilliant and entrancing, her hair a radiant cloud atop her head, her flesh buttery in its softness, like a fluffy French pastry. Coquettish, she could be whisperingly seductive when it served her purposes, but she could be hectoring, too, full of trumpeted exclamations—*every* other *word* somehow *italicized*—that carried to the far corners of any room.

Yet she had a crippling disability—a withered left arm. Accidentally snapped at birth by a clumsy obstetrician's forceps, it remained oddly babyish, a short, pudgy, useless appendage dangling at about breast height, slightly uplifted, as if seeking attention. Sally was outraged that, of all people, *she* should be deformed. Histrionic by nature, she was convinced that if it hadn't been for her arm, she could have been a brilliant stage actress, another Fanny Kemble. Her brother Eliot had been a decent character actor on Broadway. Sally thought she could be more. A dewy-eyed Ophelia when she was young, a rasping Lady Macbeth later—all the great roles were within her reach, *if only*. As it was, she created drama wherever she could. Eventually she taught theater at Milton Academy, entrancing generations of Milton girls with her canny remarks and showy provocations. At a faculty meeting one sultry spring evening, she unbuttoned her blouse to cool herself—blithely baring a breast. But the cleverness was there, too. She took one shy thespian aside and told her she mustn't be embarrassed onstage. "Just think of the audience as if *they're* the ones performing for *you*."

Even before her William Ellery died, Sally had started engaging in outrageous romances, starting with her husband's friend Genot Sluder, a professor at MIT. Ellery had been having affairs of his own, but when he learned about Sluder, he flew into a rage and demanded Sally give his friend up, which she tearfully did. But the moment William Ellery was lowered into the Pie, she flew back to Sluder, only to be dumped when a hysterectomy made it impossible for her to give Sluder the children he longed for. She took solace in the arms of Ted Spencer, a Harvard English professor who had also been close to her husband.

But then, early in the summer of 1946, Minturn appeared on her door-step.

HE HAD A plan to take the Old House off her hands. He had discussed the matter with his uncle Ellery, long retired now from the *Atlantic,* and had come up with the idea of putting the house in trust for the whole family—all the living descendants of the Judge—to own collectively. He'd also gone to some trouble to determine just who those descendants were. Harry II had made his stab at it, years before. When Minturn got back from the war, he worked with a Hubert Sedgwick, a former *New York Times* reporter who was retired and living in New Haven, to create a definitive genealogy. It was an exhausting business. Where, for example, was the exotic Ernst Franz Sedgwick "Putzi" Hanfstaengl—a collateral descendant of the Civil War's General John Sedgwick—who'd been an early supporter of Adolf Hitler, then fled Germany when Hitler tried to have him killed, and finally been detained by the Allies somewhere in North America? Or the celebrated Finnish architect, Eero Saarinen, who had briefly married into the family; or the writer Nathalie Sedgwick Colby, who married Woodrow Wilson's secretary of state? Thrilled with the luster such celebrities added to the family, Minturn was desperate to hunt down every last one. Hubert Sedgwick had just turned eighty and was in frail health, but, lashed on by Minturn, he rounded up five hun-dred living descendants of the Judge's father, that Benjamin Sedgwick who had died of an apoplectic seizure when Theodore was so young. (They'd started one generation back partly to increase the pool of possible subscribers to such a project.) In taking on such a task, though, Minturn was also establishing himself as the one who determined who was family.

Another psychological factor: the Rector had finally died, at age eighty-seven, while Minturn was off at war, freeing him from a father figure—and further, freeing him to *become* one.

For the money to maintain the house, he'd turned to Uncle Ellery, who'd emerged as the major moneymaker among the living members of Minturn's branch of the family. He'd sold the *Atlantic* to an heir to the John Deere fortune, and then used a good portion of the proceeds to buy stock in the companies that invested the most heavily in research and de-velopment, thereby capitalizing on the MIT-spawned electronics compa-nies that were just sprouting around Route 128. Eager to secure the Old House for the family, and happy to have Minturn as its titular head, Ellery offered to start an endowment with the handsome sum of $10,000, much of it to come from the royalties of his freshly printed memoir *The Happy Profession*, which he'd dedicated to the Old House.

When Minturn arrived at Sally's that summer of 1946, he pushed open the big front door without knocking, as was his custom, and stepped inside the front hall. I can so easily picture the scene: Outside, the hot summer air was heavy with that wetness that seems to rise off the Housa-tonic, but inside, the house was cool, with its high ceilings and all its big casement windows thrown open to the breeze. He glanced at the letters of Theodore's to various heroes of the Revolution, framed over the side-board; his English shoes made quite a clatter in the hall. Sally, coming toward the sound, was relieved he wasn't a burglar, and said so, laughing. She took him into her arms, just to greet him, that was all. Feeling her against him, he held her longer than either of them intended.

Minturn had wept openly at Sally's husband's funeral. But this did not weigh on him now. In that big house, a world away from Dedham, he might be in England during the war, where he felt free to form whatever attachments he pleased. Sally's two teenage children, Alex and Sarah, were home for the summer, but they slept upstairs in the wing. Minturn enjoyed a big dinner that night with Sally as he laid out for her his plans for the house. Thrilled to see a dignified way out, Sally offered the Old House for a dollar. She was as giddy to be free of the property as he was eager to take it on, and they celebrated with more wine than they possibly should have drunk. That night, when the children were safely tucked in bed at the other end of the house, they were still flushed with happiness, with the joy they took in being allies in such a cause. They retired late to their separate, large bedrooms across from each other in the upstairs hall. It's unclear whether Sally came into his room during the night, or he into hers. But they slept together that night, tangling their bare limbs in one big bed until the first light of morning.

THERE WERE MORE details to be worked out before the Sedgwick Family Society could become an official entity, but the pact was set. And that fall Sally rented the house to a Mr. Knight, with the rent money to accrue to the society, and she took an apartment on Beacon Street, where Minturn would come by to see her on occasion for a drink, or more, after work.

Helen descended deeper into lethargy. As she neared sixty, her old sparkle was gone. Only Babbo, back from California, remained true to her. While Minturn was drawn increasingly to Stockbridge in the summers when Sally was there, Helen never accompanied him. She must have known, and it must have galled her. But she was too proud to speak of it. On this, and many other topics, silence reigned. As he drew away from her, Minturn devoted himself all the more to the "greater family," as he termed it. He was always the dominating presence at any Sedgwick gathering, spouting stories, making introductions, providing marvelous good cheer, leading the toasts. His wife, however, was rarely at his side.

In the fall of 1948 the whole clan descended on Stockbridge for the marriage of Christina and John P. Marquand's daughter Tina to a young Harvard historian, Richard Welch. Minturn had taken Helen and the children for a week at Murray Bay, and they drove down from there. Mr. Knight had kindly decamped from the Old House for the occasion, but it was jammed with Marquands. So Minturn settled his family with his cousin Symphorosa and Monty Livermore in Lenox. When the wedding day dawned, however, Helen declared that she was too tired to attend the service. Minturn wasn't inclined to inquire. It frustrated him, but relieved him, too. For Sally would be there.

Minturn and his three children attended the wedding without Helen. It was a very joyous occasion. At the last minute, John P. Marquand showed up to give away the bride; there had been some question whether he would, as he had felt so burned by the Sedgwicks' endless meddling in his marriage to Christina. He escaped in his limousine just after the ceremony, leaving the Sedgwicks to another of their gay parties back at the Old House afterward. Helen didn't attend that, either.

The next day, Helen said she still did not feel particularly well, so she and Minturn stayed on at Symphorosa's after the children left. It was not a matter for doctors, just more of the same. Or so Minturn thought. During the night, however, Helen weakened terribly. She could scarcely breathe, and when she woke Minturn, he could see her eyes were wild with panic. He frantically summoned a Lenox doctor, who rushed to the house, but it was too late. Helen died of respiratory failure shortly after he arrived, at about three in the morning.

Harry V had gone back to Manchester, New Hampshire, to spend a few days with Lennie Burroughs, his first serious girlfriend. Obviously distraught, but doing his best to keep control, Minturn reached him there by telephone that night, waking him from a deep sleep. Overwhelmed, Harry went in to see Lennie, to tell her the awful news. He climbed into bed with her, and she held him as he lay there, stunned, staring at the ceiling, and then she kissed him and, feeling for him in his sorrow, she slid a hand down his chest. She slipped off her pajamas and pulled him free of his. Just sixteen, Harry had never had sex before. He wasn't, in fact, quite sure he was ready for it with Lennie, who was spirited but brittle. Still, they made love for the first time that night, setting them on a course that led to an early, disastrous marriage when Harry was just a Harvard junior.

The Welch-Marquand wedding had only just been cleared away, and poor Mr. Knight scarcely returned to his rented house, before he was obliged to leave again. Many of the same participants returned for Helen's funeral, but not the bride and groom, who were off on their honeymoon. Minturn was resolute in his mourning, remarkably stoic. He did not cry. He'd put Helen's ring on his little finger, and after the long, sad walk behind the carriage to the graveyard, he kept twisting the ring as he stood by her grave. The gentle Babbo, now eighty-six, was far more open to his grief. He fell to his knees beside his daughter-in-law's grave. "Oh Helen, Helen!" he wailed. "It should be I!"

Minturn stayed on for a few days in Stockbridge to gather himself. When he returned to Boston, he did not go directly home. He drove instead to Beacon Street, and pressed the buzzer for Sally's apartment on the third floor of number 367. She was not in, but it was a beautiful day, bright and crisp, and he sat down on the front steps to wait for her. When at last she sauntered down the wide brick sidewalk, she smiled to find him there and, grabbing his arm, gaily took him inside with her.

THIRTY-THREE

SALLY & SHAN

Minturn's affair with Sally remained secret, but it didn't last. After the initial upheaval over Helen's sudden death, he fell into a deep despondency, taking frequently to his bed with the sorts of medical complaints that had so troubled his wife, but had never brought him low before. Sally soon lost patience with him. But his new availability may have counted against him, too. Sally could not put up with a man on a full-time basis, and a few months after Helen's death, she threw Minturn over for an even more outrageous choice: her late husband's brother, Shan. By now, he was a *New York Times* correspondent stationed in Athens, where he lived among the ex-pats on Odos Karneadou in the wealthy part of the city with his ferocious wife, Roxane, the arch-conservative daughter of a Greek academic. There was little happiness in the marriage, but neither could bring themselves to end it. Shan came back by himself every year or so to visit the Stockbridge of his childhood, where he'd tramped all over the Berkshires reciting Wordsworth.

For Sally's purposes, Shan was the perfect lover. Tubby, mustachioed, wild-eyed, he had published a couple of impassioned novels (the one he'd worked on while John P. Marquand walked Choux-fleur was never published) and had a flock of high-flown literary friends like Lawrence Durrell and the British traveler (and John Murray author) Patrick Leigh Fermor. He was the rare man whose passionate intensity matched Sally's own, and his romantic streak possibly ran deeper. In the many effusive love letters he addressed to "My darling love" from "your Shan," he thrilled to the uncertain nature of their affair. "Try to keep your courage up," he implored her in 1951, when the first fissures in the relationship started to appear. "In this both of us are going to fail often each day and each night, especially."

The affair was a tangle. To Shan's regret, Sally told her daughter Sarah all about it, just as she had told her all about the tryst with Minturn (which is how I came to know of it). Roxane herself suspected it, but she had her own attachments, which kept her from prying too deeply. But neither of them was up for a full-fledged sex scandal. So Shan's affair with Sally sputtered on, sustained by the distance between them, for almost three decades. Gradually everyone in Stockbridge came to know of it, and most of the family, which bothered Shan a good deal, not that there was much to do about it. There was no shutting Sally up on any subject. Over time, Sally had taken other lovers besides Shan, most notably the Boston artist Gardner Cox. When Cox died in the mid-1980s, she kept a small dog of his to remember him by. The dog even looked like Cox a little, with its bushy eyebrows. When Shan came for a visit a year or two later, he stayed with Sally at her apartment, a colorful place on Commonwealth Avenue. By then, Shan was well into his eighties, and he had his infirmities, gout among them. One evening he left his pill box open on the side of the bed and accidentally knocked it over, sending the gout pills all over the floor. Shan didn't notice, but the dog did. He ate them, and died on the spot.

The evidence of what had happened was unmistakable, and Sally was beside herself with fury. She stormed off to Stockbridge, where she summered in a small property by the Old House called the Chicken Coop. There she railed at her daughter, Sarah: "Shan killed my dog, you know. He *killed* my dog. He did! He killed him. *He killed my dog!*"

That ended it. Shan never returned to Stockbridge. Eventually, Alzheimer's claimed them both. Sally died babbling of that early apartment where she had first taken Shan (and Minturn) in, repeating "367 Beacon Street, 367 Beacon Street." Shan died imagining he was in Stockbridge. He mistook Mount Hymettus, the Athenian mountain visible from his apartment, for a similar peak, Bear Mountain, in the Berkshires, and the dining room of his Athens apartment for the dining room of the Old House. "Move that blue china into the light, would you?" he once asked a friend who came by, thinking he was seeing some porcelain pieces he'd grown up with. In his delusion, his eyes misted up as he told friends that, gazing out from his Athens balcony, he'd seen once more "that bend in the river," meaning the Housatonic as it curved past the Old House.

THIRTY-FOUR

THE ANTI-HELEN

After Sally, Minturn was ready for someone more demure. He dated widely and spent many a night with Virginia Ellis, the imposing woman he'd "theatered" with in London. Then, in late 1949, he met a former tennis champion named Emily Lincoln. Mutual friends had set them up to play tennis at the Badminton and Tennis Club in Boston's Fenway. Singles. Afterward, they could never agree on who'd won, and not out of chivalry, either. Each claimed victory.

Emily had reason to believe it was her. A slim, girlish five-six, she was fifteen years younger than Minturn, and she played tennis with a lot of balletic grace. In her competitive days, she'd won the Massachusetts Ladies Singles championship three years in a row, a state record. Minturn, by contrast, played tennis like a boxer—stiffly but with power. Whatever the result, it was game, set, and marriage. Outwardly, Emily was the anti-Helen: not just younger, but infinitely more sprightly and appealing—especially in her clingy tennis clothes and with her shy smile. Inwardly, however, Emily might have been Helen's twin.

The grim dynamic that governed the Judge's relations with Pamela, ultimately driving her to suicide, had returned in Minturn's first marriage to Helen, as a strong-minded husband took a lively wife and ground her into dust. Tragically, that disparity of interests was the driving force in this new marriage, too, and it proved only a little less destructive. As each

end of a magnet yearns for its opposite pole, Minturn's strength sought
out Emily's weakness, and made it weaker still. One is tempted to call it
fate, but it seems more like a law of marital thermodynamics that certain
powerful Sedgwick males should select mates whom they would then en-
feeble. And initially Emily thrilled to Minturn's zest, incredulous that she
could ever catch a man with such flair. But she was easily intimidated,
and remained so, no matter how many parties he dragged her to. At first
Minturn found her charmingly adolescent, a "gorky young thing." But
her dislike of the social whirl soon frustrated him.

The Lincolns were cold-roast Bostonians, much like the Peabodys,
but considerably less regal. The son of the owner of a series of passenger
liners that steamed between Boston and Great Britain, Emily's father,
Alexander Lincoln, had put his Harvard law degree to use in Republican
politics as the state's assistant attorney general, but after getting drubbed
in a 1926 campaign to become attorney general, his professional life was
limited to dashing off indignant letters to the *Boston Evening Transcript*
and performing fitful public service on blue ribbon committees. In his
frustration, he wanted to try the psychoanalysis just coming into vogue
among a small set of Bostonians, but his proper wife, Eleanor, wouldn't
hear of such a thing. He had an affair instead. When Eleanor discovered
it, she never let him into her bed again. He was not yet forty.

Eleanor was an Ames of the Ames Shovel Company of North Easton,
Massachusetts, which was built by Eleanor's flinty grandfather Oliver
Ames. At its height in 1879, the company manufactured three out of every
five shovels produced around the world, making the family firm a colos-
sus of the industrial age. Two of Oliver's sons—Oliver Jr. and Oakes—
created the Union Pacific Railroad. When the famous Golden Spike was
driven in Ogden, Utah, in 1869, joining the Union Pacific to Leland Stanford's Central Pacific and creating the country's first transcontinental rail
line, it fulfilled a dream of the dead President Lincoln to link east and
west as the Civil War had united north and south. But the accomplishment brought only shame for the family once the newspapers revealed the
Ames brothers' role in what was termed the Credit Mobilier scandal,
named for the Union Pacific's construction company: they had won federal approval for their railroad by bribing several congressmen, including

Schuyler Colfax, the former Speaker of the House who had since risen to the vice presidency, with stock in the new venture. The *New York Sun* headline conveys the explosiveness of the story:

THE KING OF FRAUDS
How the Credit Mobilier Bought
Its Way Through Congress
COLOSSAL BRIBERY

The Ames family retreated inward and, for consolation, engaged in a spate of mansion-building that transformed North Easton into something resembling a royal seat. Various family members put up great towering Anglophiliac piles, ten of them altogether, bearing names like Langwater, Spring Hill, and Wayside that would also have warmed the hearts of those Lenox plutocrats of roughly the same era; also an Ames public library, a town hall, and a railway station. Five of these edifices were designed by the society architect H. H. Richardson, making tiny North Easton the holder of the world's largest collection of Richardson buildings. Richardson, in turn, arranged to have the family's private landscapes all done by Frederick Law Olmsted; much of the sculpture by Augustus Saint-Gaudens; and many pieces of ornamentation by Stanford White.

Eleanor's own grandfather, William Leonard Ames, a brother of the two railroad-building Ameses, had sought his fortune in the Midwest as a cattleman in St. Paul, Minnesota, on 2,000 acres his father had bought for him. William in turn gave his son, another Oliver Ames, a farm on Grey Cloud Island in the middle of the Mississippi near where it passed St. Paul. That Oliver raised a family there, but he died at thirty-three when he was thrown from his horse, leaving a widow and three children, of whom the youngest was Eleanor, just three. Two years later, her mother decided to send Eleanor back east rather than let her grow up fatherless in the still-untamed Midwest. So a pair of spinster Ames aunts raised Eleanor in their luxurious Clarendon Street apartment in Boston's Back Bay. When Eleanor arrived in 1885, one Ames cousin, yet another Oliver, was occupying the State House as governor, and another cousin was erecting the world's tallest building, at fourteen stories, on Court Street downtown. The two aunts sent Eleanor to finishing school in Vevey, Switzerland, arranged for her to study the piano with Anton Rubinstein, and once had her travel in the private railway car of James J. Hill, along his Burlington Northern railroad. They cultivated her, in short, into a perfect Boston lady.

THIRTY-FIVE

MY MOTHER'S DIARY

To tell Emily's story properly, I have to jump ahead to a time after she became my mother, in order to convey the full dimensions of the secrecy in which she operated, one that both nurtured and manifested the depression that ate away at her until it devoured her completely. Her secrets, you see, were my own. I realized this in 1999, when a nightmare came back to me from my early childhood—age seven or eight, I'd guess. It was a recurring nightmare, one that I'd had so frequently, and that had frightened me so deeply, I dreaded falling asleep. It had eventually faded, and I had completely forgotten about it for almost forty years, until I finished writing that first, crazy novel, *The Dark House*—the one about the wealthy loner who follows people at random in his car—and set myself up for my cataclysmic depression. When the book was done, the dream returned to me with stunning clarity.

In the dream, I was lying in bed in my room, which was on the other side of a bathroom from my parents. Out of the bookcase on a near wall a dusty volume of my encyclopedia would slowly pull itself off the shelf and float toward me, growing ever larger and more threatening as it approached, accompanied by the sound of my parents arguing more and more loudly, although I could never make out any of the words. The sound, though, grew just as the book did. I tried to fend the book off with my hands; the surface was spongy, and it gave a little where I touched it, trying to push it away as it swelled up over me, like some strange planet that I had somehow attracted. I would wake up in a fright, my heart racing, before the book smothered me, my parents roaring at each other.

There is a lot to unpack in that dream—the sexuality of that growing book; the scary nature of knowledge; the crackling marital tension; and the vulnerability of the little boy lying there. I incorporated many of these ingredients in the novel unknowingly, or perhaps I should say subconsciously. Born into a tense marriage and surrounded by secrets, my lonesome protagonist turned to his lubricious pastime to find answers he was not aware he was seeking. When the novel was done with my dream, it returned it to me.

What strikes me now is the part about the arguing, for my parents almost never argued, at least in my hearing. The big Dedham house was largely silent by then—they'd been married for a dozen years, each parent removed to his or her separate sphere. And yet, of course, they were arguing viciously in every other way their entire marriage, for they disagreed about everything—politics, manners, style, scale. I almost never saw them touch beyond a perfunctory peck on the cheek on greeting. They lived largely separate lives, with separate friends and separate interests. When I moved out of that frightening bedroom to the former guest room on the third floor at age ten, my mother took my place there, leaving my father alone in his twin bed in the room they had always shared.

Tense as all of this now seems, my mother had been silent about it. Angry and frustrated as she had reason to be with my too-oblivious father, she never once spoke to him of her distress. And I had always assumed that she had gone through the rest of her life silently as well. After she died, though, I went through the desk she had always used in our living room, at the other end of the room from where my father had his office after he retired. The desk was a formal antique, made of swirling walnut in an English style, that she'd inherited from her mother. She had always been very possessive of it, which was unusual, since she normally didn't appreciate elegant furniture. The desk had a series of cubbyholes on the top toward the back, and I noticed that each one was thick with yellowing papers—spiral notebooks, jottings, paper-clipped sheets. When I pulled them all out and started to go through them, I could see they comprised nearly a lifelong journal. Her desk had been honeycombed with her secrets.

The very first diary, labeled "My Diary by Emily Ames Lincoln" in large flowing script, is covered with advisories—directed, presumably, to her two older brothers—"Keep Out" and "Please do not read this—if you do you are a big cheater." That one dates from 1927, when she was in the ninth grade at the very proper Winsor School in Brookline. The first entry, on Saturday, January 1 of that year, describes an outing with her mother to see a "very amusing" play called *Tommy,* about a "young unmarried girl with two lovers and she does not know who to take." The journals continue, year by year, until 1941, when, by a quirk, she finds herself in an identical predicament, which she finds anything but funny. She is torn between the well-born but "boring" Dick Prouty, who had been a friend of her brother William's, and a warm-hearted, gangly dipsomaniac named Charlie Balch, whom she knew from her summers in Chocorua, New Hampshire. By then, her chipper account of life at Winsor and then at Smith College, where she'd been a star athlete, has descended into a fretful chronicle of her days as a Red Cross volunteer during that early war year and her nights with her distant parents on Codman Road, a lonely private drive in Brookline.

Her father's scrapbooks, which my mother also saved, were filled with newspaper clippings of her exploits as a tennis champion. But her own diaries barely mention any of that. Instead, for page after page, she broods about her life. She records a daunting variety of psychosomatic complaints—exhaustion, headaches, upset stomachs, insomnia—that had her dragging herself off to the doctor every few weeks. The handwriting, at first a freewheeling schoolgirl scrawl, tightens over time into a jiggly script that scarcely rises up from the line, as if it, too, was crushed by her many burdens. (So different from Minturn's jaunty script on his unfeeling aerograms to Helen during this same period.)

In 1937, Emily had devoted months to a cheery Harvardian named Ken Brown, who'd "fooled around" with her at the beach in Nahant, making her heart soar, she was so sure he was her "one and only." They'd been secretly engaged, and she'd tried on her mother's wedding dress in front of the mirror, melting at the image of herself as Mrs. Brown. A photo album has snapshots of each of them one warm afternoon behind the Lincoln house, first her shooting Ken—stiff-looking, in a rumpled ill-fitting suit—and then him shooting her, looking a little lost in Newbury Street haute couture. But then all of that simply faded away. Brown took a job in Southport, Connecticut; Emily stayed in Brookline. For a while, they exchanged letters, but the letters became more halting and more remote, and then she didn't hear a word from him ever again.

My mother wasn't all that sorry, although I don't know why. Dick Prouty moved in to fill the vacuum, and then Charlie Balch displaced him. Toward the end of 1941, though, Prouty had won out once more. But it was a pointless romance. At least the affair with Brown had had some passion—kisses and, as Emily wrote, "going too far," although not really very far at all, and some breathless talk of love. She never even kissed Prouty; and he never once tried to put a hand somewhere he shouldn't. The whole business had a damp, draggy quality, much like her mood.

> *Oct. 4, Sat.*
> *Another day of uncomfortable excitement. Dithered all day till 6 doing nothing having slept practically none last night. Tried to get some exercise at the tennis court. Then at 6 after I had finally got my clothes for the week end decided upon, Dick arrived. It was nice to see him but he does bore me badly. Was glad he was here when Charlie arrived, but I suppose it hurt him. O dear!…*

Ten days later, she decided to marry Prouty after all. But the prospect was so upsetting, she couldn't climb out of bed until nearly three the next afternoon. So she held off. The tepid affair continued, neither side committing or breaking it off, until April, when she finally sat down to write him a long letter closing things off. But she set it aside, and before she could take it up again, she received a letter from him ending it. He got engaged to another Smith girl, Ann Jenkins, a month later. Emily pasted the engagement notice in her scrapbook, along with the wedding invitation that came a year later. The ceremony was at the bride's hometown in Illinois; Emily, needless to say, did not attend.

The journal breaks off seven days later, to be continued only with occasional scraps—brief notes about a few days here or there—for the next two decades. So there is no interior account of meeting my father, or marrying him. Most likely, she wrote those diaries, but destroyed them. At the end of September in 1949, she'd finally summoned the courage to move out of her parents' house and into an apartment of her own on

Beacon Hill. She began a tiny datebook at that point, and practically from the day she began her independent life, it's filled with appointments with my father, often for tennis, sometimes for dinner, and a few times for a "night" with him in Dedham.

After they'd declared their engagement in the winter of 1949, just a year after Helen's death, Minturn took Emily for a weekend together in the Old House, just the two of them, and a test run of the dream outcome that my father had been pressing for ever since he acquired the house from Sally. But my mother did not see the attractions of a town that had by then become a community largely for retirees and weekenders. Only Austen Riggs was much of an employer. So there would be few playmates for their own children, which Emily was very much hoping for, and the fall was already so cold, Emily could tell that winter would be interminable.

Minturn walked taller in that big house, Emily could see that. But she also could feel herself tightening up. That portrait of Pamela with Catharine, the one that now hung on the wall in the living room—Emily could see the age lining Pamela's face. The deepening cold outside, the lure of the warming fires in the fireplaces—they only scared her. And so when Minturn showed her around the house, and asked her if she would be willing to live here with him, she told him no.

That was bitter. Later that winter, before the wedding, Emily asked him if he'd mind taking a drive to see her Ames haunts in North Easton, where many of her relatives still lived. Any one of the ten Ames mansions dwarfed his blessed Old House, just as her North Easton overwhelmed his Stockbridge. North Easton was still almost entirely an Ames affair, whereas Stockbridge operated quite independently from the Sedgwicks, and had for over a century. The Ameses even had an impressive graveyard, too, a sprawling hilltop affair that Olmsted had designed.

Minturn said no, he'd rather not go to North Easton.

STILL, THE WEDDING went on as planned that March of 1950. It was held in the Lincolns' wide living room, with a view of the overgrown tennis court through some trees out the back. At thirty-six, Emily thought herself too old for bridesmaids; instead of a wedding gown, she wore a cocktail dress with short sleeves, and a large corsage. Minturn was resplendent in a double-breasted suit, with his green-and-white-striped

Porc tie. Minturn's cousin Ellery, the son of the *Atlantic*'s editor, was his best man. There was a large contingent of Peabodys, and Helen's brother Malcolm, then the Episcopal bishop of eastern New York, performed the ceremony. Afterward, for the reception under a tent in the backyard, Emily wore a huge, relieved smile, but a cigarette—always a token of her anxiety—burned between her fingers.

After a brief honeymoon on Francis's ranch in California (where Minturn obtusely gave a toast to *both* his wives), they settled into the Yellow House in Dedham. Emily had carpenters in to open out the downstairs into sunny bay windows. She took Helen's seat, with Babbo on one side and May, the sharp-tongued eldest, on the other. Harry and Fan were off at college. From the other end of the dining room, Minturn regaled his bride with war stories, generating conversation as he never had with Helen, but leaving Emily with little to add. Soon it was just the two of them, for Babbo remarried, remarkably, at age ninety-one; Harry, at twenty, married Lennie; and May, at twenty-four, followed suit with Erik Osborne, the gangly scion of an upstate New York newspaper-publishing family.

SEX FOR MY mother was not a pleasure, but the children came along quickly—his namesake, called Rob, and then hers, nicknamed Lee, thirteen months later.

Frustrated to be up to his elbows in small children once again, Minturn persuaded Emily to fly with him to London, just the two of them, so he could show her off to the many friends he'd made in London during the war. The trip was to last six weeks. With Lee just one, and Rob two, Emily didn't want to leave them. But Dad had been left at a similar age by Babbo and May, and he assured her it would be fine. Her parents, and their maids, would take the children.

In London, Dad trotted her around to absolutely everyone, from "Aunt Angela"—as Emily, too, soon learned to call her—to John Martin, Churchill's wartime secretary. But they spent much of the time with the Murrays in their snug Cannon Place house in Hampstead Heath on the hilly outskirts of London. The newlyweds were encamped on the third floor, with a view of the garden. On one of their many long, exhausting rambles across the vast country park there—while Dad swapped stories with the garrulous Jock over highballs in the garden—my mother blushingly asked Diana for some vinegar for a douche, which made them instant friends.

OVER TIME, THAT trip became a pivotal event for the family, the dividing line between a paradisiacal Before and a ruinous After. As my mother often put it in solemn, guilt-stricken tones, Rob had been an "angel" when they left, just the sweetest little boy, and when they returned, well, he was a "devil"—short-tempered, sleepless, irritable, prone to tantrums. And Lee, a jolly little thing, had turned silent, withdrawn, almost unreachable. The transformation was so dramatic that Emily asked her mother if anything had happened. Her mother told her, Why no, nothing had, of course not, nothing at all. But as the months passed, the two children's personalities did not improve. Rob's features, once so radiant, remained clouded over. Eventually dark circles hung under his eyes, and his mouth, which had quickly curled into a smile, now tightened into a mean, scary smirk. Lee was so little, it was hard to tell, but she seemed less responsive, less her.

It troubled Emily enormously, but she had no time to focus on it. While she was in England with the Murrays, she'd become pregnant once more.

THIRTY-SIX

A LITTLE SCRATCH
ON THE CHROMOSOMES

Well before dawn on Wednesday, May 5, 1954, Emily's water broke, soaking her twin bed upstairs in the Yellow House while Minturn snored beside her. She telephoned her obstetrician, a Dr. Gould, who told her to hurry in to the hospital. He'd been planning to induce her in a few days anyway. With his stiff neck, Minturn was a terrible driver, so she drove the two of them to the Boston Lying-In Hospital, in the Longwood medical district downtown.*

The pains started at seven, and tightened at eight, sending Emily into the labor room for painkillers. A "corker"—as she later put it—of a contraction a little before nine brought her into the delivery room for ether. I appeared fifteen minutes later. When she was capable of registering me through the ether haze, my mother appraised me as "sweet" and "very lovable" with a "square face" but "big jowls," a "cunning little whorl" of hair at the crown of my head, and "a lovely peaceful contented expression." She set these impressions down in my baby book, a bound volume

---

* Now part of Brigham and Women's Hospital, it was founded in 1832 by Dr. Walter Channing, brother of the Unitarian minister William Ellery Channing, as one of the nation's first maternity hospitals.

labeled "Record" on its spine. It took over from the assorted small spiral-bound notebooks she'd poured her heart into for years, and reveals a more confident hand.

She named me John for my grandmother's favorite brother back in St. Paul. My father added that illustrious Shaw, and threw in the nickname Jock for Jock Murray, the jolly Englishman he'd thought seriously of trying to cuckold during the war. If the name alleviated his shame, it added to mine, for I was teased mercilessly by neighborhood kids as "Strap" throughout my childhood, until I finally succeeded in dropping the name.

ON THE FOURTH day of our hospital stay, my mother made what she termed the "horrible discovery." She found a "horrid looking spot" in the small of my back. About the size of her thumbnail, it looked like a deep wound that hadn't healed, merely been papered over by a filmy membrane. In a panic, my mother yelled for the nurse, but she was unable to raise a doctor until the next morning when Dr. Gould and an unnamed "house doctor" came around to remove the bandage the nurse had placed on the small of my back, and peer at the strange hole over my spine. Neither of them had ever seen such a thing. It wasn't until my pediatrician, Dr. Gallup, arrived from Dedham later that morning that the mystery was solved.

I had spina bifida. In her distress, my mother simply could not take in the words. So he wrote them down for her, capitalizing the S and the B, and then underlined the two words. And then, when she still didn't understand, he sketched out a diagram for her, showing my "hole" in relation to the vertebrae and spinal cord underneath. Whatever calming effect that had, he terrified her by reciting the possible dire outcomes—clubfoot, retardation, "uncontrolled bladder," and other unnamed "troubles from the waist down," all of which my mother nervously jotted down, too, and then scratched out as if they were simply not to be thought of.

Literally "cleft spine," spina bifida is now recognized as a fairly rare birth defect of the brain stem, affecting one baby out of a thousand. It stems from a genetic flaw, a little scratch in the chromosomes, that creates a particular vulnerability to a variety of environmental hazards (a vitamin deficiency, a fall) to exploit. At its rapacious worst, spina bifida can ravage the developing neural tube, leaving a newborn with a grossly underdeveloped brain inside a misshapen skull, or a dangerously exposed spinal cord, resulting in early death or a lifetime of severe retardation. Dr. Gallup tried to assure my panicked mother that nothing so horrible had happened, or would happen, to me. Of the "hundreds" of cases he had seen, mine was about the mildest. Still, there was something about his tone that did not reassure her. After the sudden childhood deaths of Halla and his infant sister Edith, and then Harry's "osteo"—as the family abbreviated it—Dad knew better than to be entirely optimistic. So he was unable to cheer her, but retreated into a stoicism that was all the more unnerving. She was convinced my condition was life-altering, she simply did not know how severely.

After driving me home to the Yellow House, she installed me in the large wicker basket that she used as a crib, on the floor by the big bay window she'd put in, and she kept a close eye on me every day until nighttime, when she handed me over to a night nurse so she could get some sleep.

Dr. Gallup came around every few days to examine my "spot," as my mother termed it. He wanted to see if the skin was starting to grow over the hole, which it wasn't. He also pulled out a set of calipers and a tape measure to check for any swelling of my head, an alarming possibility he hadn't mentioned. But hydrocephalis—fluid on the brain—is a common consequence of spina bifida, requiring a shunt for drainage. My cranium seemed fine, happily, but my mother continued to scrutinize me, carefully recording in the baby book every ounce I gained; every "wiggle"; every bout of diarrhea.

Finally, after about a month, when Dr. Gallup came by for one of his regular checkups, he confessed that he *had* been quite worried about me, but I seemed to be out of the woods now. I'd exhibited none of the "dreadful signs"—my mother's term—he'd feared, and he was finally confident I wouldn't in the future.

"Wow! What a relief!" my mother wrote.

Still, the hole did not seal up on its own, and, after checking with a neurosurgeon, Dr. Matson, in July, my mother delivered me to Children's Hospital for him to stitch me up in late September. I was twenty weeks old and weighed, by her careful reckoning, sixteen and a quarter pounds. I'd be there for ten days altogether. She was not allowed to visit, but she should feel free to call every night for an update.

"Perfectly ghastly leaving Jock there," she wrote.

The nightly reports were all encouraging, but even so, when she returned to pick me up ten days later, she expected to find "a pale, thin, sad little baby." But no.

> Instead I found a rosy, plump, smiling little fellow lying in his crib. At first he didn't know me and wouldn't smile though he smiled readily at the doctor and the nurses. Every time a nurse went by the window next to his crib, she'd knock on the glass and he'd look around and smile.

I'd been the "pet of the ward," she was told, and the doctors assured her I was fine—so fine, in fact, that I could be expected to play football. "Oh wonderful!!"

So I was okay. But, in another way, I wasn't, because my mother remained overprotective of me for years afterward. This was the nature of her love, a kind of cosseting. I craved the sunny warmth of her attention, and considered it one of the great blessings of my childhood. It took me a while to realize that her love was not, in fact, a pure gift, but rather a kind of repayment for the enormous relief I'd given her, first by surviving, and then by being so cheery. I was the first piece of unadulterated good news she'd ever received. I lit her up.

As I grew older, my own vulnerabilities made me sensitive to hers, for I felt our fortunes were somehow linked, a deep, private bond that existed independently of the family. I looked out for her, and more. I was her tiny suitor. In some notes she saved, she described a trip we made together to the Cape for a weekend together in the fall after I turned four. We ate our dinner in front of the fireplace, just the two of us. "Very exciting + cozy," my mother purred. I opened doors for her "without being asked" and, in my tiny voice, sang her all the songs I knew, starting with "How Much Is That Doggy in the Window," a tune I listened to repeatedly off a yellow-vinyl 45 on our Victrola. "He is so easy to deal with and such joyous company."

Even though my spina bifida was a genetic accident, no one's fault, my mother convinced herself it was because she'd delayed too long in having children, and her eggs had deteriorated. Rotted. If she'd only married earlier—to the zesty Ken Brown, or even that dull Dick Prouty—like a normal person, she would never have inflicted such a disease on me. Mistaken as this belief was, she could never be shaken from it. Sure that her eggs had gone bad, she decided to stop at three children, even though she'd hoped for one more to square off the family. So I was the last of my father's line.

Despite Dr. Gallup's assurances, my mother remained on the lookout for any aberrations in my physical development, convinced they portended a slide into drastic retardation. When I ran with a peculiar, galumphing gait in grade school, my mother was convinced I'd be hobbling soon. And it troubled her I was frequently sick as a child. I remember these as phantom illnesses, excuses to stay home. But my mother dutifully brought Dr. Gallup around each time, and then kept me in bed almost the entire day. I attended kindergarten so infrequently, in fact, my mother put a sign in the front window to alert the carpool when I *could* go, since that was rarer. I never thought of myself as sick, though. I loved being able to stay home with her, bathing in her concern. On fall afternoons, she would sit me at the kitchen table and peel a Bartlett pear for me, her knife curving around the fruit, and then quartering it and scooping out the seeds before placing all four sections, glistening with juice, on a yellow plate before me.

WHEN I THINK back on my childhood, I think primarily of my mother. But my parents had a couple of servants, a young black couple named Lucille and Willy, who lived in the wing of the Yellow House. It's tragic that no one now remembers their last names. Handsome, bespectacled Lucille cooked for the family, and the languid Willy did odd jobs around the house, like laying fire logs or trimming hedges. Around the time I was born, Lucille gave birth to a little boy of her own named Michael, and the three of them became a kind of shadow family for me in the house. Michael was my first friend, and as soon as I could walk, albeit unsteadily, we were inseparable. "He and Michael are very cunning together. Though Michael is bigger and seems older they have great fun together," she wrote when I was just eighteen months old. "Much conversation and noises—very loud. If Jock is walking around and Michael confined to a chair, Jock will give him a toy. And vice versa. One so blond and the other so dark." In a photograph taken after a snowstorm that winter, we're standing side by side in the white driveway in our snowsuits, striking identical poses as we lean on our snow shovels.

We were just a couple of toddlers, of course, but even so, it was the first genuine friendship between a Sedgwick and a black person in the record since the Judge's children fell in love with Mumbet in the 1780s. Despite the gallantry showed by Colonel Shaw, and the idealism of the many abolitionists who supported him, Boston had evolved into a highly racist city in the century since, and blacks and whites lived a more racially divided existence than in the South. As a devoted liberal, Dad was open to black people in a theoretical way, as a part of his political beliefs. But the only black man he actually knew was George Washington Lewis, the very formal steward of the Porcellian Club. Aside from Willy and Lucille, there were hardly any black families anywhere in Dedham, and precious few in the surrounding suburbs, either.

By the time I was three, my mother decided that we were too isolated there on the outskirts of Dedham, out by the shore of the Charles River. Once again, my father pleaded with Mum to move the family to the Old House, but to no avail. She scouted out houses closer to town, and set her heart on a small gray house with a cozy backyard. My father considered it a comedown after the palatial Yellow House, with its acres of woods. But my mother said it would be better for the children, so we moved. I never saw Michael again.

DAD CONTINUED TO work at Scudder's. He still had not made partner, and never would. He had a small office toward the back, where he kept pretty much to himself. As his frustration at Scudder's mounted, he turned his attention to developing a different kind of stock investment, one that would challenge the conventional mutual funds that had been a Boston—and Scudder—specialty since the 1920s. "The Twenty Largest," he called it, since it involved investing only in the twenty most highly capitalized industrial stocks—Ford, General Electric, and IBM were all on his list then, as they would be now—strictly in proportion to their percentage of the total. Instead of relying on the genius stock pickers who ran mutual funds, it went purely by the numbers. If a stock ranked as one of the twenty largest, it was automatically in, regardless of what any money manager thought of it. It required none of the expensive investment advisors provided by firms like Scudder's at all, just a calculating machine. It was a revolutionary idea, unheard of, when he first explored it in the early 1950s. Now, of course, it has emerged as the "indexed fund," a concept that has taken over much of the mutual fund industry.

Initially, Dad thought of it as a kind of capitalist populism, an attitude that grew out of his dedication to FDR. His system would "bring the common man to common stocks for their mutual benefit," as he often put it. When he first wrote of it in a 1953 *Harvard Business Review* article, he thought of it just for pension plans of workingmen, to supplement their social security benefits. But gradually he realized that the Twenty Largest system would be ideal for investing the endowments of colleges, hospitals, foundations, and the holdings of private investors as well.

As an investment strategy, the Twenty Largest idea was brilliant, which is why today's indexed funds are so popular. But since this was an investment scheme that ran by itself, without the high-priced help of an investment house like Scudder, Stevens & Clark, the partners at Scudder's were appalled. When Ronald T. Lyman talked about my father, he did so in utterly exasperated terms. "Duke's just crazy," he'd growl. "Just crazy."

Dad's scheme was such a glaringly obvious affront to Scudder's, it is hard to imagine that Dad didn't intend it solely to stick it to the men who had stiffed him. Indeed, later on, he would pointedly include the results of Scudder's own leading mutual fund, Scudder, Stevens & Clark Common, in a list of the best efforts of twenty-four "distinguished investment firms," all of which had been soundly beaten by his Twenty Largest over the last quarter century.

In his retirement, Dad devoted himself increasingly to his trusteeships at Groton, Garland Junior College, the Massachusetts Society for the Prevention of Cruelty to Children, and the Austen Riggs Center, which had been his salvation in 1939, largely in hopes of persuading them to adopt his investment plan. Of them, he could persuade only Riggs. And one other concern: the family graveyard, turning a $5,000 endowment in 1934 into a very handsome $80,000 by 1969, even after considerable expenses were taken out.

The results were impressive, but Dad was intimidated by the powerful investment executives who would decide his fate. The Judge would not have countenanced such timidity, nor the *Atlantic*'s Ellery. But Dad would not push his ideas in the relentless, all-out, total-war fashion such a campaign would require if it was to succeed, for he was taking on the entire investment world, as John C. Bogle ultimately did in pushing the indexed fund, and creating an investment empire, the Vanguard funds, that now vies with Fidelity as the world's biggest investment fund company. Beside that first article in the *Harvard Business Review,* Dad published just one more two decades later, in 1973, in an obscure publication called the *Financial Analysts Journal.* He gave up that year, resigned himself to failure, and started to sink into a depression that was born of his defeat.

Floated by my mother's money, ours was a comfortable life, all the same. We belonged to The Country Club in Brookline (that *The* pronounced with telling emphasis), where my mother whiled away many a winter afternoon playing tennis in the hangarlike indoor courts. We had Sunday lunch at my grandmother's, in Brookline and then on Commonwealth Avenue, where she was waited upon by aproned Irish maids, who silently appeared at the press of a buzzer under the dining table. We spent a month in the big ramshackle house in Murray Bay every summer, paid weekend visits to Stockbridge, and in the late 1950s added a shingled cottage in Barnstable that looked out over a hayfield to the blue water off Cape Cod.

Comfortable as all of this should have been, it was never relaxed, for Rob had emerged as near-constant trouble. He looked like a nice enough little boy, with short-cropped hair and a generally amiable expression, but he was a hyperactive child, easily provoked if he didn't get enough exercise, and there was a disturbing woundedness in his eyes, their hollows ringed with brown, as if they'd been bruised. They were always on the lookout for slights, each one reminding him, no doubt, of being abandoned so wrongfully as a toddler.

Ever since the London trip, Rob had been prone to tantrums, but they grew over time into tornadoes of rage that sent everyone scrambling. He was a particular fiend for candy, forcing my parents to keep any sweets padlocked in a cupboard. But when he felt deprived, anything he desired—TV time, the front seat of the car—was a kind of candy to him. Before dawn one Christmas morning when he was quite young, he snuck downstairs and ripped open all the presents my mother had painstakingly selected and wrapped. Mum was very sentimental about Christmas, seeing it as the most prominent emblem of family spirit, just as Catharine would have wanted. Rob just wanted to make absolutely sure he'd received his fair share. The first I heard of the calamity was the sound of my mother wailing in the kitchen.

I'd thought I was the one who was most keenly aware of my mother's emotional frailties. But my brother was sensitive to them, too, for his own reasons. In his frustration, he'd pound her with his small fists, but his words were far more wounding. Inevitably, she'd falter, her guilt undermining the little willpower she possessed. And while we often played together quite contentedly, Rob was always checking to see if I'd wangled rewards he had not received. When I followed in my mother's footsteps to play in tennis tournaments and to win a few, he snuck into my room to scratch my name off all the trophies with a house key, then scribbled on the wallpaper for good measure.

Much of this was standard sibling rivalry, accentuated by the closeness of our ages and the favoritism my mother showed me as her injured darling. Rob is an infinitely kinder, happier, and more considerate person now, and for years I have regarded him as one of my very best friends. But my parents' response made matters worse, running a line down the household between the mad and the sane. (As was common, *mad* was our word for being angry, but we meant more than usual by it.) Dad took over Rob, while Mum took charge of Lee and me. Dad enlisted professional help, hiring child psychologists to try to understand the sources of Rob's anger and putting him in special art classes to soak up his extra energy before he could turn it to more destructive uses.

But ever after, ours was a divided household. Rob continued to test Dad's patience. My father styled himself as a stoic, enduring figure—the omnipotent Rector of his household, not one to give in to anything so petty as anger. But Rob could routinely get Dad to explode, and I found that terrifying, sure that my aged father would collapse, grabbing his heart. And lose it Dad would, stamping his foot, pounding a pillow on the sofa, raging at Rob with a stream of oaths. He took his belt to Rob's bare backside one time; I was sure their screams—one low-pitched, the other high—would bring down the house.

It was brutal, but it did forge an intimacy between them that made me jealous, just as I'm sure that my closeness to my mother elicited similar feelings in him. Put to bed early, on summer evenings when it was still light out after eight, I used to hear them gaily playing badminton on the lawn below my window. Dad drilled Rob for his spelling tests, read to him at bedtime, and then would sit beside his bed deep into the night if Rob had trouble sleeping, as he often did.

In my entire childhood, I can remember only one moment of genuine intimacy with my father. He'd taken me skating one winter on The Country Club pond, just the two of us for once. In the log house, by the big fire where we put on our skates, though, I dug into my coat pockets and realized I'd forgotten my mittens. My father gave me his socks to warm my hands, while he skated barefoot.

BECAUSE I HAD so little to do with him, it took me a while to realize how unusual my father was. Fifty-five when I was born, he was over sixty when I entered first grade. But the years didn't mean much to me; I was struck more by his manner, which seemed so much stiffer and more proper than the fathers of any of my friends. He was still the English country gentleman, drawn to that feudal sense of order, regularity, permanence, and above all class, just as the Judge had been almost two centuries before. He felt somehow above it all. He wore his double-breasted Groton School blazer so often that he wore out the inside of the left sleeve where it brushed against the crest's prickly golden threads as he reached across it for his billfold in his inside jacket pocket. So he had his tailor stitch on a protective leather patch. (That was the coat I recently found moth-eaten in New Hampshire.) Kipling's hymn to manhood, "If," was his favorite poem. He had his secretary type it out, and then framed it for his bedroom. And his manner of speaking seemed foreign even to me. In his English accent, birthday was birthdee, and Anglicisms littered his speech, making a flashlight a "torch" and our garbage can a "swill pail." It seemed fitting he should call himself Duke. My little next-door neighbor Katie Woodworth thought he *was* a duke.

As the 1960s progressed, the times ran against him. The Beatles songs I loved left him howling about how I could ever listen to such "noise." He himself preferred Broadway show tunes, which he sang off-key. Every morning, he played "Old Man River" while he did calisthenics from the Royal Canadian Mounties handbook. He never got the hang of anything electronic. To operate the hi-fi, he'd get down on his knees to eyeball the needle as he set it down on the record by hand, unable to believe that this could happen automatically. He termed the contact lens he had to wear for a cataract his "eye," making him cry out "I can't find my eye!" every time it slipped out of his hand by the sink, which happened often. And such a bundle of mannerisms! If a football game was on TV, he'd always ask, "*Who* is playing *whom?*" emphasizing his correct grammar. When anyone set the table, he'd proclaim: "Fork, or fork and spoon, but never spoon alone." Ask him how he was, and he'd reply: "Fair to middling, not choice to prime."

VERY EARLY ONE morning in the fall of 1962, he had a heart attack. When the medics came, they carried him down the stairs in a sturdy chair from my bedroom. He looked like royalty as he descended on this seat, his scarlet wrapper over his Brooks Brothers pajamas, a pair of leather slippers on his feet. The attack was not severe, and he recovered in a month or two. Then, two years later, he suffered an aneurysm, and nearly died. He was in the hospital for months. We weren't allowed to see him for the longest time, he was so frail. When I finally did go, I remember coming up to his bed and scarcely recognizing him. He was all bone. His hospital ID tag hung off his wrist. Three years later, he had another heart attack, and he was never the same again.

THIRTY-SEVEN

LITTLE DUKE

Rob's rages were a strain on my father. But he was facing other family stresses, too. Harry's marriage to Lennie had broken up a few years before, in 1956, after the couple had three children. Two years later, though, Harry married again, to Patsy Rosenwald. That ended in 1968 when Harry fell madly in love with a stewardess on a transcontinental flight and moved in with her, leaving Patsy and their three children, the youngest, Kyra, just three years old. And Fan was a wreck. Repeatedly frustrated in love, she had taken refuge in politics, working for Adlai Stevenson, then the Peace Corps, but that gave way to less satisfying jobs, and men were put off by the intensity of her moods. She took out her distress on Dad, frequently tearing into him on the phone. In San Francisco later in the 1960s, she fell in with Leonard Orr, the founder of the Rebirthing movement, which had its followers reimagine their passage down through the birth canal in order, supposedly, to free them of any lingering traumas related to it. Fan tried that repeatedly, but never did find peace.

But the biggest issue for my father was his brother Francis, who by the late 1950s was turning into a monster all of his own. Still in Santa Barbara, he moved his family to a series of ever-larger ranches, all of them with the gorgeous olive trees and the beautiful Mediterranean light that Babbo had adored. The children had come rapidly, eight of them in fourteen years, even though Alice had increasing difficulties with the births. Sensitive to the dark moods of these western Sedgwicks, my mother had tried to discourage contact. But Francis and Alice visited occasionally in Dedham. Alice frightened me. A beauty in her youth, she'd had an operation to clear up a middle-ear infection, but it had gone awry and left half her face drooping hideously. Francis I took for a British admiral—trim, proper, and extremely stiff. Lee got a different impression, though, one afternoon when she was dispatched upstairs to his guest room to alert him to a phone call. Only ten, she timidly knocked on the door. "Come in!" came a booming voice. She gently pushed open the door and found Francis in his bikini underwear, his bare skin glistening with oil, doing his exercises on the rug. He seemed to take an erotic thrill from her startled gaze.

Nicknamed "Fuzzy" by his family, Francis styled himself the most loving and proper parent, donning a jacket and bow tie to read Greek myths to the children at bedtime before the fire. But that was all show. Harry was out there one summer when the oldest boy, Bobby, got in trouble with Francis for neglecting his chores. Harry had never seen such fear as he saw on his cousin's face, or such fury as on his uncle's. It seemed to Harry that Francis might actually try to kill his son.

It was typical of Francis that he could not forgive Babbo for remarrying. To most people, the December-May romance made the most charming story. To Francis, it was a personal affront. Babbo had met his new wife at a dinner party, where he was regaling one of his guests with quotations in Greek from Anacreon, and he stumbled over a line. At the other end of the table, the Greek floated toward him. It came from the joyous Miss Gabriella Ladd, who, at forty-four, wasn't quite half his age of ninety-one, but was smitten by this learned and charming old gent who walked now with two canes. She wooed him assiduously, and soon they were exchanging daily letters, hers in a gorgeous, looping caligraphy, his in a tight, steady hand. "I have a new name for you," he wrote in one of them. "It is Great Heart." At his age, Babbo was certainly not considering marriage, but Gabriella was, quite determinedly, and to that end sent him a letter from Bermuda in the spring of 1953 saying that she'd met an absolutely delightful young man... Aflutter at the news of sudden competition, Babbo flew straightaway to Bermuda, and shortly afterward he sent Minturn a cable: ALL IS WELL. SHE ACCEPTS.

The wedding ceremony was at Gabriella's church in Philadelphia. As Babbo ambled down the aisle with his canes, he wore a lily of the valley in his buttonhole, a rose-pink cravat, and, on what he called his "Cinderella feet," wonderful soft, shiny English leather shoes that he'd owned for nearly half a century and polished every morning. Everyone's heart was alight at the sight of him. Minturn was his best man; Francis, conspicuously absent. He remained in Santa Barbara, smoldering, refusing to dignify what he viewed as an outrageous embarrassment.

The minister, a gentleman well into his sixties, gently asked Babbo to repeat after him, "Thereto I plight thee my troth"—but he pronounced the last word "trawth,"sending Babbo into a fury of indignation. He stamped one of his canes on the floor and brandished the other one menacingly. *"Troth,* young man, *troth!"* he cried in a booming voice. "Not 'trawth'! *Troth! Troth!"*

Delightful as others found this wedding tale, Francis could not get over it. He blamed Gabriella for luring Babbo into matrimony, and was aghast that she could contemplate having children by him, as she did. "It ain't no good no how," Francis complained to Minturn in one letter. "The idea of having that horror as my stepmother makes my hair wiggle like a serpent's."

Babbo and Gabriella had four delightful years together in Stockbridge, where Babbo felt he had a home at last, even though he was in a wheelchair by the end. When he finally died there in 1957, Gabriella stayed on in the Old House, taking the tiny bedroom next to the big corner room she'd shared with her husband.

Francis did not relent. He sent the widowed Gabriella a stream of vituperative letters:

> *Because you came like a wall between my father and myself, regardless of personal qualifications, the mere sight of you makes me sick. It has nothing to do with either reason or justice, but I loathe you. I will never willingly go near Stockbridge because you are there. Pray be so kind as to stay away from Santa Barbara, where I live.*

Francis's own particular vein of madness, unchecked, had led to something profoundly disturbing. If his illness had started out as the family's standard-issue manic depression, it had by now turned pathological, the mania drawing on deep reserves of frustration and anger to forge a personality that was unspeakably malicious. It was the culmination of the generations of madness that had begun with Pamela, passed through Harry, Arthur George, and so many others, and then, just as it finally seemed to be on the verge of petering out into the eccentricity of Harry II, picked up an extra stock of malignancy from the Shaw-Parkman line. And in Francis's formative childhood, it took on something more potent—an insidious, gnawing despair that was born of all the deaths in the family, the dispossession, the drift, and caused him to lash out in a frenzy at everything that, in fact, he held dear. While previous bearers of the Sedgwick disease had ended up torturing only themselves, Francis turned his anger outward. If Minturn was the paterfamilias of this generation, Francis became his nemesis, trying to wreck the edifice that Minturn was trying to build up. Unimaginably brutal as he was to his children, to Gabriella, and to my father, he also ended up attacking some of the essential values of the family, singling out in particular the idealistic ones pertaining to race, as if he was determined to bring down whatever honor remained in the name of Sedgwick.

The battleground for this internecine war was not Stockbridge, where the Old House was now safely in the hands of the Sedgwick Family Society, but the new family locus, the Groton School where Halla died, and whose Rector still reigned, even in death, as the spiritual father of Minturn and Francis both. To them, Groton stood as a kind of idealized society, where money and principle combined to represent the height of gentlemanly aspiration. The school had always been exclusive, its ranks confined to the sons of the industrialist class. And, needless to say, all white. But in the 1950s, the Rector's successor, a thoughtful Episcopal minister and Groton graduate named John Crocker, realized that the school was not living up to its Christian ideals if it persisted in limiting its offerings only to white children of privilege.

So at Crocker's direction, in the fall of 1952, two years before the *Brown vs. Board of Education* desegregation decision, Groton School courageously admitted its first black student, a Washington, D.C., boy named Roscoe Lewis, making it one of the first of the elite northeastern boarding schools to desegregate. It was difficult, however, for the school's admissions department to find other black students who could handle Groton's competitive, high-church atmosphere, and the school had added only two more blacks by February of 1957, when the racial issue drew Francis's attention.

That month every graduate of Groton School—several thousand men altogether—received in the mail a stunning letter, typeset, on the subject of the school's plans for further desegregation. With the crisp formality of a press release, the letter declared:

> *In consistence with Christian doctrine and the teachings of the Bible and in consistence with the human beliefs of two of Groton's most eminent graduates, New York's Governor Averill Harriman and the late President Franklin Delano Roosevelt, Groton announces its irrevocable intention to increase the number of negroes from a few students to not less than one quarter and not more than one third of its total enrollment.*

The notice went on to pledge that the school would, if necessary, devote to the cause "the full use of its entire endowment fund."

Although the broadside was quickly determined to be a hoax, newspapers pounced. To limit the damage, Groton announced that the school's policy on "acceptance of Negro boys...remains unchanged." The question was: Who had sent such a hateful, underhanded letter? To identify the culprit, the school prevailed upon specialists at the post office in Washington, D.C., to examine an envelope that had been addressed by hand to Roy Wilkins, then the head of the NAACP, and compare it to handwriting samples of seven individuals it suspected. The specialist needed only a glance at the distinctive handwriting to determine his answer: Francis M. Sedgwick of Santa Barbara, California, Groton class of 1922.

Not wishing to inflame Francis any further by accusing him publicly, Crocker delegated the delicate job to Minturn, who served the school as a trustee. He flew to Santa Barbara, where Francis met him at the airport, then drove him back to the ranch. "They have any idea who wrote that letter?" Francis asked nonchalantly as they drove along.

"Yes, they know," Minturn replied. "It's you."

It should have been a chilling moment, but Francis expressed no embarrassment whatsoever. He later claimed his brother was the startled one. Francis was about to come clean in a second letter to all Grotonians—he'd gotten, he said, down the alphabetical list "to the V's"—that would declare himself the author of the previous one, and then inflame the situation further by asking, "Why are there negroes at Groton? And, Why are there *so few* negroes at Groton?"

Psychologically unbalanced as Francis was, he had an unerring instinct for the vulnerabilities of those he wished to antagonize. Groton did have a tight limit to its good intentions: there were practical and political constraints in a tradition-minded school of limited means. In his mean-spirited delusions, Francis had no such compunctions. He was a vulgar racist of the sort that breezily compared black people to chimpanzees. Plus, he was agonized by his unacknowledged grief over Babbo's death, which occurred just a month before the first letter. Already furious to see his father "lost" to Gabriella, Francis had pointedly not attended the funeral. But his face was noticeably ashen, and he had taken to walking with a cane.

His seemingly irrational attack on Groton had its own sick logic. As he later revealed, the "real" reason he was so upset by the prospect of black Grotonians was that he wanted to spare them the frustration of not being able to marry the upper-class white women their white classmates would pair off with. What Francis really meant was that he was afraid that blacks might make off with his own daughters, a common enough racist preoccupation. In Francis's case, there was one extra detail. Francis himself had already made an incestuous pass at his oldest daughter, Saucie, after he saw some college boys flirt with her at a party at his mother-in-law's on Long Island. "Don't you think I understand how all these men feel about you?" he purred, as he sat beside her on her open bed. "After all, I'm a man, too." She rejected him with disgust, and declared years later that she'd found him a "filthy old creep."

The fact that Minturn—the "Duke" who made Francis, as he was often called, "Little Duke"—was a Groton trustee only added to the familial complications. Having seen Minturn take over the Old House, and muscle in on the Rector they both adored, Francis's own sibling rivalry was already operating at a fever pitch. Francis was now vying with his brother for the "correct" interpretation of the family's legacy, with Francis setting himself up as the one who truly knew what the Rector would have wanted for his school. It must have galled him that, instead of fighting back, Minturn turned fatherly. He sent his brother a blatantly paternalistic letter expressing his dismay about the whole hoax business. "It is always a matter of regret when a graduate loses confidence in the School and particularly so when he is one like yourself who has had three boys there and cares deeply," Minturn began, in a measured, judicious tone that must have added to the insult. He then turned more forceful, registering his disagreement with Francis's "premises" and his "conclusions," and asserting that, as the Rector's son-in-law and a trustee of his school, he could speak far better for the Rector than Francis could—a statement that must have stung. He ended on a reconciling note that only Minturn could strike: "As you see we are poles apart but that is life." He signed it only with his initials, like a memo.

Francis had indeed sent all three of his sons to the school he was now trying to destroy. The oldest, Bobby, had graduated without incident in 1951, and the youngest, Jonathan, was a junior at the school while Francis was waging war upon it. His second son and namesake, Francis Minturn Sedgwick Jr., nicknamed Minty, had graduated just the year before, and his experience there may be another key to the mystery of Fuzzy's wrath. For Minty had done poorly at Groton and, to Francis's bitter disappointment, been turned down by Harvard. He had been forced to attend Berkeley that fall—in Francis's mind, about the most ignominious development imaginable. For months before the hoax letter, Francis had been sending long, angry letters blaming the school for this outcome, which is what alerted Groton to suspect him of the crime in the first place. Pathetically, his letters of denunciation overlapped with Minty's own tender expressions of gratitude to the school for helping him. More distressing for Francis, Groton had come to see that the source of most of Minty's problems was Francis himself.

Minty was a gentle, artistic, sensitive child in a rough-and-tumble household—a ranch, after all—presided over by a man who, disturbed by his own homosexual inclinations, regarded gentleness, artistry, and sensitivity as dangerously effeminate qualities. The tensions left Minty riddled with phobias, chiefly of water, horses, and flying. Exerting relentless pressure on the boy, Francis made Minty a swimmer and a champion horseman, and sent him to a boarding school that was a plane ride away. There, Minty developed into a troubled, underachieving student.

To puzzle out the matter about why a boy with such obvious talent should be floundering, Groton sent him, in February 1955, to be examined by a psychologist named Frances B. Holmes. After a series of interviews and Rorschach tests, she concluded that Minty suffered from "a real lack of feelings of self-acceptance and personal worth" and asked to interview the parents. Francis and Alice were loath to oblige her, but ultimately did fly east for a consultation. Dr. Holmes was clearly put off by a duo who did little but heap scorn on a son who appeared to be nothing but smart, amiable, and well-intentioned. She wrote:

> Their attitude towards Minty appeared to be quite rejecting. I don't remember that either parent said anything good or appreciative about the boy except to agree with me that he was intelligent. Their main reason for withholding their approval of Minty was that they consider him a "coward." This word was used by both parents over and over in what was almost a sadistic way. It was a rather startling and unpleasant experience to hear a mother call her son a coward repeatedly and state that he "ratted" in his refusal to play football etc.

This summation did not go down well with the parents, but it did encourage Minty. The following spring, he resolved not to go home to Santa Barbara for vacation. He went through four drafts before finding the right note of assertion in his letter home explaining his decision. Moments after receiving it, Francis telephoned Minty long-distance to tell him he had thirty seconds to change his mind, or be disowned. Happily, Minty managed to steer a middle course. Hearing about the standoff, Minturn drove out to Groton to take his troubled nephew out to lunch and hear his side. He wrote Francis that Minty had "stated with touching emphasis that he was dreadfully unhappy and frustrated at the ranch, and he would not go there for his holiday." At his own expense, Minturn put Minty on to a psychiatrist who confirmed this appraisal.

This time, Alice flew out and, after a long, tearful conversation with her son, worked out a compromise whereby he would spend part of the vacation at the ranch after all. Francis followed that up with a diatribe to the school saying that if it had only forced Minty to play football, he would have overcome his "confounded timidity." As it was, he said, "the little psychiatric lady" placed the "blame" for Minty's distress on him, Francis, poisoning his paternal relationship with his son.

That June, the Harvard rejection came. Alice laid that at Groton's door: "This proves we were not too strict, but the School too lenient, with Minty."

Minty slept through two months of Berkeley, then enlisted in the army for three years, where he did well, emerging with a medal for sharpshooting despite his fear of firearms, a "best-soldier" rating, and a near fluency in the Mandarin Chinese he'd studied at the army's language school. He bowed to his parents' wishes and applied to transfer to Harvard, which accepted him this time. The pressures proved too much for him, and he secured a medical leave a few months later. He lived with his de Forest grandmother in New York, but quickly ran through the small trust that had been established for him. Although he'd been an AA loyalist, Minty started drinking heavily again, and his behavior turned increasingly bizarre and frightening. When his oldest sister Saucie found him one morning in bloodied pajamas, maniacally waving a baton to "conduct" the music coming out of the stereo, she had him placed in the psychiatric ward of New York Hospital. Eventually he ended up at Silver Hill, a Riggs-like mental hospital in the woods of New Canaan, Connecticut, that had been founded by Dr. John Millett, the psychiatrist who'd treated Francis at Riggs. Minty was still unsteady when he emerged, and in October of 1963 he was found making loud, crazy speeches to no one in Central Park. The police took him to Bellevue, where he screamed about "helicopters in the sunlight." He called himself Francis now—but was unsure if he was Francis Sr. or Jr. His parents returned him to Silver Hill, and his father finally flew east to see him there six months later. He told him he had to stop all the nonsense and buck up, but Minty only burst into tears and told him he had fallen in love with another man. His worst fears realized, Francis screamed, "You're no son of mine!" They were his last words to his son.

One evening a few days later, the staff noticed Minty had not come down for dinner. A maid was sent to investigate, and she found him hanging off the bathroom door from a noose he'd made of his necktie.

The next morning, Francis sent a brisk letter to Rev. John Crocker at Groton. He noted that it was Minty's birthday, and after once again pressing his interpretation of the spring vacation incident, he closed: "Yesterday, Francis Minturn Sedgwick, Jr. hanged himself."

At the end of the month, he typed out two more lines for Crocker:

"Try to wash those blood stains off your hands, Judas Iscariot! They'll never come off—God damn you." In May, when he learned of Crocker's plans to retire, he wrote: "GOOD!" He followed that with his final circular, his fifth. It was headlined "In Memoriam...1938–1964," in which he ratcheted up all of his claims against Groton to come just short of accusing the headmaster of murder.

That was the fall that my father had his aneurysm.

THEN IT WAS Bobby's turn to fall. Francis's oldest son, born in 1933, he had always been the most handsome of the Sedgwick boys, the most talented, and for all that, the most complicated. Aware of his beauty's effect on people, he used to carry a pocket mirror to admire himself, as an attractive woman might. And he frequently wore an odd Cheshire-cat grin that suggested a serene detachment that was largely illusory. After Groton, he'd gone on to Harvard, where he'd majored in fine arts and made the Porcellian Club, much to his father's excitement. But none of this meant much to Bobby himself. Like Minty, he'd also been beaten down by Francis, who accused him of being "just a good-looking Airedale." He took up with a raw-boned Radcliffe girl named Randy Redfield, complaining to her about the impossibly high standards of his domineering father. His sophomore year, he finally snapped. He was taken to a psychiatric hospital in a straitjacket. On his return, he dabbled in Marxism, and hung on to graduate. After a brief stay in New York, he left to do some union organizing for the International Ladies Garment Workers Union in Kansas City. For a time he had a black fiancée, before taking up with a married Jewish woman with four children. He finally drifted back to New York, where he lived in a railroad flat and settled into a catatonic depression.

Saucie, always the one in the family to pick up the pieces, put him in Bellevue in August 1963, a few months before Minty ended up there. Lacking the money for any better care, she telephoned her parents to beg them to help, but Francis slammed the phone down in disgust. Bobby was shifted to Manhattan State Hospital, and finally to the Institute for Living in Hartford—formerly the Hartford Retreat, the psychiatric institution that had housed the first Henry Dwight Sedgwick. He was there when Minty died. It barely registered, he was so heavily sedated.

Bobby's old Harvard professor, Sidney J. Freedberg, managed to spring him to do some graduate work as a special student in fine arts. Back in Harvard Square, Bobby hooked up with a lovely Mexican girlfriend with raven-black hair, and he bought a motorcycle, a huge Harley-Davidson. He was reckless on it, though, and never wore a helmet. Late on New Year's Eve, he was roaring up Eighth Avenue in New York, trying to beat the lights as they turned, when he slammed into the side of a bus that crossed in front of him. He never regained consciousness, and he died on January 12, 1965, not quite a year after Minty. His parents did not come east. At their instructions, Saucie sent his ashes in a cardboard box to Santa Barbara, care of general delivery.

At the ranch, Francis combined Bobby's ashes with Minty's in a small wooden box. Rather than return them to the Pie, he conducted a private ceremony all of his own. Francis saddled up his favorite horse, which he decked out in full, grand Spanish regalia with *tapaderos* stirrups and splendid decorative tassels hanging from the bridle. Then he pulled on his leather leggings and his wide belt, set his Stetson on his head, and rode up to a high ridge where the boys had all galloped about, years before. And there, alone, Francis opened the box and, jerking it into the air, scattered his sons' ashes to the winds.

THIRTY-EIGHT

EDIE, SUPERSTAR

I'd met Minty only once, the summer before he died when I was nine. He spent a night with us on the Cape. He wore black pants and a long-sleeved black shirt, very unusual for a sun-splashed summer place. He seemed hesitant, spooked, unsure how to fit in with these rambunctious cousins he only dimly knew. I got word of his death when we were back in Dedham the following spring, but in memory I place it at our house on the Cape, probably because of all that funereal black he'd been wearing when we met. Otherwise, the news that he had hanged himself in a mental hospital had little weight, for I couldn't conceive of such a thing.

My sister Lee was in the room when Dad received the telephone call about the death from Francis. He'd repeated a few of the grisly details over the phone—hanging, psychiatric hospital. It wasn't until he'd put down the receiver that he realized that Lee had been listening. "I'm sorry you had to hear that," he told her gravely. But he said no more.

I never did meet Bobby, but I had one of his paintings. I got it from my father around the time that Minty died, but I didn't combine the two events; I was too young to make sense of such a sprawling family I never saw together, only in ones and twos, like refugees from a distant country. I had little use for art in those days, but I treasured Bobby's picture. It was a miniature version of the art that would later be called psychedelic, a rectilinear labyrinth done in brilliant, electric colors, painted on a linoleum tile. It was both adventurous and finely controlled, and it evoked an imaginative realm that was thrilling to me.

I don't recall being told of Bobby's death, but I have had a dread of motorcycles ever afterward.

The only one of those cousins I knew at all well in those years was the youngest, Suky. She came to live with us in our little gray house in Dedham in the fall of 1962, while she attended Winsor School in Brookline for her junior year. Silky-haired and curvy, she made a big impression on Rob and me. On Rob because she was so blithely sexy, so abundantly physical—probably the first person to live with us of whom that could be said—and on me because I'd never known anyone so keenly alert. Plus, she was a brilliant pianist. Her sheet music was absolutely black with notes, and her hands turned nearly invisible as they flew over the keys. I'd been trying to learn myself, and I couldn't imagine how she could cover the keyboard as she did. She leaned in to the piano, too, as if it were a somewhat stiff but loving dance partner she needed to steer around the room. She bit her fingernails to the quick, which I thought so sophisticated I started doing it, too. I've never stopped.

Suky moved into her own apartment in Brookline the next fall for her senior year at Winsor. It was unusual for a high school senior to live on her own, but she was unusual, and I didn't question it. She stayed in Boston to attend the New England Conservatory the following year, when her dazzling older sister Edie arrived in Cambridge.

Only twenty, Edie had already lived several lives by then. Slim, leggy, doe-eyed, lustrous, she caught everyone's gaze and held it until it hurt. Dad was always transfixed by her. She'd been in boarding school in San Francisco as a young teen, but withdrawn after she developed an eating disorder, alternately picking at her food and wolfing it down, and then vomiting up anything that reached her stomach. So she returned to the ranch, where she was the only child; all the rest were off at school. She thought herself a prisoner there. Then one afternoon she glanced into a spare bedroom and found Francis there fucking—there is no other verb—a woman from town. He swore at Edie, claimed she was seeing things, irrational, and, when she responded with hysterics, summoned a doctor to give her tranquilizers.

The whole experience tamped Edie down drastically, and it never left her mind. (She can be heard mumbling about it in the documentary-style

Ciao! Manhattan, heavy on interior monologues, the last film she made.) She was sent off to the private all-girls' school St. Timothy's in Maryland the next fall, in 1958. Francis had started to turn against Groton by now, and Edie bore the brunt of some of her father's fiercer delusions when she returned from school. The eating disorders returned with a vengeance, and the fall Suky came to stay with us in Dedham, Edie was delivered to Silver Hill, the genteel mental hospital Minty had already attended. I think my father's idea was that if Suky was with us, she might be saved.

Dad was right to be fearful about those children. Edie's story reaches past the particular terror of the stunning downfall of a great beauty to evoke something central to her era, something that is otherwise hard to capture or express. It is part of the persistent, if sometimes demented, genius of the Sedgwick family that its members have so frequently come to represent a significant aspect of their time, whether it be through the Judge's stout Federalism, Catharine's Victorian pathos, or Ellery's traditionalism. In many ways, Edie outdid them all to become the very personification of the sexy, incandescent nowness that remains the most distinctive aspect of that strange, psychedelic era. In the ranks of the Sedgwicks she stands unique in another way as well. Although Edie's style became *the* style, she could never really be said to *do* anything at all. She just *was*. A presence. That was a first for the family. And so she represents another turn of that great wheel, from active to passive to an eerie neutrality, just being there, for others to make of what they would. She was like one of the pacifist Buddhist monks who maintained a transcendent serenity as they burned themselves alive. There is the same eerie silence as Edie's flaming mania consumed her, while a nation watched, mesmerized. This was her brilliance.

SILVER HILL PROVED too indulgent for Fuzzy's taste, so he shifted Edie to the much stricter Bloomingdale, the Westchester division of New York Hospital, which had a more militaristic feel. Toward the end of her stay there, she met a Harvard student and had sex with him at her grandmother's apartment in New York. She got pregnant but had an illicit abortion—"on the grounds of [being] a psychiatric case," she whispers in *Ciao! Manhattan*. It was her first sexual experience.

That was when she arrived in Cambridge to study sculpture with our cousin Lily Saarinen, the ex-wife of the architect, Eero Saarinen, who'd designed the wonderful, floating TWA terminal at Kennedy airport. Edie spent almost a full year laboring over a single clay horse, almost full-sized. It must have evoked the best part of ranch life for her. Cambridge was just a half hour away from us in Dedham, but her world could scarcely have been more removed from ours.

To keep an eye on her, my father invited her several times to Sunday dinner, a tedious affair with roast beef, cheap sangria, and the same Churchill stories. She came only once, arriving behind the wheel of a gray Mercedes her parents, in a moment of weakness, had bought for her, the sort of car my parents would never own. It was a hot, bright day, and she wore sunglasses, a tight blouse, and short pants. I was spellbound. Edie livened the lunch up considerably, though, by skipping out on dessert to slip out through the French doors to the backyard, where she stripped down to her panties to sunbathe half naked—I remember nothing but a sense of spectacular whiteness against the green lawn—while the grown-ups sipped coffee from demitasse cups, unaware.

EDIE DEVOURED ALL there was to Cambridge back then. Like Boston generally—which had still not recovered from the decline after World War I—Cambridge wasn't much, just sandwich shops, used bookstores, and tweedy Harvard boys. She went to a psychiatrist most days, and did her sculpture, and fended off the boys—many of them proper, three-named types—who started to fawn over her. She was drawn to what there was of a homosexual scene, partly because the men would make no demands on her. She set up court underneath the Brattle movie theater in the Casablanca bar, where everybody went. She had her first legal drink there the day she turned twenty-one and came into her grandmother's trust fund. (Years afterward, when I was at Harvard, I had my first legal drink there, too.) After that, she much preferred the Ritz, where she once stood on top of a table and, in her breathy voice, crooned a Richard Rodgers show tune, "Loads of Love": "I want some money, and then some money and loads of lovely love."

Bored with Cambridge, Edie drove her Mercedes to New York, which was brighter, bigger, faster. After a stint at her grandmother de Forest's on Park Avenue, she rented a place of her own in the East Sixties off Madison. Hoping her daughter would settle down as a nice socialite,

Alice came east to help her decorate with embroidered pillows and scatter rugs. But Edie started to dress more wildly, wearing high-heeled boots and black stockings that drew attention to her legs, topped off with fox-fur waistcoats. It was an outrageous getup, but it was totally her. She developed a taste for Bloody Marys, and for the new European-style discotheques. After she absentmindedly totaled the Merc, she got around by limousines. Her trust fund money flowed like water.

Fuzzy didn't like the way things were heading, and he summoned her back to California. By a weird coincidence, she cracked up a family car there at almost exactly the instant Bobby was killed on his motorcycle, running a red light on that New Year's Eve, and breaking her knee. Afraid her father would send her back to Bloomingdale for her misbehavior, she conspired with her mother to fly back to New York, where she chiseled off her cast and went dancing at an uptown discotheque called Harlow. Men crawled all over her, but she just laughed at them.

All except for Andy Warhol. She met him at a party at the penthouse of an openly gay ad man named Lester Persky, known as the Wax Queen. A source of the New York coolness that would be a big part of the 1960s, Persky used to collect people like Truman Capote and Tennessee Williams—and Andy Warhol, who was searching for a new fashion "superstar" (a word he coined; in his perpetual-teenagery way he loved super anything) to champion, now that his old one, Baby Jane Holzer, was, as Persky told him, "running out of speed." Lester turned Andy on to Edie, who was everything Warhol was not. A former commercial illustrator who'd been much in demand for shoe ads, his fascination with brands elided into the Pop Art ideal of life as image, and then into image as icon, that place where mass publicity merges weirdly with total anonymity. For him, Edie was a brand name only, whose selling power came, aside from her cover-girl beauty, from being steeped in subliminal associations with old money, New England prep schools, European aesthetics, fine art, generations of breeding, and a certain knowingness—everything, in short, that Edie ran from as *nothing at all*. But it was everything to a bewigged, asexual, pale-faced, hollow-hearted immigrant's kid from the steel town of McKeesport, Pennsylvania, who had, along with an innovative brilliance, an uncanny feel for the class consciousness that underlay a big part of the 1960s zeitgeist. He destroyed her, but he made her first.

A year or two before, Andy had moved into a loft on East Forty-seventh Street that was called the Factory, ostensibly for the silk screens produced there. It was a glamorous wreck of a place, with drab used furniture and tinfoil wallpaper and bare floors, where the newly, briefly cool like Ondine, Viva, Chuck Wein, Paul America, and Billy Name had been prominently on display. With her glow, and Andy's blessing, Edie quickly became the Factory queen. Edie made the scene, and made it a scene. For celebrity, of course, was the Factory's primary product: those fifteen minutes Andy would later speak of, in that weird, flat, uncaring voice of his. Just moments, flashes, like the quicksilver reflections on its tinfoil walls. When it was happening, it was really happening. When it was over, it was hard to know what, if anything, was left.

Andy was starting to make his experimental movies when Edie arrived, and he wanted her to be in them. Most were mind-numbing cinematic exercises that, as Warhol himself admitted, were more interesting to talk about than to watch. The ones that Edie were in, though, were the other way around, even though there was absolutely nothing happening in them except Edie herself. I'd never seen one until fairly recently, when Harvard put together an Andy Warhol retrospective at its Carpenter Center. It was an eerie cinematic experience for me.

I saw *Outer and Inner Space* and *Poor Little Rich Girl,* Edie's two star turns. Both films have the humdrum immediacy of home movies, for they're just Edie being Edie, nothing more. No script, no story. Of the two, *Space,* shot at the Factory (you can see the silvery walls in the background), was the more considered, and it was done later, in August of 1965. It has not one Edie, but four. For the movie is two films projected side by side, simultaneously, each of them showing Edie in close-up, great peacock earrings dangling off the sides of her face and her hair done up à la Marie Antoinette as she chatters to Warhol offscreen, while another, slightly larger Edie in profile talks to her from a TV set. It's Edie as animated pop icon, a Marilyn or a Jackie. She is certainly captivating. Her skin has the creamy softness of skin that's never been touched by anything but air. And the alertness of her, the way she is so plugged in to Andy, her eyes locked onto him, her voice breathless for him, it is almost perverse, like something we shouldn't see, like sex.

Take away all the artsy, high-tech proto-Cubism, and it was a screen test of the next Judy Garland. Warhol saw that in her, and more, he saw a kittenish innocence, too, which gives all the cameras—and the sense of

Warhol himself as the man behind all the cameras—a ghoulish quality. It's the movie of a woman being devoured by her own image.

The second one, *Poor Little Rich Girl,* had been shot a few months before in March 1965, one of the first of Warhol's Edie movies, and it consists, for the first half hour, of a luminous white cloud drifting across a hazy interior. Andy had neglected to focus the camera, and then decided he liked it like that. It makes the movie a Rorschach test. At first, I saw in the soft focus a Hollywood glamour shot. But once Edie sharpens, the image turns beautiful and tawdry. It's Edie in a skimpy bra and underwear stumbling about her cramped bedroom, where she would later be burned in a fire. She's trying to get dressed, mumbling words that the microphone barely picks up, and which probably meant nothing to most of the film's few viewers, but hit me hard. "I've got to get to the party for Bobby," she says halfheartedly, referring to the brother she was closest to. This was March, and he'd died that January. She doesn't seem bothered, but everything in the scene evokes disarray, disorder, dysfunction. She's ordering coffee over the telephone, smoking pot from a porcelain pipe, and doing her exercises to the Everly Brothers—"Wake Up Little Susie" plays on an unseen record player. She's out of it, but In.

THAT WAS 1965, probably the height of Edie's fame. She was showcased in *Vogue* that year in black tights as a "Youthquaker," as she balanced with balletic grace atop a stuffed rhino in her apartment. She and Andy were *the* item that year, and they went everywhere together—his movie openings, discos, parties—her hair frosted silver to match his wig, as if they were not just paired but twinned. *Life* magazine ran a big photo spread on the "cropped-mop girl with the eloquent legs," and the gossip columns couldn't get enough of the starlet they dubbed "The Girl of the Year." Mick Jagger grabbed for her amid a mad crush of groupies at a nightclub called the Scene. The press mobbed her at the opening of the Warhol exhibit at the Philadelphia Museum of Art.

We subscribed to *Life* in Dedham, so the pictures came into our house. But the excitement somehow passed me by completely, probably because my mother disapproved. Breathing some of the reckless 1960s air, I'd acquired a drum set to start a group of my own with a few friends from school, but I had no idea that Edie was It. My father got caught up in all the Edie publicity; he was always interested in any fame that came to the family, and he talked up Edie to his Boston friends. Dad was well into his sixties now. Rob was off at Groton, which took some of the pressure off the household, and Dad had recovered from the aneurysm, but he was starting to stoop with age, which irritated him, and his hair was thinning at the crown. Still, he started to get into the new fashions, and began wearing bell-bottoms around the house, much to the embarrassment of my sister and me.

Francis allowed himself a little enthusiasm, too, and at Edie's urging, agreed to meet Andy Warhol one evening at the River Club, the old-line watering hole by the East River in New York City. It was a collision of eras and mind-sets, and it must have been a mind-bender for Edie to see the two dominant men in her life, both of them tyrannical artists, despite their opposing demeanors: Francis with his colossal equestrian statues that had been Out for a century; Andy with his oversized Campbell's soup cans that were totally In. The clash of cultures was so complete, there was no clash at all, for there was no possible point of contact. It was like two armies that have mistakenly massed for battle on different continents. Handsome, impeccably dressed, erect, Francis ordered drinks, tried to chat. Decked out in sloppy coolness, the whispery-voiced Andy barely spoke at all. The encounter was over before it began.

Afterward, Francis turned to a friend and nearly shouted with relief: "Why, the guy's a screaming fag!"

Edie thought of trying to parlay her appearances in the Warhol films into Hollywood movies, but she didn't have the energy, or the nerve, for anything so career-minded. Before long, Warhol and Edie started getting on each other's nerves; the need on each side was too great. He stopped including parts for her in his films, and she took the hint. She dated Bob Dylan, who wrote the song "Just Like a Woman" for her, played the diva in the back room of Max's Kansas City on Seventeeth Street, did singing gigs with the Velvet Underground at Delmonico's and other places, and finally decided to be the American Twiggy.

With her looks and fashion sense, that might have been a possibility, except for the drugs. Her "vitamins," she thought of them. Which they were, at first, albeit vitamins heavily laced with speed and injected directly into her buttocks. "Buzzerama," Edie called it once when she was flying, "that acrylic high, horrorous, yodeling..." Then it was cocaine, and then heroin. She took her first shot in the arm at her apartment, and it sent her running naked out onto Park Avenue. Drugs were a kind of sex for her, and vice versa. All night, she'd dance and fuck and do drugs and fuck and dance some more. Then she'd sleep all day.

In one heroin stupor, she set her apartment on fire while she slept: the stuffed rhino at the foot of her bed exploded into flames, turning the room an acrid orange. Firemen burst in and rushed her to Lenox Hill Hospital. When friends heard what had happened, they called her parents in Santa Barbara, but neither Francis nor Alice would come to the phone. With two sons already dead, they couldn't face another corpse. My father was in New York at the time, and he immediately hurried to Edie's bedside. Amazingly, she had only one bad burn, on the inside of her arm. She asked him for makeup, and he bought some for her from the hospital pharmacy, and then watched, aghast, as she blackened her beautiful eyes with it, turning her face into a death's head.

Her parents hauled her back to Santa Barbara for Christmas. She was a waif, a junkie, a shell. Stunned, Francis had her committed, but a boyfriend, an artist named Bob Neuwirth, rescued her and brought her back to New York. But by then there was almost nothing left. One night as she and Bob drove by limousine to a party, she was suddenly fed up with him, opened the back door, and tumbled into the street. "I thought I was going to explode," she says in *Ciao! Manhattan*. An oncoming truck swerved just in time to avoid crushing her. She hunkered down in the Chelsea Hotel, the artists' hangout on Twenty-third Street. But, drugged out of her mind, she set the place on fire when she tried to bake a sweet potato. She wasn't badly burned, but she couldn't have looked much worse, with long hospital bandages scrolled around her arms as if she were a mummy from a horror film. She got a bit part playing Lulu in a ghoulish film sequence to be interspersed in the Alban Berg opera about a woman who devours the men in her life. Edie was too out of it to grasp the larger themes, but she did the movie anyway. She'd gotten hooked on stardom.

EDIE RETURNED TO New York just before her father died. He'd been ill for at least a year with what proved to be pancreatic cancer. His mortality softened him slightly. At the end of 1964, he'd written a letter to Rev. John Crocker asking for forgiveness. "My profoundest apologies for the odious, swearing letters I wrote you at the time of Minty's suicide. I wish it were possible to retract them. As it is I can only apologize. I was at a

pitch of anguish and poured it out on you." Still, he could not resist a last
dig at Minty for his "public cowardice," which he insisted was "the beginning of his troubles," and a final swipe at Crocker for siding with his son
against him. He closed by "extending an olive branch—rather chewed
and measly, it is to be feared." Crocker generously wrote back, accepting
his olive branch, "and offering one...in return."

The friendship was not renewed, however, and the other relationships
in his life remained tattered as well. Francis took up with a young married woman in town, Ann Morrison, reading her Balzac and Stendhal
short stories to show off his cultivation. She was Catholic, and he converted for her. His last sculpture was of Saint Francis, his arms out-
stretched to the sky as he receives the stigmata, which was fully in keeping
with his own exalted view of himself as heroic victim. It was placed in
front of the Mission Santa Barbara, where he received the last rites from
Morrison's priest, Father Virgil. He died in October.

I'D ARRIVED AT Groton in the fall of 1967 as a well-scrubbed eighth
grader, and one morning in October one of the older masters, a wizened
creature named Corky Nichols, came down the stairs in the School House
toward me with the *New York Times* in his hand. He asked me if I was related to Francis Sedgwick. My own relations with Uncle Francis were so
remote that I had to think for a moment before I answered. Nichols told
me he'd died of cancer; he'd just read the obituary. I'd had no idea he was
sick. The exchange was brief, and unfeeling. Now that I know the tortured history of Francis's connection to the school, I can see why. Nichols
was the unofficial historian of the school, and I'm sure he felt very little
regret that Groton's great irritant was dead.

Rather than be buried in Stockbridge, Francis's ashes were scattered by
his only surviving son, Jonathan, who misguidedly released them *into* the
wind. The ashes flew back into his face, filling his mouth.

WHEN FRANCIS DIED, Edie was in Gracie Square Hospital, a psychiatric hospital specializing in drug disorders. She was trying halfheartedly to
get over her various drug addictions. A screenwriter named L. M. Kit
Carson had fallen in love with her, and he finally got her out. They lived
together for a time in New York's Warwick Hotel, but the relationship
was hardly blissful, for she wasn't clean, and never would be. He threat-

ened to kill himself if she didn't stop, but she was too crazed to care. Exasperated, Carson left for Texas. Edie flipped out on speed and barbiturates, ran down to the hotel lobby, and ended up in Bellevue for a few days, and then to Gracie Square once more.

The rest of it is a blur of hospitals—Lenox Hill, New York State Psychiatric Institute, Manhattan State, Cottage Hospital in Santa Barbara. In New York, the newspapers that had fed off her every move now published mean-spirited where-is-she-now articles. In her few moments of lucidity, Edie flipped through a photograph album she'd managed to keep. One of her last acts was to complete the movie *Ciao! Manhattan,* which a director named David Weisman had begun a few years before in New York. Half documentary, half hallucination, the film was meant to chronicle a day in the life of the former Superstar. It was Edie's *Sunset Boulevard*. For a set, the producers had rented a castle on the banks of the Hudson and got the poet Allen Ginsberg involved, and the whole thing made sense only if you were stoned out of your mind. The production fell apart, but Weisman persisted, starting in again in Santa Barbara, and Edie made the movie her life and vice versa. The film has a high-1960s depravity, with an obviously drugged-out Edie lounging about most of the time half naked at the bottom of an empty swimming pool, showing off her boob job. It's a film about a slow suicide, and when I saw it recently on DVD, I couldn't bear to watch it to the end. Much of it was improvised, since Edie was too wasted to remember many lines, scarcely even to stand up, but the filmmakers continued to grind away.

After the filming, Edie went in for shock treatments in Santa Barbara's Cottage Hospital to try to rewire her mind, but it was no use. She met a fellow addict named Michael Post there, not yet twenty to her twenty-eight, a sweet boy who genuinely cared for her. They got married in the summer of 1971. It was sudden; no one from my family went, although my father would have loved to be the one to give her away. The wedding was at the ranch and formal, with Edie in a beautiful white dress and Michael in cutaways. Afterward, the bridal party all ended up naked in the swimming pool.

The Posts lived together in a tiny apartment in Santa Barbara; Edie called Michael "Daddy." She was fading fast. One night, she attended a fashion show in Santa Barbara that was being filmed for the documentary *An American Family,* since Lance Loud, one of the family members, was there. Edie appears in a few frames; it's her last film appearance. She returned late, exhausted, but as always too wired to sleep, so Michael gave her some quaaludes. When he awoke the next morning, she looked like she hadn't moved the whole night. He gave her a shake—her shoulder was stone-cold.

The police came, and took her body away. After an autopsy, the coroner recorded her death as "accident/suicide." As for the cause, he put "barbiturate over-dose."

It was a cruel end to a gorgeous young woman whose beauty was not entirely exterior, and it marked a turning point for the family as well, since it was the last time a Sedgwick held the nation in thrall. With Edie's passing, the family went into a kind of senescence, broken only by an occasional intrusion of one Sedgwick or another into the national conversation, for an art exhibition in New York, a turn with Miles Davis, a book, a movie performance, or TV show to name a few of these media flurries in the years since. But the time of prominence for us had passed, turning the glory days themselves into something Edie-esque, a spray of light from a comet that briefly dominates the heavens but leaves precious little behind. Not all of these starbursts through the years have been manic, but Edie's was, in its frenzied, pointless sexuality, its craving for a stimulation that would substitute for genuine engagement, and its glittering emptiness. Her mania was, of course, compounded by the mania of her father's, and turned morbid by the suicides of her brothers. A mania built on mania. As such, it was a kind of full flowering of Sedgwickness—but it is the special talent of the family that our own mental fits are not solely self-destructive delusions, as is so often the case with others, but have an eery way of connecting. It sometimes seems as if Sedgwicks, like dowsers, have a weird affinity to the deep currents of their era. I feel that tug myself, but in a much smaller way, in an ability to forge a quick, often too-easy intimacy with people I have just met. I have not been able to make too much of that so far. But having trained my eyes on the stars, the better to see the Sedgwick comet trails, I now carry the family memories as few others do, leaving me with the perspective of a fish-eye lens, at once distorted and unerring.

THIRTY-NINE

"ONE LOVES TO REMEMBER BEAUTY"

After Babbo died in 1957, Gabriella continued on in the Old House, but she was never comfortable there. She slept in the little bedroom at the top of the stairs, one that was tucked between the two vast ones. A free spirit, she did dance, yoga, and meditation, and often drifted off to the cabin of a young actor named Bill Roerick in the woods of nearby Tyringham. But she was never one to perch anywhere for very long, and loved to nip about Stockbridge on that moped of hers. For longer trips, she would take the bus, gaily chatting up her seatmates. One of them was an American Indian who'd been indicted for murder. Unfazed, Gabriella put her in touch with a friend from town whose brother was a trial lawyer. She also took an interest in our dotty cousin Louis Agassiz Shaw II, Harvard '29 and a brother Porcellian, whose manic depression had inspired him one afternoon to strangle his sixty-year-old chambermaid with his bare hands at his Topsfield estate, adjoining the Myopia Hunt Club, because she was "bothering" him. Obviously deranged, Shaw was incarcerated temporarily at the state asylum at Bridgewater (he would spend much more time at McLean), and Gabriella often went to see him there, sometimes hitchhiking when her funds ran low, to lift his spirits.

Wherever Gabriella roamed, she often sent back my parents artfully calligraphed letters that might be the text of a treasured children's book. One was from Murray Bay, where she saw my moody sister Fan, with a rare boyfriend named Geoffrey.

> *One loves to remember beauty. On the evening of the second day in one another's company, they came walking by, & into the cottage, out of the mist.... Both looked very fair and young, he silent and self-possessed, studying the unique room, she up-cornered in the eyes, with a fairy and maiden air. I had never seen her look like that. It will make me happy for the rest of my life to see this young Fan in my mind's eye.*

She was often throwing off phrases like that, as if life for her was a Tennyson poem. But eventually my father began to tire of her impracticality, and in the spring of 1965, she agreed that it would be best if she didn't keep the house all to herself. Ellery had died by now, and his widow, Marjorie, paid to turn the servants' wing into an adjoining apartment for family members, with its own small kitchen and several bedrooms upstairs. Gabriella would move there, and Edie's sister Saucie, married to an art historian named Hellmut Wohl, would move into the Old House in her place. It was an ideal arrangement for them, since Hellmut could commute from there to Boston University, where he was teaching, and it gave Saucie a place that was beyond the reach of her father. Despite her efforts, Minty had died the year before, and Bobby just that winter. Saucie and Hellmut assumed they would be there just a few years until Dad was ready to realize his life's dream and move with my mother into the Old House.

He already regarded the house as his. He chaired all the meetings of the board of the Sedgwick Family Society, and he oversaw the house's finances. Even after Saucie and Hellmut moved in, he often dropped in for the night unannounced, without knocking, and then took the big bedroom at the top of the stairs. He adored parties, a joy my mother did not share, so he had Saucie and Hellmut give them, with himself as the unofficial guest of honor. He insisted they maintain the custom that Babbo had begun there of an annual St. Lucy's Day dance in mid-December, with the house decked out in greenery.

Dad also took charge of the graveyard. He always led the mourners trailing after the cart in Sedgwick funerals, and could be counted upon to serve the cocktails at the Old House afterward. Around this time, he started to ponder the future of the Pie, and after some calculations determined that, if the family continued to grow and die at the present rate, we would run out of space by 2101. I can't imagine how he settled on that particular year. To my father, the situation demanded immediate action, and in 1971 he petitioned the selectmen for the right to buy an adjoining parcel. At the selectmen's first meeting, the question was tabled as being "premature." It went down again at the second meeting, the shortest meeting, it was noted, "ever." But my father prevailed in the end, obtained that adjoining parcel and securing the Pie's future for over another century.

Gabriella took to the wing even less well than to the main house, and spent as few nights there as she could. One winter, she passed a night curled up in the back of a car rather than sleep in her bed. She freely opened the place to strangers, including the songwriter Don McLean who wrote his hit song "Vincent," a tender ballad about Vincent van Gogh's psychotic episode, there. He met Gabriella only once, but he never forgot the vision of youthful loveliness he found on her face. Years later, he could even recall the shades of the plaid shirt she was wearing.

The perfect time for my parents to move would have been the fall of 1967. Dad had finished at Scudder's, and finally closed up the little office he'd rented to push his Twenty Largest plan; he'd retreated to the far end of the living room, where he continued his lonely labors at a handsome desk on which he'd placed a large plywood board, to increase the surface area for all his stock tables. That was the fall his brother died and I went off to Groton, leaving him and my mother alone in the house, which must have seemed very empty, just as Dedham itself must have, with no children to connect him to the neighborhood. But that year came and went, and the next, and the next, and they did not move.

My father never told the Wohls that the plan to move to the Old House was off, suggesting that my mother never said no, but simply failed to muster any enthusiasm for a house that seemed to exclude her, much as it had, long before, excluded Pamela. And so the Wohls stayed, year after year, and gradually Saucie took on some of the functions that had fallen to Catharine in that first generation of Sedgwicks. She started to write a newsletter, keeping everyone abreast of family news. She generously opened the house to any Sedgwicks passing through, onerous as that sometimes was. It was again the "general depot" that Catharine described it as being in the Judge's day, but now a depot for Sedgwicks only.

FORTY

IN LOCO PARENTIS

I was a frizzy-haired senior—or sixth-former, as we said—at Groton the fall Edie died, but I still didn't know much about Edie-as-degenerate-superstar by then, and if I even heard about her death, it didn't register. We'd lost the final football game to hated rival St. Mark's by a single point the week before, blowing our chance for an undefeated season. I'd played a big part in bungling it: in the waning seconds, our quarterback tried to throw the ball to the ground to stop the clock and allow time to bring on our field-goal kicker, but as the wide receiver I'd instinctively dived for the ball and made a miraculous, but idiotic, catch. So the clock kept running, and the field-goal kicker dashed onto the field and ended up smacking the potential winning field goal off the center's rump. The referee blew the final whistle, and we had lost.

In that context, a remote cousin's death from a drug overdose on the other coast didn't quite compute. It really wasn't until the 1980s, when the blockbuster biography *Edie* came out, that I had any real sense of the tragic arc of her life. It was strange to read it then. The book was an oral history, with my father one of the many narrators who relate and comment on the story like a Greek chorus. At that point, Dad had been dead for nearly a decade, so it was as if he was whispering family secrets to me from the grave.

Pieces of the Edie story were all around me at Groton, actually, but no one ever spoke of it. One of my classmates, a wispy boy named Nicky Vreeland, was the son of *Vogue* editor Diana Vreeland, who'd cut Edie loose because of her drug-taking, kicking off her decline. And the names of her adored brothers were on the walls. But Groton was a place of so many unexpected connections for me. Another classmate, Steve Strachan, was the son of Marion Campbell, who had inherited the *Atlantic* from her father, who had bought the magazine from my great-uncle Ellery. A third, Rob Manning, was the son of the magazine's current editor, Robert Manning. And a fourth, Dan Davison, was a Peabody descendant. That seems to have been the whole point of my going to Groton, to grow into my place in the family, and into the family's place in the world. As such, Groton represented the upper reaches of the social superstructure that, in fact, the Judge had been so determined to marry into, and then have his children marry into as well. I was joining my class.

Since the Judge's day, of course, the establishment has become considerably more established. In the Revolutionary period, there were few institutions to confer status on the worthy, so individuals had to be almost entirely self-made. But since then countless institutions, mostly educational but not all, had sprung up to provide social credentials. Harvard has now done this for Sedgwicks, as it has for many old-line families, for almost two centuries. Since Henry Dwight Sedgwick II graduated in 1843, virtually every male descendant of the Judge has borne a Harvard degree, and virtually every female descendant married a Harvard man, until recent times, when she was likely to be a Harvardian herself. But for my own branch of the family, Groton School has served as a kind of pre-Harvard, providing some useful connections and social refinement before matriculation to "the College." Steeped in family associations, Groton stood for me as a kind of family temple, initiating me into the mystery of who I really was.

For years, Dad had always brought us out to the village of Groton to spend Thanksgiving with all his erstwhile Peabody in-laws at the home of Bets and Marge Peabody, his first wife's aging spinster sisters, on Peabody Row, not far from the school. Bets was a raspy-voiced little dynamo with a rheumy eye; Marge taller and more presentable, but sometimes bewildered seeming. (My "aunt" Helen, as we called her, had obviously been the beauty of the family.) The twosome continued to live in their father's house, worshipful nuns to the Rector's memory. It was here that Helen had passed the war, while Minturn frolicked in High Wycombe. Not that I knew that then. These Thanksgiving get-togethers were great, sprawling affairs, with tables set up in every room of the house to seat a vast family that reached from their brother, the Reverend Malcolm, and his better-known wife, Mary, who'd gotten jailed in Birmingham in a famous civil rights protest; to socialite Marietta Tree, the sometime lover of Adlai Stevenson; to future Vietnam War chronicler Frances Fitzgerald, and a slew of others. At the time, though, the only one of the illustrious Peabodys I could actually figure out was their nephew Endicott "Chub" Peabody, who'd been elected Massachusetts governor in 1962. He swept in with an impressive entourage of state troopers wearing jodhpurs and flat-brimmed hats.

Still, when I arrived at Groton School in the fall of 1967, the place seemed completely foreign to me, a little village of its own with mostly Georgian buildings around a circular green, and forbidding in the chilly, rule-bound manner of the English. On the drive out from Dedham, my mother had bought for me a bag of apples and some cider, and I felt almost desperately attached to these provisions as I stashed them away in a padlocked wooden locker in the vast, dusty basement of my dormitory in Hundred House.

It wasn't the food; it was my having a place to call my own, to hold a secret. Groton was like Jeremy Bentham's panopticon, the prison where a single guard can see all. This was a private school without privacy. It stripped you bare—the better, I suppose, to clothe you in its virtues. Every corner was open to public scrutiny. Even the toilets had no doors. I studied at an assigned desk in a big, echoing hall overseen by a grumpy proctor; ate every meal at a table for eight in a massive, high-ceilinged dining hall; and slept in one of a row of cubicles, just as my father had, each one a kind of horse stall off a wide hallway, no ceiling, and only a flimsy curtain for privacy. The furniture was all standard-issue, circa 1900: a saggy cast-iron bed, a pine bureau, and exactly six hooks on a side wall for my jacket, trousers, and neckties. Many nights that first fall, hot and lonesome, I sat by the tiny open window, staring out past the tennis courts to the distant trees, while the other boys tossed in their beds around me.

My father had let me look at other schools, and professed not to care whether I went to Groton or not. I'm not sure, now, why I chose it; I can only think that I was drawn to the place the way salmon are to their original spawning grounds, powerfully but mysteriously. So it took me a little while to realize the full extent to which this was in fact the family school. I recognized the many vaguely English brick buildings from the dinner plates my father brought out for Sunday dinner. And I could finally place the framed sketch of a small Gothic cathedral he'd hung on my bedroom wall. It was the Groton School chapel; he'd drawn it himself when he was a student. When I passed inside its heavy doors, I couldn't miss the names of many dead Sedgwicks inscribed on the high granite walls, including Helen, with a motto, "Of Such Is the Kingdom of Heaven," that must have troubled my mother. Halla's funeral service had been performed here, too.

Rob had been at the school for three years by then, and established himself as an ace student, decent athlete, and all-around good fellow—which threw me, since I knew him so differently. Or had I been wrong about him? But, as with Rob, so with everything. It appeared I had completely misunderstood my family. It was not the little, messed-up one in Dedham; it was part of a great, aristocratic race. While Dan Davison was a kind of pseudocousin, my formmates were all family of a sort, having come, most of them, by parallel routes of privilege to this sacred spot in an otherwise forgettable portion of rural Massachusetts. It was as if we were all conscripts in some elite army, on the go from dawn nearly to midnight, virtually every minute of every day filled with some improving activity.

In loco parentis was the Latin phrase the school used freely to describe its role: In place of parents. A sad idea, really, suggesting an orphanage, but the hope was that Groton would be a better parent than our actual ones, since it was more fully versed in the ways of the world. The school thought of itself as one big all-male family. The headmaster, a Texan named Bert Honea, was supposed to be the symbolic father of us all. Twangy and somewhat uncouth in a manner the trustees (including my father) had assumed the students would relate to, he wore out his welcome my second year, to be summarily replaced by Paul Wright, a forty-year veteran of the school, who took his paternalistic role a bit too seriously. Styling himself as the last Victorian, he even affected a straw boater on occasion, and might have been the Rector's shorter younger brother as he read to us at Christmastime from *A Christmas Carol,* just as the Rector had in his time.

Grated by Groton's many niggling rules, I started in on an adolescent rebellion that was complicated and intensified by the fact that my true parents were not around to be attacked. For a sacred studies exam, instead of the two-page essay my teacher was expecting on a book about Jesus' resurrection, I wrote a single angry sentence: "I read the book, okay?" That comment, I learned later, took up most of the next faculty meeting, but no one asked me about it directly. Like Edie with Andy Warhol, I felt alienated from a school that made presumptions about me based on my heritage, ignoring my own fairly distinctive personality, to say nothing of my ambivalence about being in such a place. I suppose I should have been happy to be taken for an insider; that was always my father's position. But I felt strangely absent in my own life.

And of course, society was enduring a widespread adolescent rebellion of its own, as the young everywhere were pitched against the old. I became that queer bird, the prep school radical, wearing my hair in a white boy's Afro and complaining about everything. I once sent Paul Wright a typewritten list of about a dozen items I thought the school should change, starting with mandatory breakfast and ending with the boys-only admission policy. Wright said that if I felt that way, there was no need for me to return, which quieted me right down, since I would desperately miss the friends I'd made there, and couldn't imagine life with my elderly parents if I stayed home. Still, I had a set-to with Wright over the school's age-old tradition of marching in white duck trousers to the village cemetery on Memorial Day. What with the debacle in Vietnam, I thought this unseemly, and told Paul Wright so. He actually sputtered in his indignation.

YEARS LATER, WHEN I first started to write fiction, I wrote a novel about a boy like me at a Groton-like place, who ends up raping his girlfriend—an import from the "sister" school—in the woods when they go off for a romp after performing together in the school play. The plot's interior contradictions made it unworkable; and I'd had no girlfriends at that age. More to the point, I couldn't get the boy actually to do it. But the bitter tale was true to my emotional experience of the school. The place was beautiful, noble, impressive, storied—and yet the dominant feeling that it instilled in me was a smoldering anger at the presumption that I should love it. And, to a great degree I did love it, but I still wanted to destroy it.

I can't blame Groton completely for my frustrations. It was bound to be strange to spend the age of Aquarius in an all-boys' boarding school. And it was a wild time. In a Groton photo album showing a hundred years of the school, the 1960s section looks like a bomb hit. Hair that had always been trimmed to a military shortness suddenly spills over shoulders; formerly clean-shaven faces sprout mustaches; muttonchop sideburns creep down just short of an illegal beard; neckties are askew; shirts unbuttoned; bell-bottom pant cuffs dusty and frayed from scuffing along the walkways. The smiles are loopy, dazed. Of the fifty-one members of my class of 1972, only thirty-six continued straight on in good standing to the end. Most of the others were caught with drugs, or caught too often, and had to be sent home for a while, or expelled. Our varsity hockey goalie played a whole game on acid, and did amazingly well. He was bounced later when authorities found a pharmacy's worth of illicit drugs in his study. But a fair number simply couldn't hack the strictures of the place. The son of a Spanish marquis, Tony Portago had an arresting European flair, and he was amused by me as a Boston boy. The summer after freshman year, when I worked as a page in the House of Representatives (courtesy of Rep. Phil Philbin, a Harvard chum of my father's), Tony visited me for a few days at my seedy rooming house north of the Capitol. He slept on a mattress on the floor and, late one night, showed me how to masturbate, which—such was my innocence—I hadn't known how to do before. Tony disappeared after junior year, and we were all astonished to read of his engagement the following fall in the *New York Times*. The marriage did not last, nor did Tony. He died, sadly, of AIDS before our twentieth reunion.

Although Bobby and Minty's deaths had occurred just a few years before I got there, and Francis's the very fall I arrived, I didn't connect with their experiences at Groton. To me, they were just a few more dead Sedgwicks on various school plaques. This was a kindness to me on the school's part; theirs might have been a heavy legacy otherwise. But my ignorance left many blank spaces in my understanding of the full picture, as in a Victorian novel where a lunatic aunt is locked up in a soundproof attic. The disparity was unfair: the school elders knew something about me that I didn't know. The big secret contributed to my disconnection from a place to which I was, if anything, too connected.

As I look back, I see that a lot of my rebelliousness was stylistic and superficial, but there was a hot core: I was invariably vicious toward the older masters who seemed to be the most tradition-bound, which is to say, most like my Dad. I toyed mercilessly with my shy, ungainly French teacher, a distant Peabody relation. And I was rudely impatient with the hidebound history teacher who doubled as my tennis coach. I was the captain of the tennis team, so it fell to me to honor him when he retired my senior year. A trophy or plaque would have been suitably gracious, but I handed him an unframed, unsigned photo of the team, adding a few sarcastic remarks about the place of tennis in all our hearts—comments that were so snide it pains me now to recall them. I wasn't mad at them so much as I was mad at my father for off-loading his paternal responsibilities onto an institution. It might have been different if he hadn't been so old, and so preoccupied with Rob, back when I was still in the house. But his seeming indifference enraged me, not that I could have said so at the time, and a lot of that anger spilled over into a disdain for the school he loved. It seemed that he loved the school instead of me, just as he fathered the greater family in place of his actual children. He seemed to take refuge in abstractions, like that Old House, which had come alive for him only long ago, and never again. Lost in his memories, he was unable to animate them for his son. At Groton, I felt cut off, so cut off I couldn't even say what it was I was cut off from. But I now see, of course, that it was from Dad, that grand, foolish Duke; from the comforting presence of a father; and from the larger, confident, all-embracing manhood he might have represented. He was a man from another century, and, for me, he remained there.

I'd seen Dad stride confidently about the Circle in the strange, Russian winter hat he wore in those years, as if he owned the place. But he never explained what the school meant to him, why he'd devoted his life to it, given it two sons. I didn't know that this was where his brother died, where he'd met his first wife at age twelve, where he'd found his own true father. That was all so long before I came along, he must have thought it irrelevant, and it was all too painful to say. Still, such details might have made Groton come alive for me. As it was, Groton seemed more like a mausoleum, haunted by the unknown dead.

Saddest of all, I did love much of what Groton offered, I did—the fellowship, bookishness, sports, drama, the Ruskinian beauty of the place, the sight of the lazy sun dropping into the woods beyond the distant playing fields on a spring evening while taking postprandial coffee on the terrace by the dining hall. Yet I had largely closed myself off from such pleasures—a joy for another time, or for another person, perhaps.

BY THE TIME of Edie's death, I was not the only child of my father's to be lost to him. Lee had joined the radical Students for a Democratic Society, better known as the SDS. She'd gone to a small all-girls' school in Franconia, New Hampshire, called St. Mary's in the Mountains, and started an SDS-like group there. When she arrived at Harvard in 1969, the fall after the occupation of University Hall, she joined the real thing. By then, there were *two* groups claiming to be SDS, one trying to build the equivalent of the Communist Party, and the other headed toward becoming the more militant, bomb-throwing Weathermen. Lee sided with the more pacifist version, which splintered a few more times to emerge as the Revolutionary Youth Movement II. The campus protests picked up after the bombing of Cambodia that spring, and she gave up on her studies to organize antiwar demonstrations. Rob was there also, but too busy with his coursework to get involved with the protests. The day of the big post-Cambodian-invasion demonstration in 1970, when rampaging students smashed in all the storefront windows in the Square, Rob was headed to the library when he was chased by police in riot gear down Massachusetts Avenue. His eyes stung with tear gas, he couldn't see where he was going, and he slammed into a street sign.

Lee flunked all her courses that spring, having skipped the exams. Mum never conveyed her anxiety to her directly, but, calling from Dedham, poured out her worries to Lee's roommate.

DAD WAS AGAINST the war, too, but more conventionally, as a liberal Democrat. He supported organizations like Americans for Democratic Action, a group founded by Eleanor Roosevelt, and Common Cause. He told Lee he didn't see what good it did just to have "all these parades," as he called the student demonstrations. He thought it more effective to work within the existing political system, but Lee would have none of that. To her, the existing system was corrupt. That summer, she moved with a half dozen other Harvard radicals to Lynn, Massachusetts, to try to bring the antiwar movement to the working class. They called their group the Red Fist, and they set up a storefront to distribute antiwar literature, much of it with a Communist slant.

I thought that her activism was cool, and I'd won points back at Groton by talking up my desire to join the Cuban revolutionaries Lee had told me about and hack sugarcane in solidarity with them over vacation, not that I ever did. Any feeling for class struggle I had derived from my own ambivalence about Groton. I knew little about working-class life, and didn't have much of an idea about Lee's until Mum and I made a field trip to see her in Lynn.

In her early childhood, Mum had spent her summers in a big shingled cottage in Nahant, a Waspy enclave on a spit of land that shoots out from the Lynn shore, but that was another world from the sweltering, treeless mainland. Lee's place might have been a hippie commune. With her comrades in the Red Fist, she'd set up house on the top floor of a crumbling triple-decker, but there was scarcely a stick of furniture in any of the rooms beyond the rabbit-eared TV, just sleeping bags rolled out all over the floor, which was otherwise littered with pizza cartons, empty but for the dried crusts. There were heaps of trash in the weedy backyard; it looked like the communards had simply tossed their rubbish out the window.

I was curious about the sleeping arrangements, though; the way the sleeping bags were all jumbled together seemed suggestive. One of the sleeping bags belonged to Jack Carr, a burly, bearded high school junior at Lynn High, who was Lee's new boyfriend. And I saw a bit of him, a shy, smiling lumberjack in flannel, over the next few months at family gatherings. One of the other members, Miles Rappaport, later resurfaced as Connecticut's secretary of state. But most of them went back to Harvard in the fall, leaving Lee and one other true believer. Soon it was only her. She stayed on for six years in Lynn altogether.

Lee returned to her Harvard classes in the fall of 1972, the year I arrived. She hitchhiked to Revere, and then took the subway into Harvard Square from there. Initially, she majored in government, but when she realized that wouldn't teach her how to *overthrow* the government, she switched to American history. I never once saw her on campus.

HARVARD WAS EVEN thicker with Sedgwicks than Groton had been, and I saw any number of their names inscribed on the walls. The Harvard weeper *Love Story* had come out shortly before, and a few of the students lampooned me as the preppie Oliver Barrett IV, and asked if there were any Sedgwick Halls at Harvard. (There were not.) For the most part, I felt liberated by my relative anonymity. If Groton was a village, Harvard was a universe unto itself; it made room for a far wider range of personalities than Groton had, and I needed that latitude. After five years of monastic life at Groton, I was ready for girls, and I asked to live at Radcliffe, which had a far more favorable boy-girl ratio than elsewhere. There, I was assigned to a tiny double with a brainy Jewish kid from the Upper East Side, Pepe Karmel, who showed up with a guitar, a set of bongo drums, and boxes of poetry. I'd never met anyone like him before; we immediately became best friends, and gradually exchanged characters. I became the swinging Jewish artiste, and he the worried Wasp. Harvard was high school for me, a zooey mishmash of different backgrounds, attitudes, and personalities, none of them anything like my own. These were the days of coed bathrooms, and I started one relationship by peeking over the baffle that separated my bathtub from a female dormmate's. The round-the-clock presence of so many women gave me an orgiastic thrill, and I was soon involved with several at once.

Imbued with the politics of the day, much of it filtered through my sister, I did my share of protesting and got involved with an SDS spinoff called NAM, for the New America Movement, largely because I enjoyed the company of my fellow radicals, one of whom wooed me by leaving a sweet note in my mailbox addressed to "My jo, John," using a Scottish word for lover, which I had to look up. (I later learned that it was a quote from a poem by Robert Burns.) I picketed at the Harvard Club on Commonwealth Avenue when Kissinger came to speak, and angrily demanded that Harvard divest itself of its stocks in South Africa, not that such a move seems prudent to me now. (Tainted as they may have seemed to us then, I now recognize that the multinational corporations like Exxon and General Motors, both of which ended up ceasing operations there, were actually some of the more enlightened forces in that apartheid country.) In the spring, I received a mysterious written invitation to meet at the Harvard gate one Sunday morning to be "punched" for the Porcellian Club. It never occurred to me to go. I thought of it as a repetition of the Groton years that I'd had too much of already; besides, it seemed too discordant to be a student radical *and* a Porker. So I never responded. My brother had come to the same conclusion. My father, growing numb to all the political havoc, expressed no disappointment. He told me he thought the Porcellian should be turned into a museum.

FOR LEE, THE Red Fist offered a better alternative to becoming a wife and mother, which seemed to her to be the only other choice. But it also got back at Dad, who had ignored her even more than he had me. In our younger days, back when Lee wore white lipstick and fantasized about which Beatle she loved most, she and I had been close, united in our opposition to Rob. She used to play with my hair, and style it à la the tousled singer Bobby Darin, whose poster was on her bedroom wall. Dad bugged her, I could tell. At those stultifying Sunday dinners, she was inclined to slouch, and he'd sidle around behind her and drive his knuckles into her lower back to get her to straighten up. She'd spring up at his touch, then slump back again, obviously irritated, as soon as he moved away. On partings, when he leaned forward to give her a good-bye kiss, her whole body would arch away, recoiling from his lips.

Lee's radicalism challenged him the way Francis's hoax letters had, by showing the limits of his commitment to his political causes. If she had been a pro-war Republican, he could have argued with her; as a radical leftist, she left him speechless. Puffing nervously on a cigarette, my mother fretted openly to me about the direction Lee's life had taken, although she never told her. My father rarely spoke to Lee at all.

FORTY-ONE

"WE'RE RUINED"

In 1974, Dad was diagnosed with rectal cancer and had a grueling surgery that diverted the lower portion of his colon into a colostomy bag. It looked like a shower cap that attached to an outlet from his lower abdomen. Terrible with mechanical devices anyway, Dad found the bag a grotesque indignity as well, and he could never get the hang of it. The only time I ever heard him swear was when the bag slipped out of his hand, spattering the contents all over the bathroom floor. He cried out, "Oh, *shit!*"

He turned weak and pale, and got bitterly depressed. He slept a great deal, and turned cranky about small matters. He raged at me once because I'd failed to bring his milk in a proper "tumbler." He'd started that unpublished memoir of his, but was able to continue only through the war years of his forties before the disappointments mounted—the troubles with Francis, the alienation from Helen, the failure at Scudder's—and the narrative ground to a halt. He didn't mention his second family, which was us.

And, most cruelly of all, the stock market turned against him. He'd given up promoting his Twenty Largest system, having found no takers beyond Riggs. But the market was in the throes of one of the longest and most spectacular dives since the Great Depression. From the end of 1965 to the end of 1974, the Dow Jones lost half its value, falling from almost 1,000 to about 500, and much more than that in constant dollars, since inflation was rampaging in that era of "stagflation." Opening the newspaper, he'd always go first to the stock tables. When his stocks were down, as they usually were, he'd grouse, "We're ruined." He'd say it jokingly, but it must have felt as though we were. In the late 1960s, he once showed Rob the family's stock holdings. They totaled $2 million, making us, as he said, "bloody rich." By 1974, with all the tuitions taken out, they'd dropped to less than half of that.

He wasted away all that year, and by the end of 1975, he largely stayed in bed. All six of his children gathered for what we imagined would be a final Christmas together. May left her vast brood to drive down seven hours from Auburn in upstate New York; Fan borrowed money to fly up from Washington; Harry, divorced from his second wife, Patsy, took the shuttle from New York; Lee drove in from Lynn. I was still at Harvard and Rob at Harvard Law, so he and I were home for the holidays anyway.

We'd all set aside whatever family we had to rejoin the family of our birth. It was the first time that we had *ever* been together as just the six of us; we were all Dad's children again. May, the oldest, reverted to an oldest child's lordly superiority. Fan, wild as ever, talked a mile a minute. Harry, the heir apparent, was quietly attuned to Dad. Of all of us in the younger set, Rob was the only one to hold his own with the older ones. Lee, with her working-class manner, faded out. I hovered protectively over my tender, skittish mother.

On Christmas Eve, we fashioned a dinner table out of adjoining card tables up on the hallway of the second floor, to save Dad from having to negotiate a flight of stairs, and placed him at the head. He wore a crimson wrapper over his pajamas, and he looked gray. After he said the grace he'd learned at Groton—"Bless O Lord this food to our use and us to Thy service"—he fell quiet while the rest of us all carried on as if everything were fine. He retired to his adjoining bedroom before dessert, and we finished up quietly while he looked on through the open door, propped up on pillows, from his bed.

That night I was awoken by a tapping sound. Groggy, I thought someone was knocking at my door, but the sound came from farther away, in the hallway or beyond, and it was accompanied by a helpless, sobbing sort of moan. I pulled on a bathrobe, peered out into the hall. Across the way, Dad's bedroom door was open. The sounds were coming from there. I crept forward, fearful, but curious, too. Dad was in a heap on the floor beside the bed, twisting about, straining to get up. Rob and Harry were already there, leaning over him. Dad was like a felled horse—wild-eyed, struggling, his ankles thudding uselessly against the floor.

He couldn't speak, couldn't explain what had happened, but he must have collapsed on his way to, or from, the bathroom. I held back, frightened by all the disorder, the awful knocking of his feet, the nervous fumbling as my older, stronger brothers hoisted Dad back onto his bed. Something about the way Dad was gesturing suggested to Harry that Dad needed to urinate, and he fetched a bowl from the bathroom for him. He was holding my father's penis, aiming it at the bowl, when my mother arrived at the doorway. She took one glance at this bizarre scene and fled the room.

The medics came, radios squawking, and they took Dad down in a stretcher, careful not to disturb the etchings that lined the staircase, and out to an ambulance bound for Massachusetts General Hospital downtown. I returned to bed, but could not sleep. With the rest of the family, I went to the hospital to see him later, early that morning, Christmas morning. He was in Phillips House, a homelike place with wallpapered corridors for those willing to pay extra. He lay ghostly pale on the bed with his head tipped back, snoring lightly. He'd had a stroke, the doctors said, and was in a coma.

We all tried to speak to him; he didn't respond except occasionally to move his lips, but no sound came out. I imagined him dreaming of the past. I reached for his hand, held it. It was limp, but warm. I couldn't remember ever holding it before, and it felt strange, the stiffness to the joints, the elegance to the fingers themselves with their rounded fingernails.

The days went by without change, until a phone call came at dinnertime about a week later. It was Dad's doctor, calling to say that there was little hope that Dad would ever emerge from the coma, and virtually none that he would recover his faculties. Dad had had us all sign a living will in which he made clear he did not want to be kept alive artificially. He was being sustained now by the feeding tubes. Did we want the doctor to remove them?

When Harry, who'd taken the call, relayed this information to us, we all looked at each other around the table. It was clear to me that my mother wasn't ready, and I spoke up for her, urging that we hold off. But everyone else believed we should follow through on Dad's expressed intentions. Harry called the doctor back and told him to remove the tubes. My mother dashed out of the room in tears.

DAD HELD ON, though. His lips continued to move from time to time, but that was all. Cousin Harold, a bantamweight Episcopal minister, appeared one afternoon. He stood by my father's head and spoke sharply to him: "Minturn...*Minturn,*" he said, bending down and shaking Dad's shoulder. When Dad didn't respond, Harold laid his hand heavily on Dad's forehead and recited the Lord's Prayer. I wanted to shove him out of the room. And one morning when Mum was there alone, Sally burst in, all aflutter, and badgered my nearly lifeless father about something, literally yelling in his ear. My mother was in a fury when she left.

When it became clear that death would take a while, Harry joined the three younger children to drive up to Chocorua, where we played soccer in the deep snow for hours on end, eager to exhaust ourselves. When we dragged ourselves back to Cambridge, my brothers and I spent our afternoons playing pond hockey, despite the fierce cold. We'd just returned from a game, all breathless and oily, when May and my mother came home from the hospital in tears.

My father had been so old, I had been anticipating that moment, it seemed, for most of my childhood. At first, his death was not really a death at all, but a moment of exquisitely tragic feeling, and I was thrilled to set down a moving chronicle of his death for the Murrays in England. I remember spending a good hour or two crafting it, and enjoying the sensation of setting down the story, recording it, getting it right.

A few days later, we drove out with his ashes in a cardboard box to Stockbridge for the funeral. It was a slushy January day, and the mourners seemed to lack their customary order as they trailed along behind the wobbly cart, on which his coffin lay. I kept thinking it strange he wasn't there to lead the procession from the house.

Grief picks its own moment, for its own reasons, and it hit during the service at St. Paul's Church, when it came time to sing "A Mighty Fortress Is Our God." Beside me, I could hear my two brothers singing so bravely—Sedgwicks always belt out the hymns—that I was overcome.

Tears spilled down my cheeks, and I found I couldn't quite croak out the words.

Still, the raw truth of his death did not hit me until the following fall, after I met Megan Marshall, whom I would eventually marry. She lived upstairs from me in the Victorian house that was part of South House, up at the Radcliffe portion of Harvard, where I spent my four years. I could hear her overhead clomping about in her clogs while I lay in bed. A poet and musician, she was also a trim, athletic Californian. At Harvard she'd started a women's softball team, and we played catch one Saturday afternoon. I fell in love with her as the ball arced lazily back and forth between us, finding our mitts with a slap.

I already had a girlfriend, Maggie, but she'd gone off to graduate school in Chicago—a test of my affection, she told me later, that I failed. It had been an important relationship, though, the first one like that, and I was sorry to see it end, even though I'd brought the ending about myself. That breakup evoked the other, more profound one, the loss of my father. One night, I started missing him so badly as I sat alone in my room that the tears started pouring out of me. And then I slumped back into the chair and, digging my hands into my face, cried so hard, I felt weightless and saw stars.

ROB HAD A worse time. Like me, he'd been able to hold it together for a while, but that winter, his girlfriend ran off with a close friend, leaving him utterly devastated. Unable to concentrate on his law books, he withdrew for the spring semester and came home to live in the Cambridge house, where he listened to Jimmy Cliff's "The Harder They Come" at full volume on the living room stereo for hours.

That was the bottom for him, and he rose from there. Although Dad's death was a debacle for him, the pain confirmed a core truth that he was his father's son. That was why he'd felt so at home at Groton: it was his father's place. As such, Groton suited him far better than our divided Dedham household. It gave him a group of close friends—brothers, almost—that he has remained devoted to ever since; it also provided an outlet for his prodigious energy, racking up awards and honors. I did well at Groton, too: he finished first in his class; I was second in mine. But, more importantly, Groton confirmed his own values, whereas it did not quite line up with my own. Groton formed the basis for his adult identity; for me, it was a fun-house mirror, distorting my sense of myself. At Harvard, he went on to a joint degree in government and law, which led him to a stint in Washington as a deregulator in the Carter administration before moving on to corporate work in Washington, and then in New York. In Washington, he fell in love with a fellow associate; they married at Stockbridge in 1984, and they've since settled in tony Short Hills (which he teasingly persists in calling "Short Hairs"), New Jersey, with two bright-eyed children.

As Rob and I have moved beyond our own difficult childhood, we've become closer than I would have thought possible, as it became clear that we were both caught up in a war that was not of our own making, and that for a lot of it, we were more on the same side than we realized. Each of us, I think, is happy to see quite a bit of himself in the other. We're both athletic and boisterous, with a sense of humor that we can always count on each other to appreciate, even if our jokes might be lost on others. After many years as a corporate lawyer, Rob has settled into a lucrative niche negotiating the salaries and benefits of Fortune 500 chief executives, a position the Judge would have endorsed as pleasingly Federalist.

AT MEGAN'S URGING, that fall after Dad died, I saw a therapist named Rick—the one I would return to almost a quarter century later for my millennial-year breakdown—in the basement of the Cambridgeport Problem Center off Central Square. Out the window, I could see the feet of the hurried pedestrians on the sidewalk as I breathlessly recited to him all my woes, about my dead father, wasted mother, flipped-out brother, absent sister, and— He stopped me. "And what about *you?*" he asked. A stunning question; I'd truly never given that a thought, and that was the beginning of the road back, a road that led to marriage, to the writing life—and, of course, to my terrifying fall just when I imagined I had reached my zenith.

LEE WAS THE only one of the three of us who kept on her original political trajectory. If Rob was the Federalist in the family, she went the Jeffersonian route, and more. She'd left Lynn a year after graduating from Harvard and joined For the People, a more broadly based version of the Red Fist; it published a newspaper and supported union activities in a few industrial cities around New England. The organization sent her to Bridgeport, a gritty seaport in Connecticut, with another Harvard kid named Mark Warren, a bushy-haired leftist who'd dropped out of my class. Mark took a job at a steel mill, and Lee at Dictaphone. After a year, they moved on to New Bedford, where For the People had a stronger presence. Lee ended up at Cliftex Corp., where she sat at a sewing machine tacking down belt loops for the trousers produced for Sears, Roebuck (one of Dad's Twenty Largest) at its Town and Country clothing factory. Along with the other workers, most of them Cape Verdean, she entertained herself listening to *The Young and the Restless* and other soaps on her Walkman headset. She learned some Portugese, and as often as she dared, she snuck into the women's room to talk up the union, which was her real purpose for being at the plant.

In 1979, she married Mark. She joked that it was because they'd both get two weeks off, a concept that distressed my mother. A Cambridge justice of the peace did the honors, but Mum held a small reception for them on her tiny front lawn. Mark's mother was Italian, and the party had a lot of spirited dancing and spicy food. Of our family, only Harry came to swell the family beyond the nucleus of Rob, Mum, and me. From Auburn, May sent an engraving entitled *Death of Warren,* depicting the demise of the Revolutionary hero Dr. Joseph Warren, whose remains were briefly interred near the Judge's. Mark was not amused. "What's that?" he kept saying. "What's she mean by that?"

The marriage lasted only two years before Lee and Mark realized that, for all their shared politics, they didn't have much to say to each other. Mark gave up on the worker's life to return to Harvard and complete his undergraduate degree. He is now an associate professor at Harvard's School of Education, and a fellow at the W. E. B. DuBois Institute for Afro-American Research. Lee stuck with her belt-loops job in New Bedford—until the company decided to ferret out union activists and, after hiring a private investigator to do a background check, fired her for failing to list her Harvard degree on her job application. She appealed to an arbitrator, as provided in the union contract, and won reinstatement, only to have to take the case to the Superior Court when Cliftex refused to accept her. She won again and returned to Cliftex, but she tired of the work and moved back to Boston to get the necessary certification to try her hand at teaching. A year into it, she met a labor lawyer named Bob

Schwartz at a Martin Luther King dinner at Boston University, and they were married a year after that.

I gave a toast at their wedding reception at a Thai restaurant. "Most women are looking for Mr. Right," I joked, "but Lee was looking for Mr. Left." It was a gentle way of alluding to the political divide between us. With Bob and their two spirited teenage boys, she lives in Jamaica Plain now, and teaches math and history at the Jeremiah Burke High School in Dorchester, one of the more troubled schools in the city. I do admire her for it; she works harder than just about anyone I know. Although she is close by, I don't see her very much, and it pains me. The whole tangle leaves me feeling like Catharine, lamenting a rift in the family, as if it has nicked the heart of everything that matters.

AFTER DAD'S DEATH, I was eager to create a family of my own, one that I could shape to my liking. Because I'd taken a semester off to bicycle across Europe, I graduated at midyear, a semester before Megan, and I couldn't bear the prospect of living apart from her. Despite some misgivings on her part, she and I took an apartment together on Ware Street just off Harvard Square. I loved being with her, just the two of us. We shopped around the corner at the Broadway supermarket, joked with the building's gruff superintendent, Buddy, and slept naked together on a mattress on the floor. I started my magazine career there in one of the apartment's spacious front rooms. A talented Harvard friend, Anne Fadiman, and I wrote a piece about Harvard bathroom graffiti that *Esquire* published the September after graduation. I'd thought it would make my career, until I dutifully made the rounds of New York editors, pitching stories to no avail, and ended up retreating to fairly humble local publications. I spent months more banging out stories on a portable typewriter while puffing on the Balkan Sabrane cigarettes that I thought were essential to the writer's task, and, for the same purpose, ended the workday with Jack Daniel's. After a couple of years there we moved to Joy Street, on the back side of Beacon Hill, where the Underground Railroad had run in the years before the Civil War.

Megan's father had gone to Harvard as a scholarship kid from California. Brilliant, he'd finished near the top of his class after his freshman year, but then the pressures got to him. He couldn't decide on a major, flitted from one subject to another, and did worse and worse until his scholarship was finally withdrawn and he was forced to leave after two years. He was in the merchant marine when Megan's mother met him. After they married, he attended Harvard's design school but never did get a degree, and then tried a career as a city planner, but he had trouble holding a job.

They'd divorced just after our own graduation; with this example behind her, Megan was distrustful of marriage. Still, we were powerfully in love. I admired the independence that she'd won from her own troublesome family, putting down stakes on the opposite coast, and I craved the security of affection I found with her. I was impressed, too, that her head was not turned by my supposedly illustrious, high-Wasp background. That had become clear from the very first time she'd seen me, which was a year or so before we ended up together on Walker Street at South House. It was at a sherry hour for English majors, an event presided over by the Puritan scholar Alan Heimert at Eliot House. In the midst of it, Heimert asked me my name, and when I told him, he cried out, "You're not one of those Sedgwicks with the Pie, are you?" When I admitted I was, he went on: "You know why it's that shape, don't you?" Of course I did, but I couldn't stop him from addressing the others in the room. "So when the dead rise up on Judgment Day, they'll see no one but Sedgwicks!" Then he laughed uproariously, while I tried to disappear. Megan overheard the whole exchange, and silently vowed to have nothing to do with this ridiculous Sedgwick person ever.

But after we started living together, she decided I was fairly engaging after all. And one night on Joy Street, after we made love in our bed by the open window to the back alley, I asked her to marry me, and, after a heart-stopping pause, she told me she would.

The service was in the English garden at my great-uncle Ellery's place, Long Hill, on the North Shore. After Marjorie's death, it had been taken up by the Trustees of Reservations, which preserves historic estates that would otherwise go for subdivisions. Unlike Lee's, ours was a full ceremony, with a Unitarian minister from Megan's old church in Pasadena, a brass quartet playing from an upstairs porch, many Sedgwick relations in attendance, and Rob as my best man. Megan was marrying into the family, and, there at Ellery's with all my relations ringed about me, I was also.

I WAS WORKING on my first book then, *Night Vision*, about a grizzled private investigator named Gil Lewis who was based in Quincy. He'd drive by our apartment, honk, and then I'd jump in his Toyota Supra and interview him while he went out on late-night surveillance, following some errant spouse, usually, back to some hideaway apartment. As we sat together, gazing up at a bedroom window on some lonely street, he'd suck on a Muriel Coronella and recount his great cases—tailing Howard Hughes for the *National Enquirer,* investigating "hitchhike murderer" Anthony Jackson, staking out a graveyard on Halloween. He'd detail the seamier side of life with a clarity and gusto I found enthralling after such a sheltered existence in secluded suburbs and princely private schools. Gil Lewis was graduate school for me, an education in reality, and I was thrilled to learn about such a gritty world as the one he portrayed, and to make my way in it.

Better still, this was all in service to a higher calling, that of the literature that Babbo and Catharine had espoused, albeit on very different terrain. Indeed, that very part of it excited me most, in that I was both a Sedgwick and not. I had no doubt that Babbo and Catharine, to the extent that I could then imagine them, would have been appalled by the subject matter. I delighted, in one passage, in detailing the gradual drooping of an erection sported by a philandering husband caught in the act by Gil's photographer, as the poor startled man wheeled on his intruders, exposing himself to a series of flash photos that I later marveled over in Gil's office. Still, I was fairly sure that my writing reached the standard of literature. When I submitted the book proposal to publishers, Houghton Mifflin was an early contender. The offices were on Beacon Hill, just a block up from the Old Granary, and one moonlit evening I walked over to peer in the front window at the display of its leading titles, an edition of Tolkien's letters and *Peterson's Guide to the Birds,* and dreamed about joining them.

In the end a New York publisher, Simon & Schuster, swooped in with a higher bid, and I wrote the book for a veteran editor there, Fred Hills. By then, I'd breezed through a lot of magazine articles—on a pool shark named Boston Shorty, a deep-sea submersible, an Episcopal bishop, and many other topics—and I had written a regular column about Gil in a local tabloid, the *Real Paper.* But I was only twenty-five, not as confident as I might have been, and I found it forbidding to turn those stories into a book, which I considered a far grander enterprise, partly because these were the items my forebears had produced. And Hills let me know through my agent that he was not pleased with my early attempts at conveying such a powerful character as my detective. Lewis was tough and manly but also had a romantic side, an endearing softness, that I found hard to capture, possibly because it represented a kind of fatherliness I'd missed in my own dad. I was struck by his tale of a teenage girl named Felicia who'd hired him to track her runaway father after her mother died. She had only a few dollars to pay him, but Gil took the job on, worked the case for months, and finally found the man at a navy base. When the father would have nothing to do with his abandoned daughter, Gil ended up adopting Felicia himself, and later put her through college. Gil himself was divorced, and one of his sons ended up doing time for larceny, but, as Felicia had, I looked to him as a father to me.

This slowed the writing. After a turn as an editor, Megan was at work on a book of her own on women of the baby boom generation, but I turned in desperation to her, and she helped me, sentence by sentence. *Night Vision* appeared in 1982, the year the book about Edie came out, with Dad himself speaking to me in it, pulling me back toward Sedgwick subjects. Drawing on my Groton and Harvard connections, I wrote a book, *Rich Kids,* about the young rich of my era, Edie's world, which came more smoothly. After that, sick of the money culture, I spent a year at the Philadelphia Zoo, exploring the relationship between the keepers and their animal charges. Taking my title and a few stories, it became the basis of a TV series, *Peaceable Kingdom,* on CBS, a rather lame show starring Lindsay Wagner that ran for only one season but was a thrill for me all the same.

It all seems a little frantic as I look back on it, but at the time the brisk pace, and the sweep of my activities, all seemed quite necessary and stimulating, if a bit isolated, as I spent long days hunched over the keyboard of my word processor, as we called PCs in those days. I didn't quite notice how much I was living in my head, in narratives, and not in the world. Without fully discussing it, Megan and I had divided up responsibilities along the 1950s lines that my parents, if not hers, would have recognized, as I took charge of the moneymaking and Megan ran the household, each of us getting a little more frazzled than we let on. Returning to magazine journalism, I wrote a few pieces on technological subjects for the *Atlantic* and was surprised that the first one was heralded on the editor's page for my link to my great-uncle Ellery. It reminded me, once again, that I was not the first Sedgwick to head down this path. I became a regular at *GQ,* at the women's magazine *Self,* and at *Newsweek,* writing about everything from estrogen cycles to the class insecurities of the first President Bush to the uncertain future of Yale University. We had two daughters now, the beaming Sara having been joined by cuddly Josie; we'd moved to a big Victorian house in the leafy Boston suburb of Newton; and, just as my life was beginning to stabilize in my late thirties, my mother responded to widowhood by sinking into suicidal despair.

FORTY-TWO

"WSSHT"

Starting in her thirties, my mother had been seeing a psychiatrist, a Dr. Barry, recommended to her by her brother William. He was in his eighties and quite deaf by the time she finally broke off with him in the mid-1970s and turned, in increasing desperation after her husband's death, to a series of substitutes. The weight of sadness and isolation that comes to any widow fell especially heavily on her. Dad may have chosen her for her emotional frailty, but his presence girded her. When he died, she felt all the more infirm, having lost the habit of bucking herself up, and her own inadequacies were exposed, besides. Everything that Dad had taken care of now fell on her: finances, houses, Sedgwick relations. It was particularly awkward for her to carry on as the surviving queen of the dead king, the titular head of the greater Sedgwick family—not just the thirteen step-grandchildren produced by Harry and May, but all the cousins and in-laws and distant Sedgwick relations whom Dad alone had been able to keep track of. Mum wasn't up for any of it, and as everything got too much for her, she felt increasingly irked that these relations kept coming around, trying to keep up the family spirit.

In the twisted way that she started looking at life, my marriage meant another loss for her. One July weekend when Megan and I visited her in Chocorua, where she spent her summers, she beckoned me into another room. It felt a little strange to be alone with her, which turned out to be the point. "We used to be such good friends," she told me. "Don't you like me anymore?" I couldn't find the words to give her the obvious answer: I'd grown up, left home, gotten married. What did she expect? As I think about it now, though, I realize that the break had actually occurred far earlier, when I trooped off to take up my place in the male line at Groton, leaving female things, like my mother, behind entirely. At Groton, I was far more conscious of the gap between me and my father, which in fact was tantalizingly close at that Sedgwick family school, than the far wider, and ultimately unbridgeable, one between me and my mother. She had felt the loss, though, as only a mother can.

When Sara was born in 1984, we did our best to include my mother, dropping the delightfully bouncy little girl off at her house one afternoon a week. Mum was the one to teach her to throw a ball, just as she had taught me, and took a special interest in Sara as she developed into a powerful athlete, much as Mum had been herself. She did some tutoring at a local elementary school to keep busy; and she was a regular at the Friday-afternoon symphony, having inherited her mother's seat in row LL. But she took in boarders so the house would not feel so empty.

Still, Mum felt useless, a victim of her class and her temperament. Having never worked, she had no affiliations beyond her few friends. In earlier years, she'd released her tensions by digging in her garden or smacking balls around the tennis court. But as her body began to fail in her sixties, needless worries began to eat at her. In Chocorua, she became preoccupied with some brambles that had infiltrated the lawn and a sly leak under the fat new chimney my father had insisted on installing. To me, it all seemed darkly Hawthornean, this invasion of creeping malignancies, but she could not be shaken from her obsessions.

In August of 1985, when she was seventy-one, it all became too much for her. Alone in Chocorua that summer, she spiraled down into a paroxysm of worry. At night, when she lay in bed, glistening with anxious sweat, on the sleeping porch, she could almost hear the drips, the rustle. She fled to Cambridge but found no relief, and three days later her psychiatrist, Dr. Bernstein, thought it best to place her in a psychiatric ward for her own safety.

St. Elizabeth's is a sprawling hospital on a hilltop in Brighton, a noisy, congested part of Boston my mother would otherwise have avoided.

When I went to see her there, I felt somewhat overwhelmed, too. She was up on a long ward in Quinn 2, occupying one of a row of beds—a shock for someone so private and set in her ways. Jumpy, distracted, she was not at all well, and I couldn't bear to stay very long. "Racing-thoughts depression" was the diagnosis given to me over the phone by Dr. Keenly, the harried staff psychiatrist at St. Elizabeth's. It was as good as any. Inward-turning by nature, she now almost clawed at herself out of anxious self-loathing. It was a shock to see.

For the first two weeks, my mother was not allowed outside. The confinement gnawed at her, and Dr. Keenly finally released her to my care for an afternoon. Pent up, she'd been dressed and ready to go, with her sneakers on, for nearly an hour before I arrived.

I signed her out, and we went for a walk down Washington Street, the busy street that passes in front of the hospital. By then she walked like an old Yankee farmer, with her elbows out and her head bobbing with each step. She had her heart set on an ice cream soda at Brigham's down the street. In the shop, we settled into a booth, and she placed her order. In solidarity, I had an ice cream soda, too, and, as we sipped them, it was nice to be linked through a pair of straws into identical drinks. But then Mum fixed me with a weird, spectral look. "Do you think it'd really be so bad?" she whispered to me conspiratorially, scrunching up her nose.

"What would?"

"To—you know—" She glanced down, embarrassment stealing over her. She knew I knew what she meant.

"Mum, please."

She flicked her hand open and made a whistling sound—*"Wssht"*—through her lips, wetted slightly by the ice cream. "And you're just… gone." Then she brightened as if she'd seen God.

I tried to hide my distress, thinking that would just make her feel worse. Instead, I did my best to reason with her, and told her that it wouldn't set a very good example. For her grandchildren, I think was the idea. But really for me. Still, I knew it wasn't the best answer even as I gave it. I was supposed to tell her I couldn't bear the prospect of her leaving me. But at that point, with her so haggard and disturbed, there was very little left of her to love.

After three weeks, she was able to go back home to her widow's life in Cambridge and resume her usual load of worries. I'd become her primary caregiver, the one her psychiatrists always called to talk over her treatment. Occasionally I would sit in on her appointments, only to be appalled by the quality of her care. That Dr. Bernstein, the one who put her in St. Elizabeth's, wore a rabbinical beard that he used to push up toward his mouth with his hand, and then slyly nibble on during her sessions.

In 1992, tired of living alone in a big house, Mum moved to a retirement center in Lexington called Brookhaven, but she brought her anxieties along. Two years later, she started keeping a journal again. It was headed "My Probs."

2/19
Must have Albrechts for dinner
Can I stand it in Choc.
" " " " " Stockb.
Smoking—ever present prob.
Car fixed?
Played recorder. Failure.
Got too nervous

2/21
Bad A.M. Called Dr. Savage. Waited for her to call.
Hard to wait for calls—nervous.
Trying to throw bad feelings out
No reason to feel depressed
Lost confidence
Choc. probs—leak. Sofa fixed etc. this summer
Can't think

2/25
Back from Stockb. yesterday. Couldn't take it there. Too much
    confusion
Cold in bed here—couldn't sleep
Worried about everything.
Head doesn't work. Well—make it.
Look forward to bedtime—wish day was over instead of just
    starting

And on the journal went, a daily log of woes, set down in haiku. She went through a series of psychiatrists and tried every type of antidepressant, from Lithium to Prozac, each of which followed the same pattern of disappointment—slow to work, briefly effective, then useless except for the side effects, the stomach trouble, the jitters. A steady refrain in her journals is how she needed to believe in her drugs. On visits, I noticed the shelves in her bathroom were jammed with bottles and packages containing the many drugs she was trying, along with the sleep aids, upset-stomach remedies, and other palliatives needed to tolerate them.

In the summer of 1997 she entered the psychiatric ward at Mount Auburn Hospital, a drab place that reminded me of a dreary 1960s co-op and did little for her. The following winter, she had electroshock treatments at Massachusetts General. I was put off by the Dr. Frankenstein aspects of the procedure, but she was surprisingly gung-ho. On some level, I think she felt she deserved any pain it might cause her. She was supposed to have seven treatments, but stopped after four, afraid she was losing her memory. (Which she was; she remembered nothing of the treatments later, a common side effect.) Her mood lightened only briefly. Just before Christmas, with its bad associations because of Dad's stroke on Christmas Eve, she entered McLean.

In going there, she took the Sedgwick story full circle, back to Pamela, and to the place where that first Harry had gone for help. By now, McLean had long since relocated from Charlestown to the genteel suburb of Belmont, where it occupied a broad, grassy hillside that had been sculpted for the institution by Frederick Law Olmsted, and then, by a terrible quirk of fate, had been the last thing that he himself gazed upon as an inmate. But he did create a refreshingly bucolic setting, of the sort that Philippe Pinel had dreamed of. By the time my mother arrived, though, the wonderful old cottages and mansions of Olmsted's day had faded into sad decay. Vines crept up the walls of some of the great abandoned fortresses, blinding the windows and sometimes ascending to the roof, where they overran the chimneys. I found it almost electrifying to see these old buildings burn with a green flame. But my mother was in no mind to notice when I brought her for her first stay in Admitting Building II, a new brick building off the main parking lot. To get to her floor, an attendant had to insert a special key into the elevator. And when we passed through the glass door to her ward, it locked with a loud click behind us.

I was busy with my fiction now, partly to detach myself from her plight, hard at work on *The Dark House,* the novel about the man who likes to follow people in his car, absorbed in other lives because, troubled and alone, he lacks an authentic existence of his own. But these "pursuits," as my hero calls his nightly followings, take him back into his own past, where the explanation for his peculiar behavior lies. I did not realize that this was a message to the author, too.

My mother was back in McLean two years later when the book came out. Looking at me worriedly, she let me know that she had tried to read it, but found the effort too much. She didn't have to spell it out: my novel had landed her in the psychiatric hospital. In my own weakened state, that was the message I took, in any case. That was when—as a nurse did the intake exam, asking my mother once again for the particulars of her psychiatric history, a sad story I knew by heart—my attention wandered to other parts of the room, and I thought of writing my next novel about her. All I can say is, it seemed like a good idea at the time.

And that's when I began my fall.

PART FIVE

WHAT REMAINS

FORTY-THREE

OUR INTERIOR WEATHER

I had always thought I would conclude this narrative by returning once more to that ruinous September of 2000, newly armed with insights gleaned from the generations of Sedgwick history that I have surveyed. I thought of this in musical terms, actually, as a theme and variations; it struck me that the various characters from my ancestral past provided variations on the theme of me. Isn't that what genetic inheritance amounts to in families, as the genes of the parents get shuffled together, then further randomized by stray mutations, combining traits and features in ways that are both recognizable and not? In my case, Theodore emphasized the proud, ostentatious side of me, his son Harry a more impulsive dimension, Catharine the dreamier aspects, and so on through the ages to the killingly over-the-top Francis and tragic, transcendent Edie, and on to my heroic, underachieving father, my fretful mother, and all the rest. And I'd thought that, by returning to my period of crisis, I would find closure. This was like music, too.

So I worked through the rest of this book, researching and writing. It took me far longer than I had expected, years, all this thinking about the many sources of my me-ness, and in September of 2004 I was finally prepared to emerge once more into the time of that breakdown. Then something happened in my life that blew apart all my fine literary plans.

That particular something was not only horrendous, but so startlingly unexpected that it seemed to come from some other realm altogether, and in the many months that I have devoted to considering the matter, I have decided that the very unexpectedness of it was one of its signature qualities. It was like a catastrophe in Greek drama, which stems from—and thereby reveals—the hero's tragic flaw. In my case, there was further reason to think I was being mocked by the gods, for my own tragic flaw—or one of my flaws—was being maddeningly obtuse about the very subject to which I had devoted myself. I had failed to get my own point. I had started to think of my story not as being about me at all, but largely as a matter of others—of Theodore, and Babbo, and Francis and the rest, as if I was merely an observer, that snorkeler I mentioned some chapters back, and a somewhat disinterested one at that. Oh, what an amazing fish that one is! Look at the colors! But then my actual history called a halt to the literary proceedings, and summoned *me,* the putative author, before *it,* the putative subject, and thundered (or so I imagined), *No, you clod! Get with it, would you? This is not about them. It is about* you!

For what happened was, indeed, about me. As in a volcanic eruption, there was a long, slow, quiet buildup before my life actually blew. In this case, the first rumbling came with a request, early that September, from my wife that we go together into marriage counseling. This was not the first time she'd suggested it; we had gone to one professional or another at intervals over the course of our marriage. So I responded without undue alarm. But this time proved to be different, for weeks and then months of increasingly acrimonious exchanges in front of the marriage counselor (the first one, I joked to friends, possessing all the tact and wisdom of a lifetime staffer at the Registry of Motor Vehicles) did not produce the relief that previous sessions had; rather, it led only to a deepening sense of rift that soon reached crisis proportions. Late in January, Megan sat me down in the kitchen of our house in Newton and complained so bitterly about how things were going between us—or rather, how things were *not* going between us, how I seemed to have withdrawn all sympathy from her—that I had to ask her if she still wanted to be married to me. To my astonishment, she replied that she did not. And that was it. A twenty-five-year marriage was consigned to the grave.

I've wondered how I could so easily let my marriage go. On some level, presumably, I was only too willing, for I, too, felt the chill that had settled in between us. But there are as many layers to intent as there are to consciousness, and deeper down I was simply being obliging. It was, paradoxically, an expression of my love to accede to her request that the marriage end; I did not want to impose myself on a woman who would not have me.

We stewed for the next few months, each in our own separate corners of the house, approaching each other warily, like zoo animals placed in too-close confinement. I developed such tension in my neck that I could barely turn it to see where I was going when I backed the car out of our driveway, and I started grinding my teeth so hard at night that my temporomandibular joint went out of whack: my ears felt permanently stuffed up, as if I'd just dropped down from 35,000 feet (which in a sense I had), and I heard a near-constant ringing. Eventually, we decided that the conventional arrangement should apply: I would leave, and Megan would remain in the house with Josie. Megan had always been the more conscientious parent, attending to the children's needs with a loving vigilance that I, anxiously preoccupied with the moneymaking, could never quite equal. (I tried to compensate with spurts of fatherly rambunctiousness, but such displays were, to Megan, probably not enough, and somewhat irresponsible besides.) I planned to move out that spring on June 1, thinking that would allow me time to find a place to live, but I took forever to decide on the right apartment, eventually settling on one in the South End of Boston that wouldn't be available until September.

Still, Megan insisted I honor that June 1 agreement, and she helped make arrangements for me to stay for the summer in the third-floor space of some neighbors down the street. So, that bright June morning, I sullenly dismantled the marital bed while Megan packed up some kitchen items for me, and then, together, we separated out some other things—a few pieces of furniture, lamps, a rug or two. Somehow I had imagined that the sorting of our books, many of them constant companions since college, would be the traumatic part, but in fact I left most of my books behind. It was the division of the coffee cups that drove a dagger into my heart. Lots of them were mementos of places we'd been together as a family—travels, weekend visits, athletic tournaments. I took the colorful, chipped mug from Virginia Beach, she kept the fading one from Ogden, Utah…it seemed that we were not just breaking up our marriage, but the history of our shared life. And so we were.

Once we had gathered up my things, Megan helped me move. It was a weird reversal of all the previous moves we had made in our marriage. We'd moved in together, and now we were moving out together. Into separation. Together, we struggled down the sidewalk bearing most of my earthly belongings (my shirts, the empty sleeves fluttering in the breeze; a big stuffed chair from my childhood; my office chair, which she wheeled along the sidewalk) to the neighbors' house, while lines of morning commuters hurried by, gazing at the spectacle, and a few of the neighbors gawked.

Thus did I depart not only from my marriage, but from regular contact with our children, from our house, and from the very pleasant life I had enjoyed since college, nearly thirty years altogether. Once I was finally installed there, in a sweltering, airless, top-floor space, with just two narrow windows from which I could see a patch of the noon sky, I considered my daily round as I had known it just hours before: sleepy mornings over coffee and the newspaper before a quick good-bye hug to Josie as she left for school; then a bracing run; work in my upstairs office with its pale blue walls; a sandwich with Megan; more work in the afternoon; cheery greetings from Josie after she traipsed home from school, and then, and then, and then...the whole delightful daily swing between solitude and conviviality, interspersed by a few dreamy moments snatched to totter in the rocking chair off the back porch while I took in the sunshine through the maple trees at the foot of the lawn. My old house was just three houses away. My old house. I could see the backyard from the rickety back porch of my new place, if I stretched myself over the railing. But it was all gone.

Again, this is not the place for a he-said, she-said recounting of the issues between us. Neither of us would come off well, as anyone who has ever been married can well imagine. In truth, we both contributed about equally to the breakdown; the love between us simply ran out. After all, we'd spent nearly three decades in each other's company, working together in the house, eating most of our meals together, and bedding down together at night. Stresses can accumulate in such situations that are never adequately addressed for fear of upsetting the fundamental equilibrium on which both lives have come to depend. I'd draw a veil over this whole unhappy topic except for what gradually emerged as one of the key issues between us, as I came to understand it. It was a certain extravagance of mood and manner in me, one that, while it was reasonably appealing to my friends, seemed to Megan to be irritatingly self-centered; there was, besides, an innate sense of my own specialness that Megan sometimes took to be simply Too Much; and, at moments, an intensity she found intimidating.

This is not how I view myself, of course, but then, there are limits to one's self-awareness. Like Pamela with her Theodore, I feared that Megan knew me better than I did. I secretly ascribed to her an objectivity that I did not dare claim for myself. After my breakdown, I couldn't be entirely sure of my own judgment. And I had my ancestry to consider: the Sedgwickness of me; the traits I had been tracing in this book, ones I had been painstakingly trying to identify as my family inheritance. These were the very characteristics Megan was objecting to.

This is when my tragic flaw revealed itself. For even as I seemed to embrace these traits, I'd actually been distancing myself from them, perhaps protectively. I'd decided these things were in *them,* my ancestors, but I had been cured, purged. I'd wised up, gotten over them. They were no longer in me. I could survey the Sedgwick landscape with scientific detachment, the all-powerful, all-knowing author who is safely removed from the fray he is describing. Manic depression, mood disorders, mental illness of any kind—this was not the stuff *I* was made of.

Megan thought otherwise. For it turned out that she, too, had been scrutinizing the events of my calamitous fall in the year 2000, just as I had. Not the depression, though. That frightened her, as it did my daughters, then sixteen and ten. Even the dog—a vivacious gold retriever named Keeper—had looked at me strangely, cocking his head when I came down to breakfast, as if unsure it really was me. But Megan did not hold the depression against me nearly so much as the part that came after, when the Prozac hauled me out.

The celebrated antidepressant was a marvelous little pill, nearly hydraulic in its power. I went on it in late October of that year, and the effects were nearly immediate. As events proved, that should have been alarming, but to me it was nothing but a huge relief. At first, the change in mood was just a glimmering, like a recovered memory, but for a little while there I stopped feeling like a bundle of restless, self-destructive energy. As I downed more Prozac, morning after morning, and worked

my way up to a full therapeutic dose, the good hours started to string together. I thought of the little pills as a kind of anticoagulant that kept
stray worries from clotting up into the thick blob of obsessional anxiety
that had brought me down. And they were remarkably potent. I never
took more than twenty milligrams, which is at the very low end of the
therapeutic range, but I was a powerful responder. It seemed to me that I
had only to sniff the stuff to get a bounce. Still, in the early going, the
range of emotion of each day was vast. It seemed to run the gamut of all
four seasons—from the bleak, bone-chill of winter to the relaxed, flowing
warmth of summer—but never in any particular order.

My psychopharmacologist, Joan, predicted that I would first seem like
myself to others, and then to myself. To my delight, I entered phase one:
friends said I seemed "better." Gradually, I noticed that I could run my
(formerly) usual three miles without stopping to ponder my misery on a
park bench or wandering aimlessly off the trail into the trees, as I'd done
before. A little longer, and I started telling jokey stories again, laughing
with gusto and feeling happy most of the time. People seemed to relax
around me, wonder about me less.

As I ascended, I began to think quicker and more clearly than ever.
This was an unexpected delight, but one that I took to be small compensation for all I'd suffered. Words, images, ideas—they all flooded into my
head. I cranked out more chapters of the novel based on my mother's experience at McLean, and magazine articles besides. And then, more: I
became ravenous for a good time. Out and about, I was twice the man I'd
been, energetic, always on. I was the life of the party—*even when there was
no party*. I realize now I had become like that first Harry when, spittle
flying, he took on all those legal luminaries on the Greek frigate case, and
ended up being challenged to a duel, and many more manic Sedgwicks
since. I was getting in touch with my Sedgwick roots far more intimately
than I ever would have imagined possible. Without quite realizing it, I'd
sailed beyond the merely outgoing to the absolutely unstoppable. Friends
noticed I was having lots of conversations with strangers—clerks, waiters,
the guys at the deli counter. And women, look out. One called me Snake
Eyes for my predatory manner. Had Uncle Francis felt this way when he
hauled socialites into the bushes at his open-air parties? I remained true to
my marriage vows, but I freely trespassed over the normal boundaries
people draw around themselves. I'd get into deep, probing conversations

with casual dinner companions or seatmates on airplanes, ferreting out information that was no business of mine. This full-bore attention turned some of these total strangers into intimate friends in a matter of hours, and I was astounded to receive from one of them, several months later, an invitation to her wedding in Chicago. At that point, I had trouble remembering who she was.

At the time, I loved all of this. I loved my life, my newfound antics and powers, which felt like some previous unacknowledged part of my character, unexplored terrain within the world of me. Keyed up, lusting for connection, I imagined I was able to see into people's hearts, as if I possessed the psychic equivalent of X-ray vision, discerning the good or evil that lay there. When I confided all this to Megan at the kitchen table one late-winter afternoon, she responded with alarm, telling me that perhaps I was a bit *too* up. Relieved that I was no longer morbidly depressed, she hadn't wanted to deflate me, but seeing into people? This was too much.

"The Prozac," she told me. "John, I think it's doing something to you."

Doing something? That was a distressing thought, but I quickly deflected it. "Well, maybe it is," I snapped. "But all of it's so good!"

And it was: My "vitamin P" was slow-motion Ecstasy, turning my life into a party. How could there be anything wrong with it, with me? I was irked by the thought. "I've never felt better in my life," I insisted.

In a bookstore, though, I noticed a slender paperback called *Manic-Depression and Creativity,* and such was my frame of mind that I snapped it right up, and burrowed into it the moment I got home. As I read, it seemed to me I had a lot in common with all the famous artists described in the book as manic-depressive. Mania had always been just a word, nothing that had anything to do with me. But I was particularly fascinated by an early section of the book headed "Symptoms and Traits." It was two columns, side by side, contrasting the depressed state with the manic one. I glanced down the first column listing all the aspects of depression—"Slow thinking," "Pessimism," "Avoidance of people," "Fatigue," and the many other characteristics that were only too familiar to me. But then I jumped to the other column, under mania, and there were all the characteristics of my current, terrific mood: "Rapid thinking," "Gregariousness," "Charming and persuading people," and "Laughing." To be sure, some of these, like "Blindness to danger" and "Homicide," weren't so desirable, but I didn't believe they fit. In the main, I felt proud to exhibit such a swell bunch of personal characteristics. Reading further, I discovered a low-level category of mania that left out most of the negatives. "Hypomania," it was called. It seemed to fit my state to a T.

I told Megan all about this that night as we lay in bed. To my distress, she wasn't at all reassured to find so many of my behavioral characteristics on a list of psychiatric symptoms. As we talked it over, I finally began to see that maybe she was right. Maybe the Prozac had done something to me that wasn't so good. My amped-up state might be fun for me, but maybe it wasn't so great for her, or for anyone else, for that matter. Revved up in my eagerness for a good time, I could be impatient, it was true, and irritable. And, despite my longing for a deeper intimacy, I had to admit this idea of seeing inside people was a little much.

When I next visited Joan, I brought up the matter of my possibly having lurched a little too far in the manic direction. Joan replied that a condition called "Prozac-induced hypomania" was well described in the literature. She probed a little, asking if I'd been going out on wild shopping sprees, the classic manic symptom. It seemed significant to her when I said no. So she was disinclined toward a mania diagnosis. All the same, we decided it might be wise to bring down my Prozac dose from twenty milligrams to ten. A month later, when I still seemed somewhat juiced, we dropped it to five.

Gradually I stopped taking it altogether, and, more or less my former self, I carried on for about nine months, until the McLean-inspired novel came out and, dreading my mother's reaction, I started to fall apart again. Once more, I felt nervous, fragile, insulated. The depression settled over me like a deep-sea diver's helmet, insulating me so much from the world that my thoughts seemed to echo in my head. And then, far more frighteningly, it got deeper inside. I remember having to give a speech in St. Paul, Minnesota; remarkably, I was able to rouse myself out of my anxious lethargy for the performance, but when it was done, I went for a run to try and get the nervous buzz out of my system, and in my distress I started shouting at myself, "What am I supposed to do now? Kill myself? Is that it? Is it?" I imagined myself possessed by a black viper that was crawling up my digestive tract, and threatening to devour me from the inside out. My Jungian therapist nodded his head sagely when I told him about that. As with most such psychiatric manifestations, at least for me, to name them was to tame them. Just to describe this monster, which was the depression itself, released its hold on me.

I thought some medication might be in order all the same. This time, I gave up on Joan. In some desperation, I asked a psychologist friend for a recommendation, and she said that, for me, by far the best person to see was a psychopharmacologist named Dr. Gopinath Mallya.

AT McLEAN HOSPITAL. In Bowditch, a long, low building for outpatients, that was diagonally across the parking lot from Admitting Building II, where my mother had been twice at that point. I tried to screen out her experience from my own. Just on first meeting, I could tell Dr. Mallya was infinitely more learned and perceptive than Joan had been. Calm and attentive, he listened to the tale of my fall, my subsequent rise, and then the current dip that had brought me to his office. He inquired further about the "up" portion of my travails. To him, the fact that just a whiff of Prozac had an effect suggested to him a powerful susceptibility to mania. He prescribed Lamictal, a far gentler antidepressant than Prozac that had useful mood-stabilizing qualities. And he offered a diagnosis. "I think the best description for what you're experiencing is sub-syndromal manic depression," he said quietly.

There are far worse diagnoses, certainly, and the mysterious modifier *sub-syndromal* took a bit of the sting out of the idea of being permanently and unalterably sick. It seemed to mean that, given certain triggers, I am prone to the disease, but that I would never quite get to the full-blown state. Not so bad, really. Nonetheless, after I walked out of his office and back to my car, I imagined that the diagnosis was emblazoned on my forehead.

All the same, I did trust my dear Dr. Mallya, with his quiet manner and soft eyes. Perhaps I was overreacting to what I took to be an Indian background, but I fancied him a seer. Given what I'd been through—the down, the up, the down—I had to think there might be something to the idea that I was some sort of manic-depressive. My mood oscillated, no question. And as I thought back, it had before my collapse, with the occasional bluesy period usually followed by a nice rebound high. But I also recognized that my mood did not, in fact, rise and fall on its own.

Like the stock market, it responded to the news of the day—the initial enthusiastic response to my first novel made my spirits soar, and the dashing of those early hopes caused them to slump. And, inevitably, the subtle, but nonetheless heavy, pressures in the house's emotional atmosphere contributed, too, as each of us was burdened by our separate responsibilities. Still, I clung to the word *sub-syndromal* as proof that I could keep my feelings within the bounds of sanity. After all, I hadn't leapt out of that third-floor window, and the black viper had not devoured me. My reactions might have been overreactions, but who is to say? When the Dow loses 250 points on a weaker-than-expected jobs report, is that hysteria, or perspicacity? The short of it was that I seemed to have a larger emotional range than most people, or at least that is how it seemed to me. I had an irregularity inside my head like the deformity at the bottom of my spine. I registered more—more angst, certainly, but perhaps more *everything*—because I could. I had, no question, a certain intensity, a gemlike flame that lit up my writing and accounted for much of my personality. But it also had an edge, plus a degree of self-absorption, that could strain relations with those around me.

I was torn by this awareness. Psychiatric diagnoses have their own stigma, of course. But it was more than that for me. I had always resisted the prospect of being summarized in any fashion, which I took as tantamount to being dismissed. A preppie; a Harvard guy; a Wasp—I hated all of that. But it went further. One reason I never wanted a regular job, with an office and a title, was that I didn't want to be defined by it. Catholic in my tastes, I listen to just about every kind of music, favor no particular style or period of art. I like just about every sport, and have played most of them, but would never want to be thought a sports nut, or a jock. I dress largely to confound expectations, looking sharp when others are likely to dress down, and vice versa. My only absolute sartorial rule is never to wear blue jeans, since everybody else does. I am determined to be the odd man out.

And now, after all this, I get a label that sticks? What did it mean, anyway? Sub-syndromal. Was I, or wasn't I? It was a torment for me— one made all the worse by Megan's reaction. She had no doubts about this. She was convinced I was. As she said to me firmly one day, "You have a mood disorder, John." To her, sub-syndromal meant that I was the bearer of a disease that intruded into our lives together, and could only be eradicated from her life if I was. And to be fair, she had the children to consider, hard as that was for me to accept. Plus, I could not deny that I had at least a touch of the fire. Sub-syndromal is not *non*syndromal. I could only protest that I wasn't intrinsically sick, merely pushed into sickness by Prozac.

Besides, I wasn't prone to mania nearly so much as to depression, since I'd had two fairly serious encounters with it. Happily, I recovered from that second depressive episode more quickly than the first, and the Lamictal eased me out of it more gently than the Prozac had. I did not vault into any of the unbounded exuberance that had succeeded its predecessor. I kept going to Dr. Mallya for some months so he could monitor my progress, and occasionally Megan came along so she could weigh in with her own observations about me, ones that often ran contrary to my own. She would bring them up with Dr. Mallya while I listened, helpless. So the good doctor was made the ultimate judge of my sanity.

Dr. Mallya was too much of a gentleman to come down decisively on one side or the other, but acknowledged the validity of both sides, leaving Megan and me to quarrel afterward about what he'd finally ruled. Of course, in matters of sanity, one can never entirely be cleared, and when I finally agreed to leave the house, I did with the heavy heart of the convicted felon. Somehow my very act of self-exile, no matter how much I might claim that it was voluntary, amounted to an admission of guilt. And in this particular case, by now I had conducted the research on my forebears that made me recognize that these very qualities were in my blood. It was incontrovertible. I was a Sedgwick, guilty as charged. I could never peer into my own soul and find nothing, for my soul is not entirely my own; nothing of mine is. And I had peered into my soul by peering into my ancestral past, these past lives that had become, to a large degree, my past lives. And there the evidence of my guilt—if guilt is indeed the word—was overwhelming. As in the past, so in the present. As I pondered all this on the third floor of the neighbor's house, I thought of that black viper again, but this time I had a new name for it. Sedgwick.

This weighed on me, as you can imagine, alone in my bleak attic space. But after a few days of utter misery, I called a woman friend in Cambridge and begged her to take me in, which she kindly did, installing me in her guest bedroom, serving me meals, and listening to my nearly endless tales of woe. After a few weeks of that, I started to look for a place of my own, recognizing that I couldn't survive two more months of limbo before my South End apartment was available. I found a delightful carriage house just off Massachusetts Avenue, not far from my old haunts at Radcliffe, where Megan and I had first met over thirty years before. I had come full circle. The place had a second bedroom for Josie or Sara to spend the night; I bought a pull-out couch in case they both wanted to stay. Since I had never lived alone before, it took some getting used to, but I had some writer friends in the neighborhood, and I moved fairly easily into the social scene and developed a new routine of coffee shops and after-work drinks at nearby bars to replace the suburban life I had gotten used to, finding, to my surprise, that it was a good deal of fun.

As this book attests, I am inclined to brood, and I won't claim that it has been easy for me to come to terms with my new life, especially given all the elements that converged on me—the separation, the move, and, by a strange turn of fate, the completion of this book, which was intended to make sense of who I was, just at the time when so much about me was changing. Through it all, I have wondered just how much responsibility I should bear for this unexpected swerve in my life. Are my genes my fate? Or do circumstances play a role, too? Surely, the genetics alone don't determine outcome. Geneticists, after all, speak only of predispositions, not of predetermination; it invariably takes something more—an event, a change in blood chemistry—to trigger an episode. Indeed, as I survey my Sedgwick ancestors, I am as struck by how many of them did *not* succumb to any of the traits they, too, were heir to. Just in that first family, for example, Harry alone fell to the illness—and only when life seemed to conspire against him, nearly blinding him at his moment of greatest triumph with the Greek frigate case, and then dashing his hopes for Rhode Island coal. The others all made it through unscathed. I take a lesson from this. I give myself the same advice that Theodore II, anxious about the inherited aspects of his brother's disease, gave his son Theodore III as he endorsed "a conduct regulated by wisdom." I try for constancy of habits, regular exercise, loyalty to friends, realistic expectations, conviviality, and dedication to work. As my circumstances stabilize, I expect my mood to steady out, too. But life is an adventure, and I enter this latest phase in some way reborn, with a fuller understanding of myself, and a keenness for the new.

And I have thought more about our family disease, that manic depression that riddles the line, for among the few books I brought with me, one changed everything for me on that score. I had become distinctly uneasy about my pedigree, feeling embarrassed, *afflicted,* by it, but this book made me think that the condition was not necessarily a subject of shame at all. No emotional condition is, just as no physical condition is. But it went beyond that. I came to see that our supposed affliction was to a surprising degree our salvation. The disease that Francis Parkman called "The Enemy" was a friend.

For I came to the classic work of Kay Redfield Jamison, a distinguished professor of psychiatry at Johns Hopkins who is herself prone to mania, and in her pathbreaking 1993 book *Touched with Fire,* she concludes that manic depression is closely allied to the "artistic temperament"—so much so that the psychiatric condition seems to be virtually a prerequisite for the production of serious, deeply felt art. Emil Kraepelin, interestingly, made note of this association, but only in passing. And I have since come across Harvard neurologist Alice Flaherty's study of the creative temperament, *The Midnight Disease,* which bears out much of Jamison's research. But Jamison cites study after study to this effect—of thirty living American writers; of twenty award-winning European writers, poets, painters, and sculptors; of forty-seven British writers and artists. All of these, and many more, reveal an incidence of manic depression in serious artists that runs far higher, a full ten times higher, than in the general population. And Jamison's own study, the last one mentioned above, suggests that the upswings in mood—the swelling enthusiasm, growing self-confidence, quickening "speed of mental association," and heightened sense of "well-being" that I knew so well—all correlate with periods of the most intense creativity, just as they had in me. Indeed, that surge might be their inspiration.

No question, manic depression can be destructive; it has led to those suicides in the Sedgwick family, as in countless others; it accounted for my own flirtation with my demise, and figured in the breakup of my marriage. This is nothing I'd make light of. Jamison herself is only too aware of the hazards, for she attempted suicide by deliberately overdosing on the very lithium she was taking to regulate her moods. But there are many shades of the illness, and not all are so dangerous. A look at the genetics shows why. Of all the mental illnesses, manic depression is the one that is most likely to cluster in families. Jamison cites geneticists who place the gene on chromosome 11, but more recent research has suggested that at least sixteen different chromosomal regions, and possibly even more, may be involved. With so many genetic ingredients, the disease is not registered simply as "on" or "off." Like baldness, the expression of manic depression in each individual follows its own distinctive pattern, from mild to severe. The full-blown cases are often markers for family lines with other bearers of the disease who would otherwise go undetected, since their conditions are relatively mild. Jamison cites a Harvard researcher, Dr. Ruth Richards, who postulates that for creative purposes, it is better not to have the full-blown disease, but instead possess something closer to what I seem to have—namely the "genetic vulnerability" conferred by being a close relative of a manic-depressive. After all, until my breakdown at forty-six, I never experienced any of the injurious symptoms, but maintained a cheerful and productive attitude that is consistent with the mild high of the group that Richards had identified. These "normal first-degree relatives" score significantly higher than average on a standardized measure of creativity, but they have generally been ignored, Richards and her colleagues report, "because of a medical-model orientation that focused on dysfunction rather than positive characteristics of individuals." This is manic depression's sweet spot, the upside of an otherwise grim disease, in which it yields prodigious supplies of energy, enthusiasm, industry, and personal magnetism with relatively little of the debilitating depression that marks the downside. For forty-six years, that was me; and I am confident that it will be me again.

This explains another of Jamison's findings—that the families of manic-depressives tend to be more accomplished than the average. Jamison went on to review the genealogies of notable bipolars like the composer Robert Schumann (who sometimes imagined himself clawed by devils, and finally starved himself to death in an asylum), the poet Lord Byron (with all his unruly passions), the philosopher William James (who sometimes experienced, he once acknowledged, a "horrible fear of my own existence"), and several others. All of their family pedigrees were black with manifestations of the disease—and with remarkably industrious and creative individuals. In the James family, for example, there was William James's grandfather, likewise named William James, a prodigiously energetic Irish immigrant who, by 1832, had accumulated a fortune second only to John Jacob Astor's as the largest in the nation; in

Jamison's account, he was likely a carrier of manic depression as well. The philosopher's father, the writer and theologian Henry James Sr., was prone to nervous breakdowns. William's brother, the protean novelist Henry James, was frequently incapacitated by his "black depression." Their sister, Alice, was a brilliant but frustrated writer who underwent treatments for "nervous attacks,"and was bedridden for much of her adult life. And the mental health of their brother Robertson, a poet and painter, was so tenuous that William James thought that suicide the "only manly and moral thing" for it. Of the five siblings, only Garth Wilkinson seems to have been immune; he was also the only member of the family to have no artistic aspirations. But the James family was, for all this manic depression, *because* of all this manic depression, dazzling in its many artistic, financial, and philosophical triumphs.

While Jamison focuses on the arts, she does not limit herself to them. Other investigators have concluded that, as in the James family, manic depression can lead to a more broadly defined success, as it provides many of the characteristics—energy, vitality, daring, and charm not the least of them—that society has always prized.

As I look again at the Pie, and see the monuments of Pamela and Theodore at its center, I've started to see them differently—not separately, but paired, as if their union is the physical expression of that founding disease. Theodore was not manic, but with his relentless energy, he may have had a touch of the fire that Jamison mentions; in any case, he was Pamela's counterweight, a lift that countered her gravity, an up for her down. It is as if that first founding marriage was not just a linking up of his strength and her weakness, his drive and her status, his thought and her feeling, as I had imagined, but a joining of his dynamic exuberance to her passionate sensitivity to forge the manic depression that has been the dynamo powering the family through more than two centuries. And this has proved to be more central to the Sedgwick experience even than the house or the graveyard. The manic depression is so much a characteristic of the family, it seems to be the place we're from. Indeed, the disease gives a new cast to Stockbridge itself, with its bright, festive summers, when there's a party every night, and lonely, frigid, endless winters, when the sun stays so low to the horizon, its wan, bone-white light barely clears the snow-strewn hills right into March. This is our interior weather. Still, for all the troubles Sedgwicks have gotten into over the years, they have also thrown off a disproportionate amount of broadsides, pamphlets, political tracts, novels, biographies, articles, movies, TV shows, public appearances, and commentary. All of it is a pure product of the family disease.

MORE LARGELY, I see the oscillation of mood, the up-down-up-down, of the family as its quavering vitality. Nothing in life remains static, or it is not life at all. It is like music: the sound requires vibration; the music dies when the string is still.

FORTY-FOUR

A GUIDE TO LIFE

Families are like rivers: they go on and on, pooling here, tumbling there, as they flow toward a distant sea. And, also like rivers—from the royal Nile to the humble Housatonic—families have their distinctive personalities.

Ours was set down by Theodore, who instilled it in his children nearly as forcefully as he drilled a vision of his house into those poor Berkshire carpenters he hired to build it. It was a message of aspiration, of bettering themselves, embodying his Federalist convictions about the value of an establishment for future generations. The children developed their own political ideas, but they all recognized that they did indeed have a stake in the outcome of the young country, and they took their obligations seriously. They struggled to live up to the Judge's accomplishments, that first Harry dying in the attempt. But they did carry forward the Judge's standard, and my father passed it along to me.

Shortly after he died, I discovered among his papers an open letter he had written to Rob, Lee, and me about what he expected us to do with the modest inheritance—ravaged by the 1970s market decline—he was leaving us. The copy I found was on crinkly onion paper, typed by his secretary and signed with his initials. It was titled "A Guide to Life," and it went beyond the money to the larger questions of purpose in the head-on fashion that was typical of him. It began:

Fundamentally, I believe there are but two basic goals to life (1) personal happiness and (2) a determination to help make the world a better place to live in. It is my observation that long-term happiness cannot be obtained by seeking it directly, but only as a by-product of success in the second objective. Of course, immediate happiness can be secured and rightly by a good dinner, a good tennis match, an interesting trip, but if one spends one's entire life doing that I suspect, though I never tried it, that no matter what the variety of amusements eventually a life devoted to them would become boring.

I had to smile. Only my father could reduce all possible human objectives to two broad categories, and then load the deck in favor of one of them. (Interestingly, in deciding to be an epicurean, Babbo had taken the other choice.) But in his directive to make the world a better place, he might have been the Judge speaking, although the Judge himself had been content to consider only the welfare of the country. (Of course, he cared for the country in part that the country might care for him.) Dad went on to recommend specific careers like being "a good teacher, or … a useful public servant, a writer, an actor or an historian." It's sad he didn't include his own "investment adviser" on the list, but I find it significant that several of these are lines of work that had, in fact, been chosen by Theodore's own children, just as they would, for the most part, be selected by Dad's.

It is impressive that, over such a long span of time, the family has still retained the means to make such career choices without much regard for the financial consequences. Again, a series of profitable marriages largely accounts for that in recent generations. And over the generations, the Harvard gloss has been helpful, too. But the manic depression may supercede all the rest, an aspect that gives another meaning to the socioeconomic direction of "up." This was another lesson from my calamitous millennial year, when I discovered firsthand what it was to be a Sedgwick.

Like memory itself, a memoir is a nebulous form that dictates its own meaning, rather than blindly obeying its authorial master. It's a book that teaches its writer how to compose it. And while I undertook this memoir to better understand myself, I have come to see that a secret purpose was to understand my father. He embodies that past self, the one I thought I was looking for. And now that I am reaching the end, I am bracing myself for his dying all over again.

As it is, I treasure the things of his he left behind. The relics. For years, I wore his Omega wristwatch, and I revere his photograph of Robert Gould Shaw, which I have brought with me to my new digs. I brought my father's Groton School dinner plates, with the Rector's signature on the back, too. And his massive Porcellian Club glasses, although the distinctive P.C. boar's-head insignia has long since faded. I have placed Polly Thayer Starr's portrait of him over my desk in my new house, so that he can watch over me. I wish I could have brought with me the high bench that he salvaged from Harvard's Sever Hall because a lecture-weary student had whittled into it "H–7, Oregon–6," the score of his Rose Bowl game, but, alas, I don't have room for it.

It has taken this book to discover how much I miss my father, have always missed him. I miss my mother, too, but at least I had her for a good length of time. It is that longing for my father that has led me back, and back. For he did not exist entirely in his time, either. More than most people, he located himself in eras gone before. And so did his forefathers before him. This is the pull of family history, if not all history: a quest for understanding, and for reunion. A desire to hold our parents in our arms once more, and know them as they really were, that we may better embrace ourselves. The dead live in me, and I in them. Fathers are sons. I built the Old House, and I dwell in it. I am buried in the Pie even as I stroll about its stones.

CONFIDO IN DOMINO

ACKNOWLEDGMENTS

A family memoir is not easy on a family, and I'd like to thank my fellow Sedgwicks for their loving forbearance in this long and sometimes arduous project. It takes courage to willingly submit to such an exhaustive inquiry, and I am deeply moved and grateful that so many members of the family have helped me out so graciously and with such sustained interest. I am thinking first of my personal family—my two dear daughters, Josie and Sara Sedgwick—but also of the surviving members of my birth family, my brother Rob Sedgwick, sister Emily Sedgwick, and half brother Harry Sedgwick, all of whom I love deeply, and all of whom have helped me hone my memories and my interpretations of events into a far truer truth than if the job had been left to me alone. By virtue of his age and openness to my lines of inquiry, Harry in particular has been of tremendous assistance to me in compiling this narrative, as his memories go back a full generation beyond mine, and has allowed me a close-up view of family events that I did not witness and family members I did not know.

I am also immensely grateful to members of the greater Sedgwick family. Arthur and Ginger Schwartz, the current occupants of the Old House, were wonderfully hospitable to me over many visits and invariably receptive to my many rounds of questions about the house. I'd like to salute Ginger's mother, Virginia Glynn, who lives with them there, as she nears her hundredth birthday. Sarah Sedgwick Ginocchio and her brother, Alexander Sedgwick, likewise were tremendously forthcoming in helping me understand their mother, the tempestuous Sally Cabot Sedgwick—and offered many useful insights into the larger Sedgwick family in the bargain. My cousin Ellery Sedgwick generously read an early draft of the manuscript and made many helpful suggestions in the most kindly and loving tones. (Would that all such editorial comments were delivered thus!) His brother Tod Sedgwick, a trustee of the Sedgwick Family Society and Trust, was likewise an early, helpful reader; in the long run-up to publication, he was endlessly patient and supportive of this project, often directing me to research materials that I had not known about and making himself available for long chats about family history. My niece Betsey Osborne, a writer herself, also read an early version of the manuscript and provided some excellent advice and observations.

Of course, I have reached outside the family, too, in my search for understanding of my past. Of particular help to me has been Timothy Joseph Kenslea, who labored for many years to produce a detailed, carefully explicated history of the marriages in that first Sedgwick family in Stockbridge, *Awakening the Heart: Courtship, Engagement, and Marriage among the Sedgwicks of Berkshire County in the Generation after the Revolution,* for his Ph.D. dissertation at Boston College. In edited form, it has since been published by the University Press of New England as *The Sedgwicks in Love.* He also wrote an earlier paper on Harry and Robert's Greek frigate case. Tim has been extremely generous to me in answering my many questions about that first family, both in person and over the phone, and even shared many of his painstaking transcriptions with me. His work also directed me to other letters in the voluminous Sedgwick archives that I have been able to scrutinize for my own purposes. I cannot stress enough how indispensable his research has been to me, and how grateful I am to him. Peter Drummey, the marvelous librarian of the Massachusetts Historical Society, has been magical in his ability to direct me to the Society's many Sedgwick holdings and to help me to make sense of them. Likewise, at the Stockbridge Library, Barbara Allen, the director of the Historical Room, was helpful to me in bringing out a trove of Sedgwick materials over many visits. I am deeply indebted to the remarkable Lion G. Miles, a former airline pilot who settled near Stockbridge and has devoted himself to the history of the Stockbridge Indians, and many related matters, sharing his deep knowledge with me, and even spending several hours in the Pittsfield Registry searching for original Sedgwick land deeds to aid me. Gary Wolf, a wonderful architect, carefully explicated some of the more abstruse architectural details of Theodore Sedgwick's original building agreement for me. Rachel Wheeler, an associate professor of religious studies at Indiana University/Purdue University at Indianapolis, has kindly helped sort out the many confusing strands of the English settlers' complicated history with the Stockbridge Indians. Matthew P. Frosch, M.D., Ph.D., of the C. S. Kubik Laboratory for Neuropathology at Massachusetts General Hospital, generously helped me interpret the comments of Dr. Flint, who performed the autopsy on the brain of the first Henry Dwight Sedgwick. David Taylor, Don Armstrong, and Pagan Kennedy all read drafts of this book, offering much-needed reassurance and many useful comments.

My agent, Kris Dahl, has handled the business side of this project with her customary dispatch and has been a good friend to me in the bargain. My original editor at HarperCollins, Dan Conaway, was a dream in shaping this project from its earliest beginnings and carrying it through to its final editing before passing it on to the gifted Courtney Hodell, who diligently added many of her own keen insights before handing it off to the brilliant and charming Jill Schwartzman who saw the book into print.

Beyond my many obligations to the living, I owe a good deal more to the dead, to my many ancestors who have made me who I am. Foremost among them are my late parents, Robert Minturn Sedgwick and Emily Lincoln Sedgwick, to whom I have dedicated this book. I do so with many feelings, but the greatest of them is love.

A NOTE ON SOURCES

Since this is a memoir, much of the information about recent generations of my family comes from my own memory—as restored and complemented by notes, letters, and conversations with other family members who have been in a position to know. But there are, as well, plenty of historical facts pertaining to earlier generations. So that readers might be able to investigate the facts further, to discover, perhaps, their own truths in them, I will provide a short guide to where they all came from.

Chapter 3: A Man of Property

In reconstructing Theodore Sedgwick's ride up from Sheffield to Stockbridge, I have relied on several descriptions, some of them near-contemporary, of the local landscape, customs, and characters. *Sheffield: Frontier Town,* a bicentennial history written by Lillian E. Preiss in 1976, provided an excellent political and social overview. A more recent work, *Early Life in Sheffield, Berkshire County, Massachusetts: A Biography of Its Ordinary People from Early Times to 1860,* by James R. Miller, was wonderful for quotidian details. Great Barrington has its own fact-filled town history, written by Charles J. Taylor in 1882. The fullest account of Stockbridge's history was written by my cousin Sarah Cabot Sedgwick, and her sister-in-law, Christina Sedgwick Marquand. That book is titled *Stockbridge: 1739–1939; A Chronicle.* And I have turned to the standard history of early Stockbridge, *Stockbridge, Past and Present; or, Records of An Old Mission Station,* by Electa F. Jones, from 1854.

To evoke Theodore Sedgwick in this and later chapters, I have depended heavily on the work of his sole biographer, Richard E. Welch Jr., a Lafayette College historian who married into the family. Helpful as the published version of his *Theodore Sedgwick, Federalist: A Political Portrait* has been, I have found the original, much-longer Ph.D. dissertation that he wrote at Harvard University to be even more so, and I have turned to it again and again, as it is leafed out with personal and familial information that was pruned for the hardcover volume. As noted in the text, I have also relied on Theodore's daughter Catharine Maria Sedgwick's own memoir, initially published as *The Life and Letters of Catharine M. Sedgwick* and edited by Mary E. Dewey. The modern edition, *The Power of Her Sympathy,* edited by Mary Kelley, has been invaluable as well.

For the history of the Stockbridge Indians, there are innumerable sources, but I mostly relied on *The Mohicans of Stockbridge,* by Patrick Frazier. George M. Marsden's recent magisterial biography of the towering figure Jonathan Edwards has also been a godsend in helping me straighten out the often-tangled accounts of early Stockbridge. I have also consulted the brilliant Puritan scholar Edmund Sears Morgan's 1962 biography of Ezra Stiles, *The Gentle Puritan.* To see how all the Williamses' and Dwights' dealings with the Indians affected the Sedgwicks, I have turned to Karen Woods Weierman's article "Reading and Writing Hope Leslie: Catharine Maria Sedgwick's Indian 'Connections' " from the *New England Quarterly,* September 2002. In many generous conversations with me, the researcher Lion G. Miles provided the details about the Stockbridge Indians' likely appearance and demeanor. I have also relied on Daniel R. Mandell's *Change and Continuity in a Native American Community: Eighteenth-Century Stockbridge,* an unpublished master's thesis from 1982 at the University of Virginia. For information on what became of the Stockbridge Indians, I turned to Dorothy W. Davids's "Brief History of the Mohican Nation Stockbridge-Munsee Band," published by the Stockbridge-Munsee Historical Committee.

Chapter 4: Mr. Sedgwick Builds His Dream House

To understand the place of the Federal style, I am chiefly indebted to Wendell Garrett's *Classic America: The Federal Style and Beyond,* with its splashy photographs and enlightening text. Theodore's remarkable building agreement is among the holdings of the Massachusetts Historical Society, which houses by far the largest single collection of Sedgwick family correspondence. I have quoted from this one so extensively, and it is so rich and unusual, I will direct interested readers to hunt it up in Box V of the Sedgwick papers.

Chapter 5: A Friend of Order

Again, I have relied on Catharine's memoir, *The Life and Letters of Catharine M. Sedgwick,* for a sense of Theodore's worldview and for an account of his first wife's death. For a fuller understanding of the ravages of smallpox, I consulted *Pox Americana: The Great Smallpox Epidemic of 1775–82* by Elizabeth A. Fenn. The information on that first Robert Sedgwick comes largely from a learned article written by my great-grandfather, Henry Dwight Sedgwick II. His article, published by the Colonial Society of Massachusetts in 1896, is posted on the Sedgwick.org Web site organized by Dennis Sedgwick that offers a great deal of fascinating material on the Sedgwick family on both sides of the Atlantic. Henry Dwight Sedgwick II's article also detailed the Puritans' sumptuary laws. Richard Welch has diligently dug out the facts of Theodore's early life from the archives but confined them largely to that unpublished dissertation. I have filled out my own understanding by turning to many of the original documents that Welch unearthed; those invariably led me to others as well.

Chapter 6: Among the River Gods

To appreciate the interlocking social networks of western Massachusetts, I have depended on a 1986 Ph.D. dissertation by Kevin Michael Sweeney, now a professor at Amherst College, called *River Gods and Related Minor Deities: The Williams Family and the Connecticut River Valley, 1637–1790.* For the more personal details regarding Pamela, I have relied heavily on another unpublished Ph.D. dissertation, this one by the extraordinarily diligent Timothy Joseph Kenslea, whom I thanked in the acknowledgments, but who deserves further recognition here. He labored for many years to produce a detailed, carefully explicated history of the marriages in that first Sedgwick family in Stockbridge, *Awakening the Heart: Courtship, Engagement, and Marriage among the Sedgwicks of Berkshire County in the Generation after the Revolution.* (It has since appeared in an abridged form as *The Sedgwicks in Love,* published by the University Press of New England.) That work directed me to particular letters, and clusters of letters, in the voluminous Sedgwick archives, which I have then been able to scrutinize for my own purposes. For the larger context of Puritan marriage, I have turned to Laurel Thatcher Ulrich's *Good Wives: Image and Reality in the Lives of Women in Northern New England, 1650–1750,* and the aforementioned Edmund Sears Morgan's *The Puritan Family: Essays on Religion and Domestic Relations in Seventeenth-Century New England.*

Chapter 7: The War Within the War

Welch provided a thorough account of Theodore's role in the American Revolution and its internal conflicts. To piece together how Theodore's war stories fit into the larger picture, I consulted widely but found that Bruce Lancaster's very readable *The American Revolution* gave the clearest and most useful account, although it lacks the fresh detail of the more recent, Pulitzer-winning account of the early part of the war, *Washington's Crossing,* by David Hacket Fisher. For Colonel John Ashley's role in the matter, I relied, again, on Lillian E. Priess's *Sheffield.* Readers curious about the extraordinary Peter Van Schaack and his family might want to consult the letters of his collected as *The Life of Peter Van Schaack,* edited by his son Henry C. Van Schaack and published in 1842. To tutor myself in the interior social alignments and realignments of the Revolutionary period, I have depended heavily on Gordon S. Wood's *The Radicalism of the American Revolution.* For the story of the many Tories in Massachusetts who found themselves under attack by safety committees and their ilk, see *The Loyalists of Massachusetts and the Other Side of the American Revolution,* by James H. Stark, published in 1910. Also, a 1904 work by Agnes Hunt: *The Provincial Committees of Safety of the American Revolution.*

Chapter 8: Williams Family Secrets

Lion G. Miles wrote an extremely helpful article detailing the Williams family's unscrupulous land dealings in *The Red Man Dispossessed: The Williams Family and the Alienation of Indian Land in Stockbridge, Massachusetts, 1736–1818,* which appeared in the March 1994 *New England Quarterly.* Karen Woods Weierman described Ephraim Williams's aberrant behavior in her previously cited article on Catharine Maria Sedgwick's relationship to this morally questionable behavior, "Reading and Writing Hope Leslie: Catharine Maria Sedgwick's Indian 'Connections,'" published in the *New England Quarterly,* September 2002.

Chapter 9: The Household Did Not Run of Itself

The family information all comes from letters and from Catharine's account. To get a sense of domestic life in the Revolutionary period, I have turned primarily to Jane Nylander's *Our Own Snug Fireside;* Alice Morse Earle's *Home Life in Colonial Days;* and David Freeman Hawke's *Everyday Life in Early America.* Welch tracked down all of the Sedgwick servants for his dissertation. Agrippa Hull makes it into Electa Jones's aforementioned history of Stockbridge.

Chapter 10: All Men Are Born Free and Equal

Mumbet's story has been told many times by now, but most frequently in tales for children. Of them, the most detailed is *Mumbet: The Life and Times of Elizabeth Freeman*. Catharine Maria Sedgwick's "Slavery in New England" was printed in *Bentley's Miscellany* 34 (1853; pages 417–24). It is available on Catharine's Web site, maintained by the Catharine Maria Sedgwick Society and its president, Lucinda Damon-Bach; it can be found at www.salemstate.edu/imc/sedgwick. The best single account of the court case is by Arthur Silversmit, "Quok Walker, Mumbet, and the Abolition of Slavery in Massachusetts," from the October 1968 *William and Mary Quarterly*. What exists of the trial documents can be found in the Berkshire Court of Common Pleas, in Pittsfield, Massachusetts (volume 4, August 21, 1781, pages 55–57), and on Mumbet's Web site, www.mumbet.com. For the short course on slavery in New England, I have turned to Edward J. McManus's "Black Bondage in the North," and Joanne Pope Melish's *Disowning Slavery: Gradual Emancipation and "Race" in New England, 1780–1860*, among other accounts. For the reference to Jonathan Edwards's slaves, I am indebted to Kenneth P. Minkema's article in the 2002 volume of the *Massachusetts Historical Review*, "Jonathan Edwards's Defense of Slavery."

Chapter 11: The Proper Object of Gibbets & Racks

There are many accounts of Shays' Rebellion, starting with the official history written by Theodore Sedgwick's friend George Richards Minot in 1788. For my purposes, I found the most rewarding to be *Shays's Rebellion: The American Revolution's Final Battle*, by Leonard L. Richards, and *Shays' Rebellion: The Making of an Agrarian Revolution*, by David P. Szatmary. Welch and Kenslea each provide a detailed account of Theodore's part in the conflict. Again, Catharine has provided the best account of its effect on the Sedgwicks generally in that article of hers on Mumbet in *Bentley's Miscellany*.

Chapter 12: Bottled Lightning

To see Theodore's role in the nascent federal government, I have turned to Welch. But I have been guided by several scholarly accounts, but most particularly the recent, comprehensive *A Leap in the Dark: The Struggle to Create the American Republic*, by John Ferling. For biographies of Alexander Hamilton, I favor the brisk one of Richard Brookhiser, *Alexander Hamilton, American*, although

I fully respect the titanic work of Ron Chernow, *Alexander Hamilton,* and its more ample references to Theodore's influence on his subject and vice versa.

Chapter 13: A State of Widowhood

This material is all drawn from family letters at the Massachusetts Historical Society, many of them ones that Kenslea guided me to in his own careful explication of the marriage of Pamela and Theodore.

Chapter 14: A Disorder of the Blood

The details of the formation of the medical society are in the *Berkshire Book,* published by the Berkshire County Historical and Scientific Society in 1892. To appreciate how mental illness was treated in the early days of the Republic, I have come to see the essential texts as *The Mentally Ill in America: A History of their Care and Treatment from Colonial Times,* by Albert Deutsch, and *Changing Faces of Madness: Early American Attitudes and Treatment of the Insane,* by Mary Ann Jiminez. But the sweeping compendium *Three Hundred Years of Psychiatry, 1535–1860, a History Presented in Selected English Texts,* edited by Richard Alfred Hunter, provided many useful nuggets as well. Readers interested in a close-up view of the "maniacs" on display might turn to the Reverend Manasseh Cutler, who recorded his tour in his diary, edited by William Parker Cutler and Julia Perkins Cutler in a two-volume compendium, *Life, Journals and Correspondence of Rev. Manasseh Cutler, LL.D.,* published in 1888.

Chapter 15: Among the Maniacs

La Rochefoucauld-Liancourt included his observations of Theodore Sedgwick in his *Travels in the United States of North America,* published in 1800.

Chapter 17: Colonel Lovejoy's Methods

The material on James Otis came from an early biography, *The Life of James Otis, of Massachusetts,* by William Tudor. Andover's early history is best collected in Sarah Loring Bailey's *Historical Sketches of Andover,* published in 1880. Dr. Willard's methods are detailed in Albert Deutsch's aforementioned *The Mentally Ill in America.* Besides recording Pamela's deterioration, and Theodore's dedication to his political career, Kenslea also details the ill-fated love affair of Frances and Loring Andrews, as does Welch. Ferling has the most thorough modern account of the machinations that led to the election of Thomas Jefferson and Theodore's role in the affair. Welch guided me through Theodore's own maneuvering.

Chapter 18: It Can Not Be Told

The Sedgwick and Marquand history of Stockbridge provided many of the details of this turn-of-the-century portrait of the town. Kenslea tracked the twists of Frances's romance with Ebenezer Watson and its terrifying outcome later. He also examined Pamela's final breakdown and performed that diligent search through the archives to find the first reference to Pamela after her death. That last letter of Hamilton's is collected in the twenty-seven-volume edition *The Papers of Alexander Hamilton,* edited by Harold C. Syrett. The original is in the possession of the Massachusetts Historical Society. Mary Bidwell's descriptions of Pamela's final bouts with suicidal depression are collected in the Historical Room of the Stockbridge Library.

Chapter 20: The Third Mrs. Sedgwick

Kenslea's account, useful in itself, guided me to the key letters recounting the details of Theodore's remarriage, decline, and death. Catharine's memoir offered her perspective.

Chapter 21: The Will

The will is archived at the Massachusetts Historical Society in Sedgwick V. Kenslea detailed the legal battle over it.

Chapter 22: To Worship the Dead

For by far the best account of burial practices in New England and their transformation with the rural cemetery movement readers should turn, as I did, to *Silent City on a Hill: Landscapes of Memory and Boston's Mount Auburn Cemetery* by Blanche Linden-Ward. My account of the development of the Sedgwicks' burying ground is drawn from the scanty records in the letters and from a careful examination of the Stockbridge cemetery itself. Indeed, the burial ground is the best "document" of all, as its dimensions testify to the children's intentions for the space.

Chapter 23: Catharine in Silhouette

Sadly, there is as of yet no proper biography of Catharine Maria Sedgwick. I have made much use of a compendium of scholarly essays, among them *Catharine Maria Sedgwick: Critical Perspectives,* edited by Lucinda Damon-Bach and Victoria Clemens. Mary Kelley's *Private Women, Public Stage: Literary Domesticity in Nineteenth-Century America* devotes a considerable attention to Catharine, and its insights and references have been key to my own understanding. It also takes up from where Seth Curtis Beach left off with his sadly outdated 1909 volume *Daughters of the Puritans: A Group of Brief Biographies* (Lydia Maria Child and Harriet Beecher Stowe were two of the others). Kenslea dug into Ebenezer Watson's complicated finances and laid out the history of his abusive marriage to Frances; he also delved deeply into the development of Harry's romance with Jane Minot. Kelley's article "A Woman Alone: Catharine Maria Sedgwick's Spinsterhood in Nineteenth-Century America," from the *New England Quarterly,* was especially good in detailing Catharine's feelings for her brother Robert.

Chapter 24: I Have Located My Heaven

Edward Halsey Foster's slender *Catharine Maria Sedgwick* stands as a kind of guidebook to her work; it was published in 1974 as part of Twayne's United States Author Series. I have also found that Charles Henry Brown's biography *William Cullen Bryant* provided some helpful views of Harry, Robert, and Catharine, who were such prominent members of his circle. As before, Kelley's *Private Women,* and the other previously cited books on Catharine, like the collection edited by Damon-Bach and Clemens, were tremendously useful in detailing her literary career as well. By far the best modern account of Eunice Williams's travails is John Demos's *The Unredeemed Captive: A Family Story from Early America.* Stephen Nissenbaum's justly celebrated history of Christmas, *The Battle for Christmas,* was my source for Catharine's significance in that arena.

Chapter 25: The Great Central Fire

Kenslea recounted Harry's arrival in New York City, and his unpublished paper "La Malheureuse Histoire Des Fregates: The Hope and The Liberator: H.D. and R. Sedgwick, Edward Everett, and the Greek Frigate Case," was my introduction to this extraordinary episode in Harry's life. I also consulted Harry's two broadsides on the subject, his marvelously titled "A Vindication of the Conduct and Character of Henry D. Sedgwick" and "A Refutation of the Reasons Assigned by the Arbitrators for Their Award in the Case of the Two Greek Frigates." Both were published by Henry D. Sedgwick in 1826 and can be found in the rare manuscripts department of Harvard's Houghton Library. For the early history of McLean, I turned to the classic text *Crossroads in Psychiatry: A History of the McLean Hospital,* by S. B. Sutton. I was also helped by Alex Beam's more recent, delightful social history of McLean, *Gracefully Insane: Life and Death Inside America's Premier Mental Hospital.* For equivalent information on the Hartford Retreat, I relied on the recent history *Mad Yankees: The Hartford Retreat for the Insane and Nineteenth-Century Psychiatry,* by Lawrence B. Goodheart. A useful overview of both institutions, and many others, comes from *Mental Institutions in America: Social Policy to 1875,* by Gerald N. Grob. For a description of Harry's decline, I depended on his wife Jane's posthumous recollection, which she set down in about thirty pages of her longhand, and included a copy of Dr. Flint's autopsy. That can be found in Box V of the Sedgwick papers.

Chapter 26: In the Country Burial Place, Would I Lie

Catharine's essay "Our Burial Ground" originally appeared in *The Knickerbocker* (volume VI, November 1835). It is reproduced on her Web site. I am also grateful to Beth Rheingold, Ph.D., of the Department of Media Study at SUNY-Buffalo, for the insights in her unpublished paper "On Common Ground: Sedgwick and the Nineteenth-Century Cemetery."

Chapter 27: The Great Wheel Turns

Many of the details of Theodore Sedgwick III's life appeared in *Gotham: A History of New York City to 1898,* that wonderfully replete volume by Edwin G. Burrows and Mike Wallace. Others were dredged from an anonymous, handwritten, lengthy appreciation of him in the Sedgwick papers at the New York Public Library's Rare Manuscript Department. The Libby Prison journal of Arthur George Sedgwick is there, too. Other details of Theodore and Arthur came from their Harvard class books. Of the many accounts of the rise of the Berkshire mansions, I favored *The Berkshire Cottages: A Vanishing Era,* by Carole Owens. *Lenox: Massachusetts Shire Town,* by David H. Wood, was helpful as well. He provided the lengthy quotation from Mrs. Burton Harrison's journals, *Recollections Grave and Gay.* Cleveland Amory's charming *The Last Resorts,* with its many passages about Lenox and the Berkshires, remains the last word on this heyday of plutocratic frivolity. The best evocation of this phase of the life of the Old House is from an unpublished reminiscence of the Reverend Theodore Sedgwick IV, which is collected with the Sedgwick papers in the Stockbridge Library. Natalie

Sedgwick Colby wrote a wonderful reminiscence of her deeply troubled, wonderfully romantic life, *Remembering,* that adds a few details to my account of this era and much else besides. And, as noted, Ellery Sedgwick's *The Happy Profession,* and Henry Dwight Sedgwick's *Memoirs of an Epicurean* have served as essential texts for me.

Chapter 28: Babbo & Ellery

I have relied primarily on Ellery's and Babbo's own accounts, as above. Ellery Sedgwick III, grandson of the *Atlantic*'s Ellery, led me to the wonderful exchange with Amy Lowell, which he recounted in the *New England Quarterly* (volume VI, number 4, December 1978).

Chapter 29: Marrying Up

Again, Babbo's own memoir is my main source for this portion of his life. For a few other details of the Minturns, I have depended on the marvelous *Edie: An American Biography,* by Jean Stein, edited with George Plimpton.

Chapter 30: Good Night, Sweet Prince

For the best contemporary account of the Parkman murder, and of Francis Parkman's manic depression, readers might want to see Simon Schama's *Dead Certainties: Unwanted Speculations.* Interestingly, one of Parkman's many biographers was Babbo, who passes very lightly over his subject's mental state. That book, heavy on extracts from Parkman's letters, was titled *Francis Parkman* and was published in 1904 as part of Houghton Mifflin's American Men of Letters Series. The account of Halla's death is drawn largely from Babbo's privately published volume *In Memoriam,* which he had printed in 1918. The material on Endicott Peabody is taken mostly from Frank Ashburn's biography *Peabody of Groton, a Portrait.*

Chapter 31: The Family Patriarch

Readers interested in a little more on the Rose Bowl game might turn to my account of it, "The Big Game," in *GQ* (October 2001). Much of the rest is taken from my father's unpublished memoir. I have written about all three of the clubs my father belonged to: "Sunset at the Somerset Club" in an undated issue of *Boston Magazine* (which I believe was from 1977); "The Country Club" in *New England*

Monthly (July 1989); and "The Brotherhood of the Pig," in *GQ* (September 1988). On the subject of Riggs, I drew heavily from *The Riggs Story: The Development of the Austen Riggs Center for the Study and Treatment of the Neuroses,* by Lawrence S. Kubie, and from a profile, "Dr. Austen Fox Riggs," by Donald Culross Peattie in the *Atlantic Monthly* (1941). His advice on worry comes from what Riggs thought of as his ten commandments, which are reproduced in the article. I also consulted a number of Riggs's "little green books," or *Talks to Patients,* as they were formally titled, on such subjects as Efficiency, The Individual, and Sensation and Emotion. The material on John P. Marquand comes mostly from Millicent Bell's biography *Marquand: An American Life.* As mentioned in the text, Emil Kraepelin's pioneering study of 1921, *Manic-Depressive Insanity,* remains indispensable to an understanding of the disease and of science's earliest attempts to recognize it as one. The rest of the story is filled out with interviews of various family members.

Chapter 32: Evasion and Escape

Information on my father's war career comes largely from his unpublished memoir, as buttressed by interviews with a few surviving eyewitnesses, and from family letters that have been gathered in the holdings of the Historical Room of the Stockbridge Library. The details on the deterioration of my father's first marriage, his affair with Sally, and the sad events surrounding the Welch-Sedgwick marriage were all told to me by family members in a position to know about them. The material on the development of the Sedgwick genealogy comes from a file that my father kept on the matter, which has come into my hands.

Chapter 33: Sally & Shan

The details of the rise and fall of Sally and Shan's relationship were confided to me by family members and friends of the couple. Some of my understanding of them, of course, comes from my own vivid memories of the two of them, whom I used to see together at Sally's apartment in Boston's Back Bay.

Chapter 34: The Anti-Helen

Readers interested in a fuller account of the Ameses of North Easton might want to turn to a more detailed source, "North Easton, Massachusetts," that I published in *Yankee Magazine* (March 1992). The best recent rendering of the building of the railroad is Stephen E. Ambrose's *Nothing Like It in the World: The Men Who Built the Transcontinental Railroad, 1863–1869.*

Chapter 35: My Mother's Diary

As I explained in the text, I am relying here on my mother's various diaries and memos, and on several scrapbooks and photograph albums that were assembled by her, or, in one case, by her father, Alexander Lincoln, which are in my possession.

Chapter 36: A Little Scratch on the Chromosomes

Most of the material pertaining to my birth and my mother's anxieties over my spina bifida comes from the baby book I mentioned. I have filled out my understanding of the condition by consulting the Spina Bifida Association, among other standard medical sources.

Readers curious about my father's stock plan might consult "A New Pension Plan," by R. Minturn Sedgwick, in the *Harvard Business Review* (January-February 1953), or his article in the *Financial Analysts Journal,* "The Record of Conventional Investment Management: Is There Not a Better Way?" (July-August 1973).

Chapter 37: Little Duke

The material pertaining to Francis's hoax letter, his war on Groton School, and the school's response can all be found in the Groton School archives. Besides my own recollections, and those of other family members, I have turned to the accounts recorded in *Edie: An American Biography.*

Chapter 38: Edie, Superstar

The information about Edie that did not come from family interviews, and my personal observation, is taken from *Edie: An American Biography.*

Chapter 39: "One Loves to Remember Beauty"

The full story of the court case involving Louis Agassiz Shaw II, better known as Louis Two, is told by David Moran, the Massachusetts state trooper who investigated the matter, in his *Trooper: True Stories from a Proud Tradition.* Gabriella's letter was saved by my father. I have told the story of expansion of the family graveyard more fully in an article, "The Eternity Club," in the *New England Monthly* (September 1989).

Chapter 40: In Loco Parentis

I have written about Groton and my experiences there in "World Without End," an article published in the *New England Monthly* (September 1988). The other material was drawn from various interviews with family members and my own recollections.

Chapter 43: Our Interior Weather

The books I refer to here are *Manic-Depression and Creativity,* by D. Jablow Hershman and Julian Lieb, and *Touched with Fire,* by Kay Redfield Jamison, whose memoir, *An Unquiet Mind,* I also consulted.

INDEX